Advances in Space AstroParticle Physics: Frontier Technologies for Particle Measurements in Space

Advances in Space AstroParticle Physics: Frontier Technologies for Particle Measurements in Space

Guest Editors

Matteo Duranti
Valerio Vagelli

Basel • Beijing • Wuhan • Barcelona • Belgrade • Novi Sad • Cluj • Manchester

Guest Editors

Matteo Duranti
Italian National Institute for
Nuclear Physics (INFN)
Perugia
Italy

Valerio Vagelli
Italian Space Agency
Rome
Italy

Editorial Office
MDPI AG
Grosspeteranlage 5
4052 Basel, Switzerland

This is a reprint of the Special Issue, published open access by the journal *Instruments* (ISSN 2410-390X), freely accessible at: https://www.mdpi.com/journal/instruments/special_issues/44O4E85G84.

For citation purposes, cite each article independently as indicated on the article page online and as indicated below:

Lastname, A.A.; Lastname, B.B. Article Title. *Journal Name* **Year**, *Volume Number*, Page Range.

ISBN 978-3-7258-2675-9 (Hbk)
ISBN 978-3-7258-2676-6 (PDF)
https://doi.org/10.3390/books978-3-7258-2676-6

Contents

About the Editors

Matteo Duranti

Matteo Duranti is a staff researcher at the Italian Institute for Nuclear Physics, INFN, in the Perugia (IT) Department. He obtained his Ph.D. in Physics from the Università degli Studi di Perugia (IT), in 2012; his thesis investigated experimental astroparticle physics based on the analysis of the data collected on ground, before the launch to space by the AMS-02. In his research activity at INFN, University of Perugia, and CERN, he has mainly been contributing to experimental astroparticle physics research, focusing on searches for new physics in space. His activities included technological and instrumentation developments and scientific data analysis for space-based cosmic ray experiments, such as AMS-02, DAMPE, PAN, and HERD. Since 2017, he has been a staff researcher at INFN.

Valerio Vagelli

Valerio Vagelli is a staff researcher at the Italian Space Agency (ASI) headquarters in Rome (IT). He obtained his Ph.D. in Physics from the Karlsruhe Institute of Technology (DE) in 2014, discussing a thesis on experimental astroparticle physics based on the analysis of the AMS-02 experiment onboard the International Space Station. In his research activity at KIT, INFN, University of Perugia, and ASI, he mainly contributed to experimental astroparticle physics research, focusing on searches for new physics in space. His activities included technological and instrumentation developments and scientific data analysis for ground- and space-based cosmic ray experiments, such as, AMS-02, DAMPE, and CTA. Since 2019, he has been a staff researcher at ASI, where he mainly participates in programs for technological R&D for space-based particle and radiation detectors.

 *instruments*

Editorial

Advances in Space Astroparticle Physics: Frontier Technologies for Particle Measurements in Space

Matteo Duranti [1,*] and **Valerio Vagelli** [1,2,*]

1 Istituto Nazionale Fisica Nucleare (INFN)—Sezione di Perugia, IT-06123 Perugia, Italy
2 Agenzia Spaziale Italiana (ASI), IT-00133 Roma, Italy
* Correspondence: matteo.duranti@infn.it (M.D.); valerio.vagelli@asi.it (V.V.)

Citation: Duranti, M.; Vagelli, V. Advances in Space Astroparticle Physics: Frontier Technologies for Particle Measurements in Space. *Instruments* **2024**, *8*, 45. https://doi.org/10.3390/instruments8040045

Received: 30 September 2024
Accepted: 5 October 2024
Published: 12 October 2024

In the last decades, breakthrough advances in understanding the mechanisms of the Universe and fundamental physics have been achieved through the exploitation of data on cosmic rays and high-energy radiation gathered via orbiting experiments, in a synergic and complementary international effort that combines space-based instrument data with ground-based space observatories, accelerator, and collider experiments.

Cosmic ray space-borne detectors, such as PAMELA [1], AMS-02 [2], DAMPE [3], and CALET [4], have been providing high-precision data on cosmic ray spectra and composition, uncovering unexpected features that could not be explained by the consolidated models of the origin, acceleration, and propagation mechanisms of cosmic rays. These achievements have been indicating the need to revise the pillars of the established theories. In reaching this results, high-precision experimental measurements from the most recent generation of space-borne particle detectors have been a game changer.

Today, advancing our understanding of cosmic ray physics mainly requires to concurrently investigate several observational frontiers.

Measurements of all cosmic ray species with energies larger than TeV to shed light on the origin and composition of the "knee" of cosmic ray fluxes is the main objective of the planned HERD experiment [5], which for the first time will operate in space an innovative isotropic 3D-imaging calorimeter to maximize the faint statistics of supra-TeV cosmic rays. Extending the precision and range of cosmic-ray nuclear and isotopic composition measurements is also one largely pursued scientific target, currently tackled by the HELIX [6] and TIGER [7] balloon-borne instruments, that requires relevant improvements in the techniques for high-dynamic range charge measurements and high-resolution velocity measurements, and that will be possibly implemented in future space-borne extensions of the previous missions [8].

Finally, one of the most ambitious objectives of space-borne particle physics is to precisely measure the faint components of antimatter in cosmic rays, possibly providing striking evidence of heavy anti-nuclei in cosmic radiation and revealing signatures of dark matter annihilation, thus helping to narrow down the origin and properties of dark matter particles. To take over the legacy of the PAMELA and AMS-02 detectors, the scientific community is envisioning large-area, high-precision magnetic spectrometers to be operated in deep space, such as the proposed AMS-100 [9] or ALADInO [10] instruments. The technological roadmap for operating such instrumentation in space combines what already being developed for high energy and high precision charge and velocity measurements with the need to develop high-temperature superconducting magnets for space applications. Besides this approach based on magnetic spectrometric measurements of particles in space, independent techniques to identify nuclear antimatter in space has been pioneered by the GAPS instrument [11], to be soon flown on a balloon, and further investigated by other proposed instruments.

Improving the understanding of the physical properties of low-energy and highly penetrating cosmic radiation is also a major requirement for enhancing radiation risk models and developing countermeasures to minimize the risks posed by radiation exposure to humans and instrumentation in space, ultimately enabling safe space colonization [12]. Most recent advances leverage on readapting particle detector technologies and approaches from the heritage of general-purpose high-energy space detectors to MeV-GeV cosmic-ray measurements (such as, e.g., the HEPD detectors onboard the CSES satellites [13]) and to instrumentation for dosimetric energy releases onboard and outside spacecrafts (such as the ALTEA-LIDAL detector onboard the ISS [14]). Novel scientific results are achieved by taking advantage of technologies developed for complementary applications and optimized for applications in space, as demonstrated by the successful experience of the Timepix hybrid pixel detectors [15]. Fundamental new information can also be extrapolated by monitoring long-term and short-term disturbances in cosmic ray fluxes due to their interaction with the dynamic heliospheric environment, with the prospects of improving the predictive models of cosmic-ray flux intensities for space weather forecasting and developing prompt alarm systems in case of abrupt increases in flux exposures.

Our understanding of the high-energy sky has also been completely revolutionized by the successful operation of gamma-ray observatories, such as Fermi [16] and AGILE [17], whose data have been providing accurate information on the most energetic and exotic phenomena in the Universe, from very powerful distant sources such as blazars to the local Galaxy. The potential of gamma-ray observatories has been additionally enriched when, in 2017 and 2018, the very-high-energy emission from extreme sources has been concurrently observed from space in correlation with the ground-based detection of gravitational emissions [18] and neutrino emissions, starting the era of multi-messenger astronomy. Very recently, the IXPE telescope [19] finally demonstrated the feasibility of precision X-ray polarimetry astrophysics measurements, de facto opening a new window for investigating most classes of high-energy events in the Universe.

Ultra-high-energy cosmic rays and neutrinos may provide information about the most energetic phenomena in the Universe, and several technological developments are ongoing to enable their observation from space for the first time, where a large-field telescope, such as, e.g., the proposed POEMMA space observatory [20], could complement and integrate the measurements made by ground-based observatories.

In the field of cosmic ray and high-energy detection in space, the reach of current observations is, in general, largely constrained by the limitations in technology and approaches. Historically, state-of-the-art level ground-based particle detectors established since several years have been adapted and optimized for space operations. Building on the experience gained from operating the current generation of instrumentation, conceiving follow-up opportunities has become largely demanded by the scientific community. In view of this, the scientific community is pushing research efforts and resources in developing novel approaches to step forward beyond the current paradigm of radiation measurement in space, possibly re-inventing the experimental investigation strategies. New frontiers in astroparticle physics can be tackled only with breakthrough advances in technological solutions and observational approaches. The development of novel technologies and techniques for the measurement of particles and high-energy radiation in space shares common heritage and approaches in detection techniques and data handling, despite being probes that historically differ in terms of observational strategies and in the scientific information they provide. The success of particle and radiation detectors operated in space is also rooted in the common, strong heritage of particle detectors first developed for ground-based accelerator or collider experiments, subsequently specialized and optimized for the harsh space environment, although this paradigm could gradually move towards an approach in which space applications will drive novel technological solutions.

Most importantly, current opportunities in the new space economy era for next-generation astroparticle and high-energy radiation observatories have significantly diversified from the previous decade. Access to small- and nano-sat platforms has become increasingly diffused, while the prospect of accessing novel platforms, such as new orbiting laboratories, planetary gateways, or bases, is driving the community to rethink the standard paradigm upon which current operating space instrumentation has been developed.

Nonetheless, the stringent requirements on mass, volume, power budget, and operation safety strongly define the constraints on mission instrumentation.

The international conference "Advances in Space AstroParticle Physics: Frontier technologies for particle detection in space (ASAPP 2023)" has aimed in reviewing the progresses in design, development, integration and test of instrumentation for measurement of particles and high-energy radiation in Space, gathering the scientific community to pave the road to future astroparticle missions for investigations of fundamental physics and the Cosmos, applications for monitoring of the space radiation environment, and investigations of the impact of low energy ionizing particles on instrumentation, Space Weather, and Earth sciences. The contributions included in this Special Issue provide a snapshot of many arguments discussed during the conference. The collection of papers covers and reports on the most recent technological developments for high-energy and low-energy cosmic rays as well as high-energy X and γ radiation, including proposals, developments, and tests of novel sensors or measurement approaches, front-end electronics and data acquisition systems, and space mission payloads and instruments. This overview largely demonstrates the advances achieved in recent years by a vivid community that targets ambitious scientific objectives to breach the frontiers of current knowledge on the origin and mechanisms of the Universe and to provide cutting-edge solutions for precise monitoring and diagnostic of the radiation environment in space. The pioneering technological solutions that are being investigated, developed, and consolidated today through collaborative international research efforts will form the foundations for next-generation space instruments, paving the way for future groundbreaking discoveries.

Conflicts of Interest: The authors declare no conflicts of interest.

References

1. Adriani, O.; Barbarino, G.; Bazilevskaya, G.; Bellotti, R.; Boezio, M.; Bogomolov, E.; Bongi, M.; Bonvicini, V.; Bottai, S.; Bruno, A.; et al. The PAMELA Mission: Heralding a new era in precision cosmic ray physics. *Phys. Rep.* **2014**, *544*, 323–370. [CrossRef]
2. Aguilar, M.; Ali Cavasonza, L.; Ambrosi, G.; Arruda, L.; Attig, N.; Barao, F.; Barrin, L.; Bartoloni, A.; Başeğmez-du Pree, S.; Bates, J.; et al. The Alpha Magnetic Spectrometer (AMS) on the international space station: Part II—Results from the first seven years. *Phys. Rep.* **2021**, *894*, 1–116. [CrossRef]
3. Chang, J.; Ambrosi, G.; An, Q.; Asfandiyarov, R.; Azzarello, P.; Bernardini, P.; Bertucci, B.; Cai, M.; Caragiulo, M.; Chen, D.; et al. The DArk Matter Particle Explorer mission. *Astropart. Phys.* **2017**, *95*, 6–24. [CrossRef]
4. Marrocchesi, P.S.; for the CALET Collaboration. CALET on the ISS: A high energy astroparticle physics experiment. *J. Phys. Conf. Ser.* **2016**, *718*, 052023. [CrossRef]
5. Betti, P. The HERD experiment: New frontiers in detection of high energy cosmic rays. In Proceedings of the XVIII International Conference on Topics in Astroparticle and Underground Physics (TAUP2023), Vienna, Austria, 28 August–1 September 2024; p. 142. [CrossRef]
6. Coutu, S.; Allison, P.; Baiocchi, M.; Beatty, J.; Beaufore, L.; Calderón, D.; Castano, A.; Chen, Y.; Green, N.; Hanna, D.; et al. The High Energy Light Isotope eXperiment program of direct cosmic-ray studies. *J. Instrum.* **2024**, *19*, C01025. [CrossRef]
7. Rauch, B.F.; Link, J.T.; Lodders, K.; Israel, M.H.; Barbier, L.M.; Binns, W.R.; Christian, E.R.; Cummings, J.R.; de Nolfo, G.A.; Geier, S.; et al. Cosmic Ray Origin in OB Associations and Preferential Association of Refractory Elements: Evidence from Abundances of Elements ^{26}Fe through ^{34}Se. *Astrophys. J.* **2009**, *697*, 2083. [CrossRef]
8. Rauch, B.F.; Zober, W.V.; Abarr, Q.; Akaike, Y.; Binns, W.R.; Borda, R.F.; Bose, R.G.; Brandt, T.J.; Braun, D.L.; Buckley, J.H.; et al. From SuperTIGER to TIGERISS. *Instruments* **2024**, *8*, 4. [CrossRef]
9. Schael, S.; Atanasyan, A.; Berdugo, J.; Bretz, T.; Czupalla, M.; Dachwald, B.; von Doetinchem, P.; Duranti, M.; Gast, H.; Karpinski, W.; et al. AMS-100: The next generation magnetic spectrometer in space—An international science platform for physics and astrophysics at Lagrange point 2. *Nucl. Instruments Methods Phys. Res. Sect. A Accel. Spectrometers Detect. Assoc. Equip.* **2019**, *944*, 162561. [CrossRef] [PubMed]
10. Adriani, O.; Altomare, C.; Ambrosi, G.; Azzarello, P.; Barbato, F.C.T.; Battiston, R.; Baudouy, B.; Bergmann, B.; Berti, E.; Bertucci, B.; et al. Design of an Antimatter Large Acceptance Detector In Orbit (ALADInO). *Instruments* **2022**, *6*, 19. [CrossRef]
11. Aramaki, T.; Hailey, C.; Boggs, S.; von Doetinchem, P.; Fuke, H.; Mognet, S.; Ong, R.; Perez, K.; Zweerink, J. Antideuteron sensitivity for the GAPS experiment. *Astropart. Phys.* **2016**, *74*, 6–13. [CrossRef]
12. Fogtman, A.; Baatout, S.; Baselet, B.; Berger, T.; Hellweg, C.E.; Jiggens, P.; La Tessa, C.; Narici, L.; Nieminen, P.; Sabatier, L.; et al. Towards sustainable human space exploration—Priorities for radiation research to quantify and mitigate radiation risks. *npj Microgravity* **2023**, *9*, 8. [CrossRef] [PubMed]

13. Scotti, V.; Osteria, G. The HEPD detector on board CSES satellite: In-flight performance. *Nucl. Part. Phys. Proc.* **2019**, *306–308*, 92–97. [CrossRef]

14. Di Fino, L.; Romoli, G.; Santi Amantini, G.; Boretti, V.; Lunati, L.; Berucci, C.; Messi, R.; Rizzo, A.; Albicocco, P.; De Donato, C.; et al. Radiation measurements in the International Space Station, Columbus module, in 2020–2022 with the LIDAL detector. *Life Sci. Space Res.* **2023**, *39*, 26–42. [CrossRef] [PubMed]

15. Bergmann, B.; Gohl, S.; Garvey, D.; Jelínek, J.; Smolyanskiy, P. Results and Perspectives of Timepix Detectors in Space—From Radiation Monitoring in Low Earth Orbit to Astroparticle Physics. *Instruments* **2024**, *8*, 17. [CrossRef]

16. Atwood, W.B.; Abdo, A.A.; Ackermann, M.; Althouse, W.; Anderson, B.; Axelsson, M.; Baldini, L.; Ballet, J.; Band, D.L.; Barbiellini, G.; et al. The Large Area Telescope on the FERMI Gamma-Ray Space Telescope Mission. *Astrophys. J.* **2009**, *697*, 1071. [CrossRef]

17. Tavani, M.; Barbiellini, G.; Argan, A.; Boffelli, F.; Bulgarelli, A.; Caraveo, P.; Cattaneo, P.W.; Chen, A.W.; Cocco, V.; Costa, E.; et al. The AGILE Mission. *Astron. Astrophys.* **2009**, *502*, 995–1013. [CrossRef]

18. Abbott, B.P.; Abbott, R.; Abbott, T.D.; Acernese, F.; Ackley, K.; Adams, C.; Adams, T.; Addesso, P.; Adhikari, R.X.; Adya, V.B.; et al. Multi-messenger Observations of a Binary Neutron Star Merger*. *Astrophys. J. Lett.* **2017**, *848*, L12. [CrossRef]

19. Soffitta, P. The Imaging X-ray Polarimetry Explorer (IXPE) and New Directions for the Future. *Instruments* **2024**, *8*, 25. [CrossRef]

20. Olinto, A.; Krizmanic, J.; Adams, J.; Aloisio, R.; Anchordoqui, L.; Anzalone, A.; Bagheri, M.; Barghini, D.; Battisti, M.; Bergman, D.; et al. The POEMMA (Probe of Extreme Multi-Messenger Astrophysics) observatory. *J. Cosmol. Astropart. Phys.* **2021**, *2021*, 007. [CrossRef]

Article

Hadronic Energy Scale Calibration of Calorimeters in Space Using the Moon's Shadow

Alberto Oliva

Istituto Nazionale di Fisica Nucleare, Sezione di Bologna, 40127 Bologna, Italy; alberto.oliva@bo.infn.it

Abstract: Calorimetric experiments in space of the current and of the next generation measure cosmic rays directly above TeV on satellites in low Earth orbit. A common issue of these detectors is the determination of the absolute energy scale for hadronic showers above TeV. In this work, we propose the use of the Moon–Earth spectrometer technique for the calibration of calorimeters in space. In brief, the presence of the Moon creates a detectable lack of particles in the detected cosmic ray arrival directions. The position of this depletion has an offset with respect to the Moon center due to the deflection effect of the geomagnetic field on the cosmic rays that depends on the energy and the charge of the particle. The developed simulation will explore if, with enough statistics, angular, and energy resolutions, this effect can be exploited for the energy scale calibration of calorimeters on satellites in orbit in Earth's proximity.

Keywords: Moon's shadow; cosmic rays; calorimeters

Citation: Oliva, A. Hadronic Energy Scale Calibration of Calorimeters in Space Using the Moon's Shadow. *Instruments* **2024**, *8*, 7. https://doi.org/10.3390/instruments8010007

Academic Editor: Antonio Ereditato

Received: 21 October 2023
Revised: 23 January 2024
Accepted: 26 January 2024
Published: 27 January 2024

1. Introduction

Cosmic rays (CRs) are energetic particles and nuclei coming from outer space from all directions with an approximately isotropic flux. The Moon, with an average diameter observed from Earth of about $0.52°$, causes a directional depletion in the observed CR flux observed from Earth's vicinity. The shape and magnitude of the deficit observed in CRs provide a measure of the pointing accuracy and angular resolution of the device used to measure it. This methodology was proposed originally by Clark [1] and has been used since in experiments measuring the extensive air showers (EAS) produced by high-energy CRs interacting in the atmosphere: CYGNUS [2], Tibet-ASγ [3], CASA [4], HEGRA [5], Tibet-III [6], ARGO-YBJ [7], HAWC [8], LHAASO [9], and GRAPES-3 [10,11]. Underground experiments such as MACRO [12], SOUDAN-2 [13], L3+C [14], MINOS [15], IceCube [16], and ANTARES [17] have also employed the Moon's shadow observed in high-energy muons coming from EAS for determining the detector's angular resolution.

With enough angular resolution and statistics, the Moon's shadow displacement in CRs due to the deflection in the geomagnetic field can be measured. Observations of the Moon's shadow can be used for momentum and charge-based separation of CRs, as originally proposed by Urban, et al. [18], as sketched in Figure 1. By observing the Moon's shadow as MACRO [19], L3+C [14], Tibet-III [6], ARGO-YBJ [7], and HAWC [8], the authors estimated upper limits on the $\bar{p}/p$ ratio in CRs. The Moon's shadow deflection dependence from energy has also been used by EAS experiments to estimate the absolute energy scale, such as in ARGO-YBJ [20] and LHAASO [9].

Imaging Atmospheric Cherenkov Telescope (IACT) experiments are able to distinguish EAS induced by γ-rays and electrons and hadronic showers with good accuracy. This separation has been used by IACTs such as ARTEMIS [21], MAGIC [22], and VERITAS [23] that were able to derive upper limits on the $\bar{p}/p$ and on the positron fraction $e^+/(e^+ + e^-)$ in CRs at TeV.

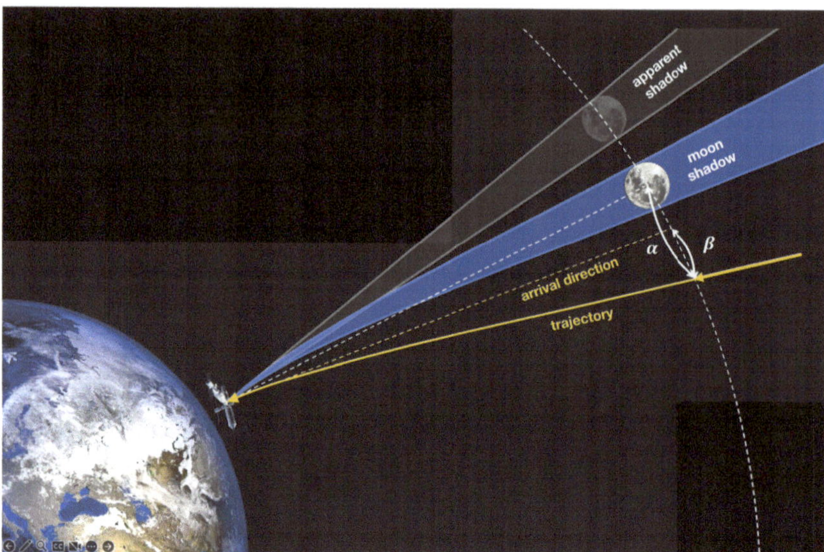

Figure 1. The Moon causes a depletion on the CR flux observed from Earth's vicinity. Furthermore, the charged CRs are bent by the geomagnetic field, causing an apparent position displacement of the Moon, as observed in the CR arrival direction at the satellite location. In the figure are defined two angles used in this work: the angle α, i.e., the angular separation between the particle and the Moon before entering Earth's magnetosphere, is derived from the scalar product of the position vector of the Moon and the position vector of the particle as observed at the entrance of the sphere defined by the Moon–Earth distance; and the angle β, i.e., the magnetic deflection accumulated by the particle traveling in the geomagnetic field from Moon distance to Earth vicinity, is derived from the scalar product of the particle direction at the satellite location and its position vector observed at the entrance of the sphere defined by the Moon–Earth distance. The angles α and β do not necessarily lie on the same plane.

The CR Moon's shadow has usually been employed in experiments measuring CRs above TeV with indirect techniques. Nowadays, direct measurement of CRs from space is reaching TeV energies thanks to large geometric factors, long exposure times, and extended energy ranges. Examples of current and future CR direct measurement experiments in space are AMS-02 [24], CALET [25], DAMPE [26], HERD [27], AMS-100 [28], and ALADInO [29]. In many cases, these experiments are based on a calorimetric approach, in which the energy of the particle is evaluated by absorbing it in thick calorimeters. The measurement of energy within calorimeters, especially for hadronic particles, requires careful calibration of the energy scale [30]. Several techniques have been employed by space-born calorimeters in space. A non-complete list includes:

- Particle Beams: the energy scale can be measured directly by studying the calorimeter response to particle beams of known energy. This calibration is limited by the maximum energy achievable for particle beams, currently set at most at a few hundred GeV [31].
- Geomagnetic Cutoff: in Earth's magnetosphere, trajectories of particles with rigidity below a given geomagnetic cutoff are forbidden. The geomagnetic cutoff depends on satellite location and particle direction and can be calculated by tracing particles in the geomagnetic field. However, the use of this effect to calibrate the energy scale is limited by the maximum geomagnetic cutoff achievable in space, of the order of tens of GeV [32].
- Cross-calibration: including in the experiment design other devices able to measure particle momentum would allow for energy scale cross-calibration. Some possible so-

lutions may include a magnetic spectrometer or a transition radiation detector [27–29]. This solution usually involves a higher degree of design complexity, often not viable for experiments in space.

In this work, we will determine whether a space-born detector with an adequate angular resolution and enough energy resolution would be able to observe the CR Moon's shadow and use it to evaluate the energy scale for protons in calorimeters in the multi-TeV region.

In Section 2, the simulation of realistic CR particles bent by the geomagnetic field and detected by a realistic device is discussed. In Section 3, the effect of the Moon's shadow on the observed angular distribution is characterized using a high-statistics simulation. In Section 4, the log-likelihood method used to estimate the energy scale is discussed and applied to a simulated 5-year data-taking period, followed by a discussion in Section 5.

2. Simulation

For the sake of a realistic case, we consider a detector in orbit similar to the high-energy cosmic radiation detection (HERD) facility [27]. HERD is a large-field-of-view, high-energy cosmic ray experiment planned to be installed on the China Space Station (CSS) in 2027. The HERD mission is based on a novel design of a highly segmented, homogeneous LYSO calorimeter of 55 X_0 (CALO). The CALO segmentation in 3 cm side LYSO cubes allows for the reconstruction of 3D showers from all incoming directions. The large thickness of the CALO will allow for measuring the energy of CRs up to 1 PeV. The CALO is complemented by a series of detectors for particle identification installed on five sides (top and four sides), enclosing it in a box. Its outermost detector, the Silicon Charge Detector, can identify the absolute charge of the entering cosmic ray, separating precisely all cosmic ray chemical species [33]. Globally, the experiment will have a field of view covering almost an entire hemisphere with a geometric factor approaching 1 m²sr for protons [34].

Let us then assume for the following a test detector located on the CSS orbit, taking data in a at least a 5-year mission, measuring particles arriving from the full hemisphere tangent to the orbit, with an acceptance of 1 m²sr, an energy resolution of about 30%, and an angular resolution of 0.1° [25–27].

To study the CR Moon's shadow observed from this test device, a realistic simulation has been constructed. The orbit of the CSS around Earth has been simulated using the Two-Line Elements of the CSS orbit in the last two years (`Predict`: https://www.qsl.net/kd2 bd/predict.html accessed 30 March 2020, and `CelesTrack`: https://celestrak.org/ accessed 22 September 2022). The Moon orbit is simulated using analytic calculations [35].

For every second of real elapsed time, protons of different energies were generated at the detector location on the CSS orbit isotropically inside the detector's full field of view. Each proton has been back-propagated, inverting the sign of its electric charge and arrival direction, effectively simulating a particle moving backward in time in Earth's magnetic field up to the Earth–Moon distance. Back-traced protons falling into Earth or the Moon are then removed. This procedure effectively simulates a flux of particles incident to our detector isotropically generated at the Moon distance, including the depletion caused by the CR Moon's shadow.

The tracing is realized by solving the relativistic Lorentz equation of motion $(\vec{x}(t), \vec{p}(t))$ for a particle with mass m and charge q in a constant magnetic field $\vec{B}$:

$$\begin{cases} \frac{d\vec{x}}{dt} = \frac{\vec{p}c^2}{E} \\ \frac{d\vec{p}}{dt} = \frac{qc^2}{E}\vec{p} \times \vec{B} \end{cases}, \tag{1}$$

where $E = \sqrt{m^2c^4 + p^2c^2}$. Solver is based on a classic Runge–Kutta method [36] with adaptive control of the time step size. During tracing, crossings of Earth or of the Moon are checked with linear interpolation between the steps. The employed magnetic field is the latest International Geomagnetic Reference Field (IGRF-13) [37] that describes the main component of the geomagnetic field caused by sources primarily inside Earth. The IGRF is a set of spherical harmonic coefficients that are used in a mathematical model that represents an accurate description of the geomagnetic field based on current and historical data. A slow secular time variation is present in IGRF, but this variation is completely negligible in the time scale of the particle tracing. The back-tracing code has already been validated by comparison with available online tools such as SPENVIS (https://www.spenvis.oma.be/ accessed 8 March 2022) and has been successfully employed in other applications [38].

The deflection angle β, defined in Figure 1 as a function of energy for the developed simulation, is displayed in Figure 2. The average deflection has been fitted as a function of E^{-1}, following the expected behavior at the first order [39]. Long tails of the distribution towards low values of the deflection angle correspond to the polar passages and can be removed by requiring equatorial passages of the CSS within latitudes of $|\lambda| < 20°$.

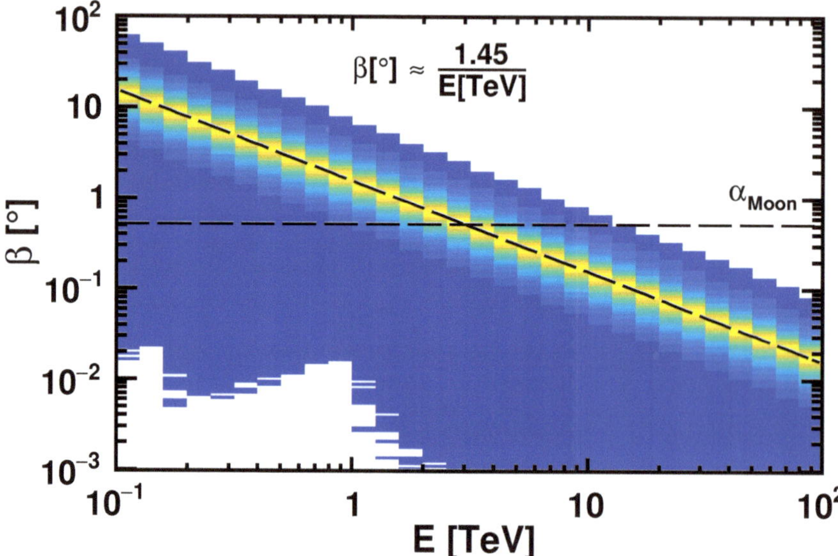

Figure 2. The deflection angle as a function of the proton energy for the simulated satellite orbit. The average deflection is inversely proportional to the energy. In particular, at about 3 TeV, the deflection is similar to the Moon's apparent size. The asymmetric tails towards small values of deflection are due to satellite polar passages.

For each proton not shadowed by the Moon, a measured arrival direction and energy are generated to simulate the particle as measured by our test detector. The arrival direction is extracted around the true arrival direction with a Gaussian smearing of $0.1°$, while energy is extracted with a Gaussian smearing of 30%. The "measured particle" is back-propagated using the same procedure described above for the "true particle", and the measured deflection angle $\tilde{\beta}$ and the measured angular distance to the Moon $\tilde{\alpha}$ are also determined and used for the analysis. With this approach, the effect of uncertainties in the calorimeter angular and energy resolutions are propagated to the angles of interest of this analysis.

3. Analysis

To study the Moon's shadow, the angle between the particle to the Moon after deflection subtraction $z = 1 - \cos\alpha$ has been used. With a perfect angular and energy resolution, the distribution of z should follow a Heaviside function with a step at the Moon's angular radius $z_{\text{Moon}} \sim 10^{-5}$. Instead, the observed distribution of the measured angle $\tilde{z} = 1 - \cos\tilde{\alpha}$ is more similar to a sigmoid due to the combined effect of angular and energy resolutions.

To obtain the distribution of $\tilde{z}$ as a function of energy, a simulation in 30 energy range bins from 100 GeV to 100 TeV has been carried out, with approximately 8 billion protons generated for each energy bin. All particles have been back-propagated successfully to the Moon–Earth distance. The events have been weighted with a weighing factor dependent on the true energy E to reproduce the CR proton energy spectrum following the parameterization developed in [40] using the latest CR measurements. With this high-statistics sample, it is possible to observe the distribution $f_i(\tilde{z})$ for each i-th measured energy interval $(\tilde{E}_i, \tilde{E}_i + \Delta\tilde{E}_i)$. These distributions have been parametrized following an effective model provided via the convolution of a box distribution, representing the Moon's shadow, and a Gaussian, representing the effective angular resolution:

$$f_i(\tilde{z}) = \frac{A_i}{2}\left[1 + \mathrm{erf}\left(\frac{\tilde{z} - \mu_i}{\sqrt{2}\sigma_i}\right)\right], \tag{2}$$

the parameter μ_i is expected to be roughly independent of energy, with the value of z_{Moon}; however, to allow an effective better fitting, this parameter has been left free. The results of the fitting procedure are exemplified at the top of Figure 3 in three energy bins. A good fit with $\chi^2/\mathrm{d.o.f.} \sim 1$ has been obtained in every energy bin. The model based on Equation (2) with the regularized parameters as a function of energy is shown at the bottom of Figure 3.

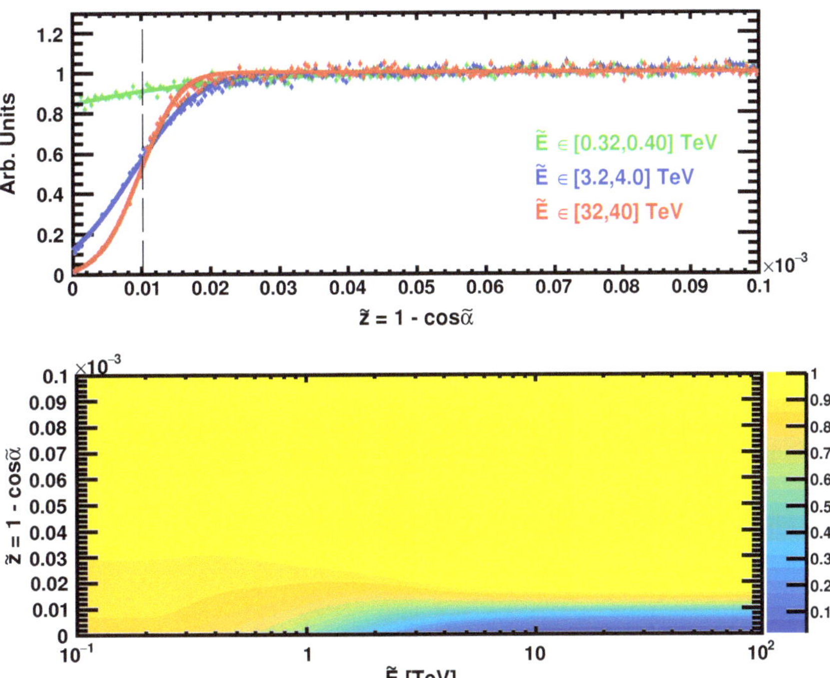

Figure 3. On the top, the distribution of $\tilde{z}$ for three energy bins: 0.32–0.40 TeV in green, 3.2–4.0 TeV in blue, 32–40 TeV in red, as obtained from the simulation. Three fits, with the effective model of Equation (2), are superimposed. The fits have $\chi^2/\mathrm{d.o.f.}$ of 1.07, 1.31, and 1.58, respectively. On the bottom is displayed the overall model that was obtained from the regularization of the parameters of the bin-by-bin fit as a function of energy.

The behavior of the distributions in Figure 3 can be understood by noticing that the $\tilde{z}$ resolution is the sum of the angular resolution plus the effect of the energy resolution that smears the deflection angle determination. In particular, above 10 TeV, the Moon's shadow becomes more defined and energy-independent since, in this range, the deflection is negligible with respect to the angular resolution. At low energies, the deflection becomes large up to tens of degrees, and the uncertainty regarding the energy causes a large angular uncertainty for the evaluation of the deflection; this causes a flattening of the $\tilde{z}$ distribution. In these two extreme cases, the dependency on energy is weak, but, around TeV, the $\tilde{z}$ distribution changes rapidly as a function of the energy. In this region, we should globally have some sensitivity to the energy scale from the observed $\tilde{z}$.

To understand how $\tilde{z}$ behaves as a function of a possible energy bias k, the parametrization procedure has been repeated, introducing in the simulation different biases on the energy scale in a range between 0.7 and 1.3. The estimated distributions of $\tilde{z}$ for each bias k are then studied employing the functional form provided in Equation (2), and the obtained fits are qualitatively similar to the ones of Figure 3. For each energy bin i, the μ_i and σ_i obtained for each different scale factor k are regularized as a function of k, obtaining the functions $\mu_i(k)$ and $\sigma_i(k)$. Eventually, the distribution of $\tilde{z}$ as a function of k in each measured energy bin, $f_i(\tilde{z}; k)$, is obtained.

4. Results

To understand the detection of the Moon's shadow with our test device, a realistic simulation of 5-year data acquisition has been developed following the recipe described in Section 2. Globally, 400 million protons have been generated between 100 GeV and 100 TeV following the energy spectrum described in Ref. [40]. For each i-th reconstructed energy bin and j-th $\tilde{z}$ bin, the observed number of events is N_{ij}^{obs}. The expected number of events that can be constructed from the Moon's shadow parameterization described in Section 3 is provided in

$$N_{ij}^{\mathrm{exp}}(k) = \frac{\int_{\tilde{z}_j}^{\tilde{z}_j + \Delta \tilde{z}_j} f_i(z; k)\,dz}{\int_0^{z_{\mathrm{max}}} f_i(z; k)\,dz} \sum_j N_{ij}^{\mathrm{obs}}. \tag{3}$$

The top picture of Figure 4 shows an example of the observed number of events and the expected ones for different values of the energy scale parameter k in a single energy bin between 1.00 and 1.26 TeV.

It is possible to create a global likelihood to measure the discrepancy between the observed number of events and the expected ones at all energies:

$$-2\log \mathcal{L}(k) = -2 \sum_i \sum_j \left(N_{ij}^{\mathrm{obs}} \log N_{ij}^{\mathrm{exp}}(k) + N_{ij}^{\mathrm{exp}}(k) \right) + C, \tag{4}$$

this negative log-likelihood can be minimized to derive the best energy scale factor $\hat{k}$.

At the bottom of Figure 4 is presented the dependence of the log-likelihood as a function of the energy scale k. The distribution leads to the result $\hat{k} = 0.99^{+0.10}_{-0.12}$. The use of a 5-year dataset allows us to measure the energy scale with the Moon's shadow with 10% accuracy, which is also the minimum detectable scale mis-calibration.

To check for possible systematic effects, several tests have been performed:

- Parameterization: possible biases due to parametrization of $f_i(\tilde{z}; k)$ have been checked by applying the minimization procedure on the same samples used to create the parametrization. Given the large statistics employed, the likelihood is conveniently approximated by χ^2. For any employed sample, the estimated $\hat{k}$ from the χ^2 minimization is in agreement with k within a percent.
- Energy Scale: the log-likelihood procedure has been tested with the 5-year synthetic data introducing different energy shifts s from 0.8 to 1.2, and applying to those samples the same minimization procedure described above. The result is exemplified in Figure 4, where the likelihood profile enables estimating $\hat{k}$ for two samples with $s = 0.8$ and $s = 1.2$. The corresponding estimated $\hat{k}$ values are, respectively, $0.84^{+0.09}_{-0.07}$ and $1.22^{+0.08}_{-0.08}$, in agreement with the imposed energy scale offsets.
- Spectral Shape: to check for a possible systematic effect due to the cosmic ray proton spectrum knowledge, data used to create the Moon's shadow parametrization have been weighted using a simple power law with index -2.7. The resulting likelihood function has been used to test the synthetic 5-year sample used before and generated

following Lipari's double-break spectral shape. The scale factor has been measured being $\hat{k} = 0.99^{+0.10}_{-0.15}$, with no sizeable impact coming from the different weighting.

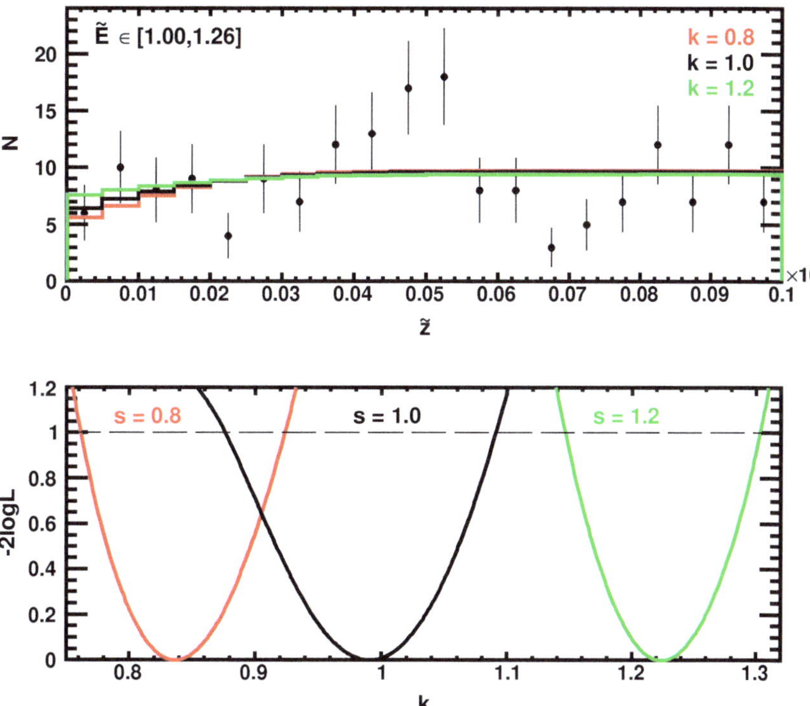

Figure 4. In the figure on the top is displayed the number of observed counts N^{obs} in the energy bin between 1 and 1.26 TeV compared with the expected number of counts N^{exp} derived for 3 values of energy scale bias, 0.8 (red), 1 (black), and 1.2 (green). On the bottom is presented the likelihood as a function of the energy scale k. In black is presented the dependency from k when no energy scale offset has been applied to the observed measurements, while in green and red are presented the likelihood dependencies in the case of two applied shifts of 0.8 and 1.2.

5. Discussion

The CR Moon's shadow position and shape depend on the cosmic ray charge, energy, and arrival direction and on the position of the measuring device, as well as on its angular and energetic resolution. In the case of our test device, the energy dependence of the Moon shadow can be used to measure the energy scale at the TeV scale. At higher energies, the effect is washed out since deflection becomes negligible with respect to angular resolution and the statistics become small. At low energy, the Moon's shadow becomes too shallow because of the energy resolution effect. In our simulation, the effect has been demonstrated to be sizeable and provided a positive result.

As stated in the introduction, while there is ample literature about the CR Moon's shadow use for pointing accuracy estimation and for matter-to-anti-matter distinction in indirect CR measurements [2–11,13–17,19,21–23], there is very little literature about the use of it for the energy scale determination or about the observation of the CR Moon's shadow with CR direct measurement experiments. This is because EAS experiments often study energies so high that there is no significant deflection to be exploited for our purposes, and, conversely, the direct CR measurement experiments study energies that are so low that the uncertainties in the Moon's position reconstruction dominate. Therefore, in this regard, this publication is exploring new possibilities that have been unveiled with the proposal of a new generation of CR direct measuring devices [27–29], with very ambitious goals in terms of energy reach and statistics collected.

We must remark that the results obtained are under the hypothesis of perfect knowledge of angular and energy resolution. In a more realistic case, those quantities may not be known. Deviations in the understanding of resolutions may cause biases in the energy scale evaluation. Various possibilities to understand the energy scale dependency from the resolution can be explored in future work: (a) studying the angular resolution using high-energy samples; (b) introducing the dependencies from variable angular and energy resolutions in the parameterization of the likelihood.

Another possible extension of the work may result from a better description of Earth's magnetic field. As discussed in Section 2, the employed IGRF-13 model describes what is known as the internal magnetic field of Earth. A weaker component, known as the external field, is produced by electrical currents in the ionosphere and can be dependent on solar activity that can have rapid variations in time. The external component is not included in our calculation since it significantly increases the computation time needed to perform the simulation. We expect a minor impact from this additional field on the deflection angle; however, in the case of application to real data, the external field should be included.

We remark that this work can be extended significantly by also employing the CR Sun's shadow. The Sun has roughly the same radial diameter of the Moon as observed from Earth's vicinity. The use of the Sun should approximately double the statistics available for the energy scale determination. However, a model of the interplanetary magnetic field between Earth and the Sun should be included and the computation investigated in more detail.

Eventually including other cosmic ray species in the study of the Moon's shadow is also of interest. Helium nuclei are quite abundant and exhibit much larger deflection than protons. The use of electrons instead could enable the study of the electromagnetic energy scale.

Other interesting extensions of this work may include the study of the energy scale time variation or its dependency on energy using restricted samples in time and energy. However, statistics is a main concern for those analyses, and only large energy scale variations with respect to the average can be detected.

6. Conclusions

The CR Moon's shadow has been employed to study the pointing accuracy, angular resolution, and derive antiproton-to-proton limits in CRs with ground and underground detectors measuring CRs indirectly. In this work, a simulation of a next-generation CR detector in low Earth orbit, similar to the wide-field-of-view calorimetric mission HERD, has been made to look for the Moon's shadow in CR directly from space. HERD is a large-field-of-view calorimeter whose purpose is to measure CR directly up to the knee. It has been estimated that, with a 5-year mission, it will be possible to observe the Moon's shadow in CRs with a HERD-like instrument and establish the energy scale for protons at the TeV scale at a 10% level. In general, this result demonstrates the feasibility of using the CR Moon's shadow for calibrating the proton absolute energy scale at the TeV scale for calorimetric space missions measuring CRs directly.

Funding: This research was funded under INFN and ASI under ASI-INFN agreements No. 2019-19-HH.0, its amendments, and No. 2021-43-HH.0.

Data Availability Statement: This work is based on the use of the publicly available resources for the calculation of the satellite coordinates (`Predict`: https://www.qsl.net/kd2bd/predict.html accessed 30 March 2020, and `CelesTrack`: https://celestrak.org/ accessed 22 September 2022) as for the geomagnetic field calculation (https://www.ngdc.noaa.gov/IAGA/vmod/igrf.html accessed 4 August 2021). The overall simulation and analysis have been developed in `C++` employing the `Boost` library (http://www.boost.org, version 1.82, accessed 1 August 2023) and the `ROOT` package (https://root.cern.ch/, version 6.28, accessed 1 August 2023).

Conflicts of Interest: The author declares no conflicts of interest.

Abbreviations

The following abbreviations are used extensively in this manuscript:

ASI	Agenzia Spaziale Italiana (Italian Space Agency)
CALO	HERD calorimeter
CR	Cosmic ray
CSS	Chinese Space Station
EAS	Extensive air showers
HERD	High-energy cosmic radiation detection
IACT	Imaging Atmospheric Cherenkov Telescope
IGRF	International Geomagnetic Reference Field
INFN	Istituto Nazionale di Fisica Nucleare (Italian National Institute of Nuclear Physics)

References

1. Clark, G.W. Arrival Directions of Cosmic-Ray Air Showers from the Northern Sky. *Phys. Rev.* **1957**, *108*, 450. [CrossRef]
2. Alexandreas, D.E.; Allen, R.C.; Berley, D.; Biller, S.D.; Burman, R.L.; Cady, D.R.; Chang, C.Y.; Dingus, B.L.; Dion, G.M.; Ellsworth, R.W.; et al. Observation of shadowing of ultrahigh-energy cosmic rays by the Moon and the Sun. *Phys. Rev. D* **1991**, *43*, 1735. [CrossRef]
3. Amenomori, M.; Cao, Z.; Ding, L.K.; Feng, Z.Y.; Hibino, K.; Hotta, N.; Huang, Q.; Huo, A.X.; Jia, H.Y.; Jiang, G.Z.; et al. Cosmic-ray deficit from the directions of the Moon and the Sun detected with the Tibet air-shower array. *Phys. Rev. D* **1993**, *47*, 2675. [CrossRef] [PubMed]
4. Borione, A.; Catanese, M.; Covault, C.E.; Cronin, J.W.; Fick, B.E.; Gibbs, K.G.; Green, K.D.; Hauptfeld, S.; Kieda, D.; Krimm, H.A.; et al. Observation of the shadows of the Moon and Sun using 100 TeV cosmic rays. *Phys. Rev. D* **1994**, *49*, 1171. [CrossRef]
5. Merck, M.; Karle, A.; Martinez, S.; Arqueros, F.; Becker, K.H.; Bott-Bodenhausen, M.; Eckmann, R.; Faleiro, E.; Fernandez, J.; Fernandez, P.; et al. Methods to determine the angular resolution of the HEGRA extended air shower scintillator array. *Astrop. Phys.* **1996**, *5*, 379–392. [CrossRef]
6. Amenomori, M.; Ayabe, S.; Bi, X.J.; Chen, D.; Cui, S.W.; Ding, L.K.; Ding, X.H.; Feng, C.F.; Feng, Z.; Feng, Z.Y.; et al. Moon shadow by cosmic rays under the influence of geomagnetic field and search for antiprotons at multi-TeV energies. *Astrop. Phys.* **2007**, *28*, 137–142. [CrossRef]
7. Bartoli, B.; Bernardini, P.; Bi, X.J.; Bleve, C.; Bolognino, I.; Branchini, P.; Budano, A.; Melcarne, A.K.C.; Camarri, P.; Cao, Z.; et al. Measurement of the cosmic ray antiproton/proton flux ratio at TeV energies with the ARGO-YBJ detector. *Phys. Rev. D* **2012**, *85*, 022002. [CrossRef]
8. Abeysekara, A.U.; Albert, A.; Alfaro, R.; Alvarez, C.; Álvarez, J.D.; Arceo, R.; Arteaga-Velázquez, J.C.; Rojas, D.A.; Solares, H.A.A.; Belmont-Moreno, E.; et al. Constraining the $\bar{p}/p$ ratio in TeV cosmic rays with observations of the Moon shadow by HAWC. *Phys. Rev. D* **2018**, *97*, 102005. [CrossRef]
9. Aharonian, F.; An, Q.; Axikegu; Bai, L.X.; Bai, Y.X.; Bao, Y.W.; Bastieri, D.; Bi, X.J.; Bi, Y.J.; Cai, H.; et al. Calibration of the air shower energy scale of the water and air Cherenkov techniques in the LHAASO experiment. *Phys. Rev. D* **2011**, *104*, 062007. [CrossRef]
10. Oshima, A.; Dugad, S.R.; Goswami, U.D.; Gupta, S.K.; Hayashi, Y.; Ito, N.; Iyer, A.; Jagadeesan, P.; Jain, A.; Kawakami, S.; et al. The angular resolution of the GRAPES-3 array from the shadows of the Moon and the Sun. *Astrop. Phys.* **2010**, *33*, 97–107. [CrossRef]
11. Pattanaik, D.; Ahmad, S.; Chakraborty, M.; Dugad, S.R.; Goswami, U.D.; Gupta, S.K.; Hariharan, B.; Hayashi, Y.; Jagadeesan, P.; Jain, A.; et al. Validating the improved angular resolution of the GRAPES-3 air shower array by observing the Moon shadow in cosmic rays. *Phys. Rev. D* **2022**, *106*, 022009. [CrossRef]
12. Ambrosio, M.; Antolini, R.; Aramo, C.; Auriemma, G.; Baldini, A.; Barbarino, G.C.; Barish, B.C.; Battistoni, G.; Bellotti, R.; Bemporad, C.; et al. Observation of the shadowing of cosmic rays by the Moon using a deep underground detector. *Phys. Rev. D* **1998**, *59*, 012003. [CrossRef]
13. Cobb, J.H.; Marshak, M.L.; Allison, W.W.M.; Alner, G.J.; Ayres, D.S.; Barrett, W.L.; Bode, C.; Border, P.M.; Brooks, C.B.; Cotton, R.J.; et al. Observation of a shadow of the Moon in the underground muon flux in the Soudan 2 detector. *Phys. Rev. D* **2000**, *61*, 092002. [CrossRef]
14. Achard, P.; Adriani, O.; Aguilar-Benitez, M.; Akker, M.v.; Alcaraz, J.; Alemanni, G.; Allaby, J.; Aloisio, A.; Alviggi, M.G.; Anderhub, H.; et al. Measurement of the shadowing of high-energy cosmic rays by the Moon: A search for TeV-energy antiprotons. *Astrop. Phys.* **2005**, *23*, 411–434. [CrossRef]
15. Adamson, P.; Andreopoulos, C.; Ayres, D.S.; Backhouse, C.; Barr, G.; Barrett, W.L.; Bishai, M.; Blake, A.; Bock, B.; Bock, G.J.; et al. Observation in the MINOS far detector of the shadowing of cosmic rays by the sun and moon. *Astrop. Phys.* **2011**, *34*, 457–466. [CrossRef]

16. Aartsen, M.G.; Abbasi, R.; Abdou, Y.; Ackermann, M.; Adams, J.; Aguilar, J.A.; Ahlers, M.; Altmann, D.; Auffenberg, J.; Bai, X.; et al. Observation of the cosmic-ray shadow of the Moon with IceCube. *Phys. Rev. D* **2014**, *89*, 102004. [CrossRef]
17. Albert, A.; André, M.; Anghinolfi, M.; Anton, G.; Ardid, M.; Aubert, J.-J.; Aublin, J.; Avgitas, T.; Baret, B.; Barrios-Martít, J.; et al. The cosmic ray shadow of the Moon observed with the ANTARES neutrino telescope. *Eur. Phys. J. C* **2018**, *78*, 1006. [CrossRef]
18. Urban, M.; Fleury, P.; Lestienne, R.; Plouin, F. Can we detect antimatter from other galaxies by the use of the Earth's magnetic field and the Moon as an absorber? *Nucl. Phys. B (Proc. Suppl.)* **1990**, *14B*, 223–236. [CrossRef]
19. Ambrosio, M.; Antolini, R.; Baldini, A.; Barbarino, G.C.; Barish, B.C.; Battistoni, G.; Becherini, Y.; Bellotti, R.; Bemporad, C.; Bernardini, P.; et al. Moon and Sun shadowing effect in the MACRO detector. *Astrop. Phys.* **2003**, *20*, 145–156. [CrossRef]
20. Bartoli, B.; Bernardini, P.; Bi, X.J.; Bleve, C.; Bolognino, I.; Branchini, P.; Budano, A.; Melcarne, A.K.C.; Camarri, P.; Cao, Z.; et al. Observation of the cosmic ray moon shadowing effect with the ARGO-YBJ experiment. *Phys. Rev. D* **2011**, *84*, 022003. [CrossRef]
21. Pomarède, D.; Boyle, P.J.; Urban, M.; Badran, H.M.; Behr, L.; Brunetti, M.T.; Fegan, D.J.; Weekes, T.C. Search for shadowing of primary cosmic radiation by the moon at TeV energies. *Astrop. Phys.* **2001**, *14*, 287–317. [CrossRef]
22. Colin, P.; Tridon, D.B.; Ortega, A.D.; Doert, M.; Doro, M.; Pochon, J.; Strah, N.; Suric, T.; Teshima, M. Probing the CR positron/electron ratio at few hundreds GeV through Moon shadow observation with the MAGIC telescopes. *arXiv Prepr.* **2011**, arXiv:1110.0183. [CrossRef]
23. Bird, R. Observing the Cosmic Ray Moon Shadow with VERITAS. *arXiv Prepr.* **2015**, arXiv:1508.07197. [CrossRef]
24. Aguilar, M.; Cavasonza, L.A.; Ambrosi, G.; Arruda, L.; Attig, N.; Barao, F.; Barrin, L.; Bartoloni, A.; Pree, S.B.; Bates, J.; et al. The Alpha Magnetic Spectrometer (AMS) on the international space station: Part II—Results from the first seven years. *Phys. Rep.* **2021**, *894*, 1–116. [CrossRef]
25. Torii, S. Highlights from the CALET observations for 7.5 years on the International Space Station. In Proceedings of the 38th International Cosmic Ray Conference, Nagoya, Japan, 26 July–3 August 2023; p. 2. [CrossRef]
26. Alemanno, F.; Altomare, C.; An, Q.; Azzarello, P.; Barbato, F.C.T.; Bernardini, P.; Bi, X.J.; Cagnoli, I.; Cai, M.S.; Casilli, E.; et al. DArk Matter Particle Explorer: 7 years in Space. In Proceedings of the 38th International Cosmic Ray Conference, Nagoya, Japan, 26 July–3 August 2023; p. 3. [CrossRef]
27. Gargano, F. The High Energy cosmic-Radiation Detection (HERD) facility on board the Chinese Space Station: Hunting for high-energy cosmic rays. In Proceedings of the 37th International Cosmic Ray Conference, Berlin, Germany, 15–22 July 2021; p. 26. [CrossRef]
28. Schael, S.; Atanasyan, A.; Berdugo, J.; Bretz, T.; Czupalla, M.; Dachwald, B.; von Doetinchem, P.; Duranti, M.; Gast, H.; Karpinski, W.; et al. AMS-100: The next generation magnetic spectrometer in space—An international science platform for physics and astrophysics at Lagrange point 2. *Nucl. Inst. Meth. Phys. Res. A* **2019**, *944*, 162561. [CrossRef]
29. Adriani, O.; Altomare, C.; Ambrosi, G.; Azzarello, P.; Barbato, F.C.T.; Battiston, R.; Baudouy, B.; Bergmann, B.; Berti, E.; Bertucci, B.; et al. Design of an Antimatter Large Acceptance Detector In Orbit (ALADInO). *Instruments* **2022**, *6*, 19. [CrossRef]
30. Asaoka, Y.; Akaike, Y.; Komiya, Y.; Miyata, R.; Torii, S.; Adriani, O.; Asano, K.; Bagliesi, M.G.; Bigongiari, G.; Binns, W.R.; et al. Energy calibration of CALET onboard the International Space Station. *Astrop. Phys.* **2017**, *91*, 1–10. [CrossRef]
31. Brogi, P.; Kobayashi, K. Measurement of the energy spectrum of cosmic-ray helium with CALET on the International Space Station. *Phys. Rev. Lett.* **2023**, *130*, 171002. [CrossRef]
32. Zang, J.; Yue, C.; Li, X. Measurement of absolute energy scale of ECAL of DAMPE with geomagnetic rigidity cutoff. In Proceedings of the 35th International Cosmic Ray Conference, Busan, Republic of Korea, 12–20 July 2017; p. 197. [CrossRef]
33. Oliva, A.; Altomare, C.; Ambrosi, G.; Barbanera, M.; Bertucci, B.; Cui, Y.X.; Duranti, M.; Formato, V.; Gong, K.; Graziani, M.; et al. The Silicon Charge Detector of the High Energy Cosmic Radiation Detection facility. In Proceedings of the 37th International Cosmic Ray Conference, Berlin, Germany, 15–22 July 2021; p. 87. [CrossRef]
34. Pacini, L.; Adriani, O.; Bai, Y.L.; Bao, T.W.; Berti, E.; Bottai, S.; Cao, W.W.; Casaus, J.; Cui, X.Z.; D'Alessandro, R.; et al. Design and expected performances of the large acceptance calorimeter for the HERD space mission. In Proceedings of the 37th International Cosmic Ray Conference, Berlin, Germany, 15–22 July 2021; p. 66. [CrossRef]
35. Meeus, J.H. *Astronomical Algorithms*, 2nd ed.; Willmann-Bell Inc.: Richmond, VA, USA, 1999.
36. Fehlberg, E. New High-Order Runge-Kutta Formulas with Step Size Control for Systems of First-and Second-Order Differential Equations. *Z. Angew. Math. Mech.* **1964**, *44*, T17–T19. [CrossRef]
37. Alken, P.; Thébault, E.; Beggan, C.D.; Amit, H.; Aubert, J.; Baerenzung, J.; Bondar, T.N.; Brown, W.J.; Califf, S.; Chambodut, A.; et al. International Geomagnetic Reference Field: The thirteenth generation. *Earth Planets Space* **2021**, *73*, 49. [CrossRef]
38. Valencia, M.; Giovacchini, F.; Oliva, A. Observation of $Z > 2$ trapped nuclei by AMS on ISS. In Proceedings of the 37th International Cosmic Ray Conference, Berlin, Germany, 15–22 July 2021; p. 1288. [CrossRef]
39. Sciascio, G.D.; Iuppa, R. Simulation of the cosmic ray Moon shadow in the geomagnetic field. *Nucl. Instr. Meth. Phys. Res. A* **2011**, *630*, 301–305. [CrossRef]
40. Lipari, P.; Vernetto, S. The shape of the cosmic ray proton spectrum. *Astrop. Phys.* **2020**, *120*, 102441. [CrossRef]

 *instruments*

Article

From SuperTIGER to TIGERISS

B. F. Rauch[1,*,†,‡], W. V. Zober[1,†,‡], Q. Abarr[2,†], Y. Akaike[3,†], W. R. Binns[1,†], R. F. Borda[4,‡], R. G. Bose[1,†,‡], T. J. Brandt[5,†], D. L. Braun[1,†], J. H. Buckley[1,†,‡], N. W. Cannady[4,6,7,†,‡], S. Coutu[8,9,‡], R. M. Crabill[10,†], P. F. Dowkontt[1,†], M. H. Israel[1,†], M. Kandula[11,‡], J. F. Krizmanic[6,†,‡], A. W. Labrador[10,†], W. Labrador[1,†,‡], L. Lisalda[1,†,‡], J. V. Martins[4,‡], M. P. McPherson[12,‡], R. A. Mewaldt[10,†], J. G. Mitchell[13,‡], J. W. Mitchell[6,†,‡], S. A. I. Mognet[8,9,‡], R. P. Murphy[14,†], G. A. de Nolfo[13,†,‡], S. Nutter[15,†,‡], M. A. Olevitch[1,†], N. E. Osborn[1,†,‡], I. M. Pastrana[1,‡], K. Sakai[16,†,‡], M. Sasaki[6,7,17,†,‡], S. Smith[12,‡], H. A. Tolentino[18,‡], N. E. Walsh[1,†], J. E. Ward[19,†], D. Washington[8,‡], A. T. West[20,†] and L. P. Williams[21,‡]

1 Department of Physics and McDonnell Center for the Space Sciences, Washington University in St. Louis, St. Louis, MO 63130, USA; wzober@wustl.edu (W.V.Z.); wrb@wustl.edu (W.R.B.); labrador.w@wustl.edu (W.L.); izabella.pastrana@wustl.edu (I.M.P.)
2 Department of Physics and Astronomy and The Bartol Research Institute, University of Delaware, Newark, DE 19716, USA; qabarr@udel.edu
3 Waseda Research Institute for Science and Engineering, Waseda University, Shinjuku City, Tokyo 169-8050, Japan
4 Department of Physics, University of Maryland, Baltimore County, Baltimore, MD 21250, USA; nicholas.w.cannady@nasa.gov (N.W.C.); martins@umbc.edu (J.V.M.)
5 SRON Netherlands Institute for Space Research, 2333 CA Leiden, The Netherlands; t.j.brandt@sron.nl
6 NASA Goddard Space Flight Center, Astrophysics Science Division, Greenbelt, MD 20771, USA; john.f.krizmanic@nasa.gov (J.F.K.)
7 Center for Research and Exploration in Space Sciences and Technology II, Greenbelt, MD 20771, USA
8 Department of Physics, Pennsylvania State University, State College, PA 16801, USA; sxc56@psu.edu (S.C.)
9 Institute for Gravitation and the Cosmos, Pennsylvania State University, State College, PA 16801, USA
10 Space Astrophysics and Space Radiation Laboratory, California Institute of Technology, Pasadena, CA 91125, USA
11 Space Coast Science, Engineering & Operations Group, KBR, Titusville, FL 32780, USA
12 Department of Mechanical Engineering, Howard University, Washington, DC 20059, USA
13 NASA Goddard Space Flight Center, Heliophysics Science Division, Greenbelt, MD 20771, USA; john.g.mitchell@nasa.gov (J.G.M.); georgia.a.denolfo@nasa.gov (G.A.d.N.)
14 National Academies of Sciences, Engineering, and Medicine, Washington, DC 20001, USA
15 Department of Physics and Geology, Northern Kentucky University, Highland Heights, KY 41099, USA
16 Enrico Fermi Institute, The University of Chicago, Chicago, IL 60637, USA; kenichisakai@uchicago.edu
17 Department of Astronomy, University of Maryland, College Park, MD 20742, USA
18 Department of Electrical Engineering and Computer Science, Howard University, Washington, DC 20059, USA; harrellangelo.tolen@bison.howard.edu
19 Spire Global, L-2763 Luxembourg-Ville, Luxembourg; john.ward@spire.com
20 Department of Astronomy and Steward Observatory, University of Arizona, Tucson, AZ 85719, USA
21 Electro-Mechanical & Systems Engineering Group, KBR, Greenbelt, MD 20771, USA; liam.williams@us.kbr.com
* Correspondence: brauch@physics.wustl.edu; Tel.: +1-314-935-6354
† These authors are SuperTIGER Collaborators.
‡ These authors are TIGERISS Collaborators.

Citation: Rauch, B.F.; Zober, W.V.; Abarr, Q.; Akaike, Y.; Binns, W.R.; Borda, R.F.; Bose, R.G.; Brandt, T.J.; Braun, D.L.; Buckley, J.H.; et al. From SuperTIGER to TIGERISS. *Instruments* 2024, 8, 4.
https://doi.org/10.3390/instruments8010004

Academic Editors: Matteo Duranti, Valerio Vagelli and Antonio Ereditato

Received: 16 October 2023
Revised: 21 November 2023
Accepted: 28 December 2023
Published: 11 January 2024

Abstract: The Trans-Iron Galactic Element Recorder (TIGER) family of instruments is optimized to measure the relative abundances of the rare, ultra-heavy galactic cosmic rays (UHGCRs) with atomic number (Z) $Z \geq 30$. Observing the UHGCRs places a premium on exposure that the balloon-borne SuperTIGER achieved with a large area detector (5.6 m^2) and two Antarctic flights totaling 87 days, while the smaller ($\sim$1 m^2) TIGER for the International Space Station (TIGERISS) aims to achieve this with a longer observation time from one to several years. SuperTIGER uses a combination of scintillator and Cherenkov detectors to determine charge and energy. TIGERISS will use silicon strip detectors (SSDs) instead of scintillators, with improved charge resolution, signal linearity, and dynamic range. Extended single-element resolution UHGCR measurements through $_{82}$Pb will cover elements produced in s-process and r-process neutron capture nucleosynthesis, adding to the multi-messenger effort to determine the relative contributions of supernovae (SNe) and Neutron Star Merger (NSM) events to the r-process nucleosynthesis product content of the galaxy.

Keywords: galactic cosmic rays; r-process; s-process; cosmic ray detectors; cosmic ray sources; high-altitude balloons; International Space Station

1. Introduction

Ultra-heavy galactic cosmic rays (UHGCRs) are the very rare nuclei above $_{28}$Ni produced in neutron capture nucleosynthesis, making them more than three orders of magnitude less abundant than those produced in stellar fusion. Measuring the UHGCRs requires the greatest possible detector exposure, which is proportional to detector area multiplied by observation time. The Super Trans-Iron Galactic Element Recorder (Super-TIGER) stratospheric balloon-borne instrument has made the best single-element resolution UHGCR measurements to date through $_{56}$Ba [1–4] with a large 5.6 m^2 detector on a record-breaking 55-day flight. The Trans-Iron Galactic Element Recorder for the International Space Station (TIGERISS) will improve upon these measurements and extend them through $_{82}$Pb [5,6], achieving comparable exposure in one year of observations following its planned 2026 launch with a $\sim$1 m^2 detector area. These measurements of the UHGCRs can address questions about the grand cycle of matter in the galaxy, depicted in Figure 1, in which material from galactic cosmic ray (GCR) sources (GCRSs) is injected into the accelerator. In a picture that has been pieced together from cosmic ray elemental and isotopic composition and energy spectra measurements, the GCRs then help energize galactic magnetic fields through their electric currents and feed back into the process of new star formation, leading to more GCRs. UHGCR measurements can provide the relative abundances of r- and s-process neutron capture elements in the GCRSs as well provide clues into how this material is accelerated to cosmic ray energies.

GCR measurements, including UHGCR abundances through $_{40}$Zr by TIGER and SuperTIGER, have implied a GCRS drawn primarily from older interstellar media (ISM) with fresh nucleosynthetic products of younger stars mixed in and acceleration by shock waves from stellar deaths. Supernovae (SNe) were long thought to be responsible for cosmic ray acceleration, and the r-process neutron capture nucleosynthesis of the heavier elements in the cycle is shown in Figure 1; however, recent evidence suggests that binary neutron star mergers (BNSMs) play a major role in r-process synthesis and may contribute to cosmic ray acceleration. Multi-messenger follow-up observations of a kilonovae identified in gravitational waves [7] provided broader electromagnetic spectral observations [8] that gave strong evidence for BNSM r-process nucleosynthesis of the heaviest elements. Extended SuperTIGER measurements providing the first single-element resolution UHGCR measurements through $_{56}$Ba show that something is missing from the GCRS model, supported by measurements through $_{40}$Zr. Superior UHGCR measurements by TIGERISS through $_{82}$Pb with unprecedented resolution will address important scientific questions about GCRSs and the cosmic ray accelerator, which are discussed in more detail in [9].

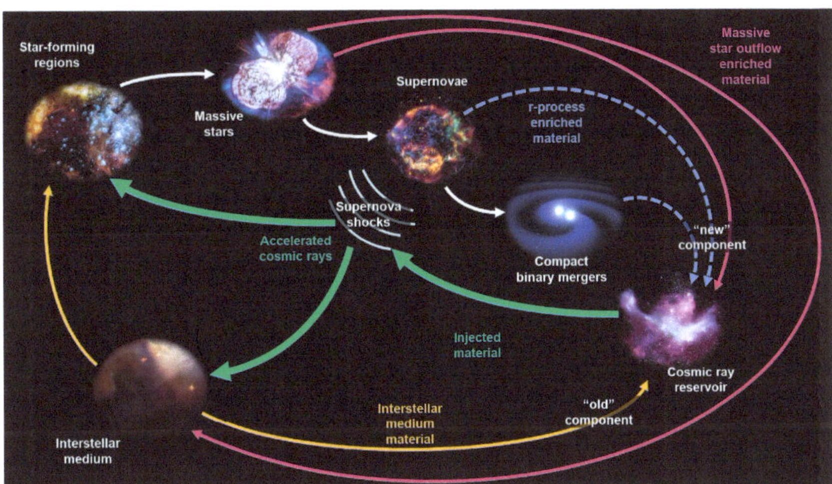

Figure 1. The grand cycle of galactic matter: massive star-forming regions give rise to SNe and NSMs, producing heavy nuclei that, along with ISM, are spread out into the galaxy by stellar winds and SN shocks.

No single instrument has been capable of measuring the GCRs from $_1$H to $_{92}$U, and their abundances must be pieced together using measurements made by multiple detectors. It is difficult to simultaneously measure the high flux of $_1$H and $_2$He that comprise ~99% of the GCRs with an instrument having the dynamic range and exposure needed to resolve the UHGCRs. Instruments like the CAlorimetric Electron Telescope (CALET) [10] and the Dark Matter Particle Explorer (DAMPE) [11] are capable of measuring abundances from $_1$H into the UHGCRs, but since they are not optimized for UHGCR measurements, they do not have the best resolution for them. Instruments designed to measure the GCRs above $_2$He can better optimize resolution and/or dynamic range for UHGCR measurements, including SuperTIGER ($16 \leq Z \leq 56$) [3], TIGERISS ($5 \leq Z \leq 82$), and the Advanced Composition Explorer Cosmic Ray Isotope Spectrometer (ACE-CRIS) ($6 \leq Z \leq 38$) [12,13], which has also made the only UHGCR isotope measurements through $_{38}$Sr. Measurements of the UHGCR abundances through $_{83}$Bi have been made by the the third High-Energy Astronomy Observatory (HEAO-3) Heavy Nuclei Experiment (HNE) [14] and by the Ariel 6 [15] satellite missions that could not resolve individual elements and measured charge groups. Passive nuclear track detectors that have measured UHGCR abundances for the heaviest elements ($Z \geq 70$) with better resolution include the TREK instrument flown on the Soviet Mir Space Station [16,17] and the Ultra-Heavy Cosmic Ray Experiment (UHCRE) at the Long Duration Exposure Facility (LDEF) [18].

2. SuperTIGER

SuperTIGER was designed to be the largest UHGCR detector that can be flown on a 39.9 million cubic foot (MCM) zero-pressure stratospheric balloon within the allowed launch envelope. The instrument was developed by a collaboration of scientists from Washington University in St. Louis (WUSTL), the National Aeronautics and Space Administration (NASA) Goddard Space Flight Center (GSFC), California Institute of Technology (Caltech), the NASA Jet Propulsion Laboratory (JPL), and the University of Minnesota. The University of Maryland Baltimore (UMBC) and Northern Kentucky University (NKU) have joined the effort under a later award, while the University of Minnesota has dropped out with the passing of Professor Cecil J. Waddington. SuperTIGER has had two successful Antarctic flights, the first for 55 days from 8 December 2012 to 1 February 2013 [19,20] and SuperTIGER-2.3 for 32 days from 15 December 2019 to 17 January 2020 [21], and a brief (~7 h), unsuccessful SuperTIGER-2.2 flight on 20 December 2018 [22]. It is the successor to the TIGER Long-Duration Balloon (LDB) payload that flew twice from Antarctica, first for 32 days from 21 December 2001 to 21 January 2002 [23,24] and again for 18 days from 17 December 2003 to 4 January 2004 [25,26]. TIGER LDB was in turn based on the original TIGER instrument that flew from Lynn Lake, Manitoba, Canada for 2.75 h on 26 August 1995 [27] and Fort Sumner, NM for 23.25 h on 25 September 1997 [28], demonstrating the instrument concept [29]. SuperTIGER uses the same two fundamental charge identification techniques demonstrated in TIGER: dE/dx vs. Cherenkov and acrylic Cherenkov vs. silica aerogel Cherenkov.

2.1. Instrument Design

Figure 2a shows a technical model of the full SuperTIGER-2.1 (2017–2018)/SuperTIGER-2.2 (2018–2019) payload comprising two instrument modules. SuperTIGER-2.3 had a 180-cell solar panel array instead of the pictured 160-cell array to support the four piggyback instruments it carried: the Advanced Particle-astrophysics Telescope (APT) prototype APTlite [30,31] and Balloon Air Sampler (BAS) [32] in addition to the Exposing Microorganisms in the Stratosphere (E-MIST) [33] and Polar Mesospheric Cloud Turbulence (PMC-Turbo) [34] pictured. An expanded view of an instrument module is shown in Figure 2b, with each module being a stack of seven detectors. Three large-area compact wavelength-shifter bar readout scintillator detectors (S1, S2, and S3) measure light production dependent on ionization energy losses ($dL/dx \propto dE/dx \propto Z^2$) and contribute to charge (Z) measurement, identification of interacting particles, and the instrument trigger. Top (H1) and bottom (H2) scintillating fiber hodoscopes provide trajectory determination for path length and areal response corrections. At the middle of the stack are two Cherenkov detectors that measure light production as a function of Z and velocity ($\beta = v/c$). Above is a silica aerogel detector (C0), with three quarters of the radiators having an index of refraction (n) n = 1.043 (KE $\gtrsim$ 2.5 GeV/amu) and one quarter n = 1.025 (KE $\gtrsim$ 3.3 GeV/amu); below is an acrylic detector (C1) with n = 1.49 (KE $\gtrsim$ 0.3 GeV/amu). The combined effective geometry factor of the SuperTIGER modules after accounting for interactions is 2.9 m^{2}sr, which is 7.2 times that of the preceding TIGER LDB instrument [35].

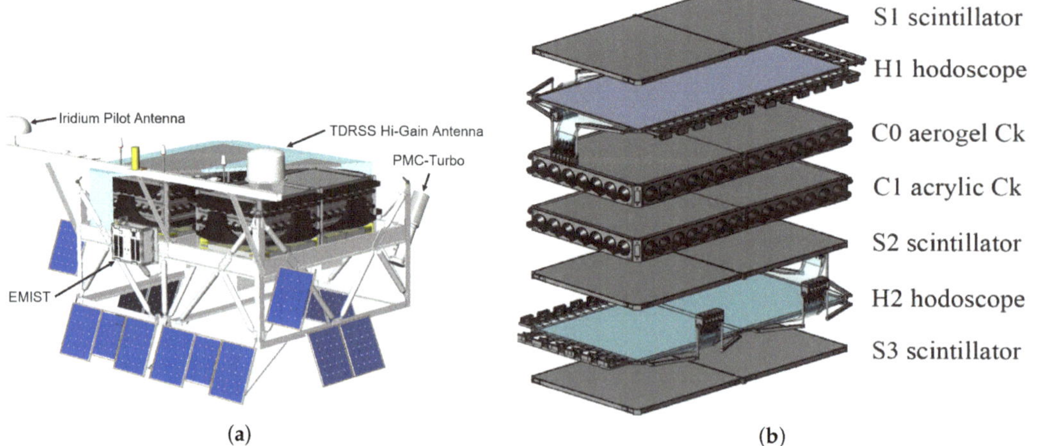

Figure 2. (a) Technical model of SuperTIGER-2.1/SuperTIGER-2.2. (b) Expanded view of a SuperTIGER module.

2.2. UHGCR Science

Figure 3a shows single-element resolution GCR abundance measurements at $\sim$2 GeV/amu through $_{56}$Ba [1,36–38] compared with Solar System (SS) abundances [39] through $_{82}$Pb, both normalized to $_{14}$Si = 1. The differences between GCR and SS abundances for the more abundant elements below $_{26}$Fe are understood to arise largely from spallation in GCR propagation from the source, a process that increases less abundant primary element abundances through erosion of more abundant ones. The GCR composition, and particularly that of the UHGCR elements not produced in stellar fusion, provides clues about the GCRS reservoirs and the acceleration mechanism.

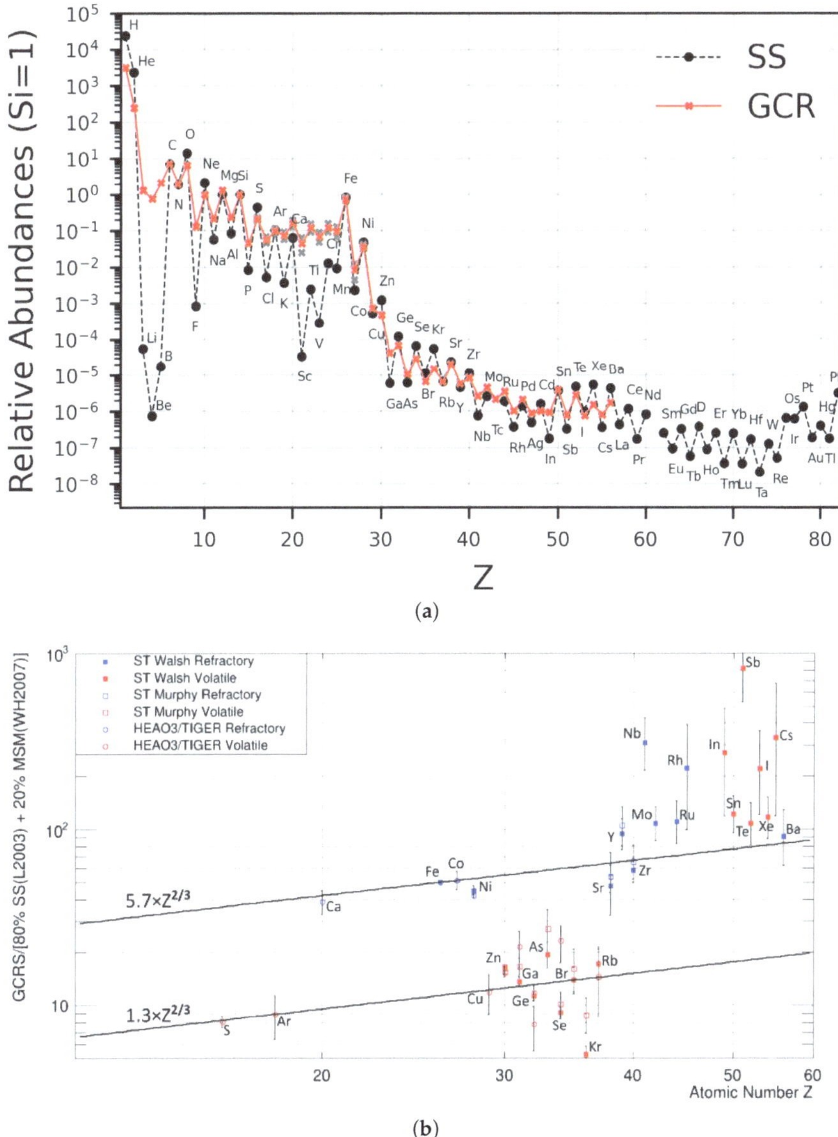

Figure 3. (**a**) SS [39] (dashed black line) and GCR at $\sim$2 GeV/amu (solid red line) relative abundances normalized to $_{14}$Si. GCR data sourced for $1 \leq Z \leq 2$ from [36], Z = 3 from [37], $4 \leq Z \leq 28$ from [38], and $16 \leq Z \leq 56$ from [1]. Gray dots depict overlapping measurements from [1,38]. (**b**) GCR measurements corrected for galactic propagation back to the source relative to a GCRS model of 80% SS [39] and 20% MSM [40] versus atomic number. Refractory elements (blue) and volatile elements (red). HEAO-3-C2 (Z $\leq$ 28) [38] and SuperTIGER (Z $\geq$ 26) [2–4,20] through $_{56}$Ba showing that the existing model is insufficient for elements above $_{40}$Zn.

TIGER made the first UHGCR measurements with single-element resolution through $_{40}$Zr [25,26], which supported a model of GCR origins with a major component from OB associations. In this model, the GCRS is composed of $\sim$80% ISM represented by SS material [39] and $\sim$20% massive star material (MSM) from OB associations, including stellar winds and SN ejecta [40]. Figure 3b shows the ratio of the GCR measurements corrected for galactic propagation to the GCRS model abundances as a function of Z, with refractory elements more likely to condense onto dust grains in blue and more volatile ones in red. GCR measurements through $_{40}$Zr fall around refractory and volatile lines, with the refractory elements being $\sim$4.4 times more abundant. The $Z^{2/3}$ slope is proportional to the nuclear cross section, which supports an acceleration model with preferential injection of elements that sputter off of superthermal dust grains [41]. SuperTIGER measurements through $_{40}$Zr [19,20] with greater statistics and improved resolution agreed with the TIGER results, but further SuperTIGER analysis pushing the UHGCR measurement through $_{56}$Ba [1–4] shows that the model breaks down above $_{40}$Zr. This hints at a potential new GCRS component, and TIGERISS will make measurements through $_{82}$Pb with superior charge reconstruction and resolution to search for new source signatures.

2.3. Future Prospects

SuperTIGER is mostly still on the high plateau in East Antarctica (71°7.53′ S, 158°35.10′ E, 6629 feet), with only a high-priority item recovery on January 21, 2020 and a data recovery on 6 November 2021. Full recovery of the payload has been delayed by the global COVID-19 pandemic, and it is now almost entirely drifted over. Recovery was initially planned for the 2022–2023 Antarctic season before being deferred to the 2023–2024 season due to limited support resource availability. With the uncertain future disposition of the payload and current backlog of Antarctic flight requests, SuperTIGER has no plans for future flights. Fortunately for the franchise, extended UHGCR analysis from the first record-breaking 55-day SuperTIGER flight hinting at new science supported a successful proposal for its successor instrument.

3. TIGERISS

TIGERISS is a UHGCR detector selected in the second round of the NASA Astrophysics Pioneers Program being developed for launch to the International Space Station (ISS) in 2026. This experiment will carry forward the UHGCR science of TIGER [26] and SuperTIGER [35] and seek an explanation for GCRS model-breaking SuperTIGER results. The TIGERISS collaboration, like the instrument, has also evolved from SuperTIGER, building on the core of WUSTL and NASA GSFC and later on UMBC and NKU additions with Pennsylvania State University (PSU) and Howard University.

TIGERISS will, in one year, measure the UHGCR abundances through $_{56}$Ba with comparable statistics to SuperTIGER, while having the extended dynamic range for the first preliminary single-element charge-resolution measurements through $_{82}$Pb by an active detector. Extended operations would allow TIGERISS to make more significant UHGCR measurements that will cover a wider range of elements produced in s-process and r-process neutron capture nucleosynthesis, adding to the multi-messenger effort to determine the relative contributions of SNe and Neutron Star Merger (NSM) events to r-process nucleosynthesis.

3.1. Instrument Concept

TIGERISS will use the same fundamental charge identification techniques used by TIGER/SuperTIGER: dE/dx vs. Cherenkov and acrylic Cherenkov vs. silica aerogel Cherenkov, as well as multiple dE/dx, but with improved detectors. Figure 4a gives an expanded view of the TIGERISS instrument stack, with pairs of orthogonal silicon strip detector (SSD) layers above and below the aerogel ($n = 1.05$, $\beta \geq 0.95$, KE $\gtrsim 2.12$ GeV/amu) and acrylic ($n = 1.49$, $\beta \geq 0.67$, KE $\gtrsim 325$ MeV/amu) Cherenkov light-collection boxes. Figure 4b shows an expanded view of an SSD layer, which will provide both dE/dx

measurements ($\propto Z^2$) and trajectory determination in place of the large-area compact wavelength-shifter bar readout scintillator detectors (dL/dx) and scintillating optical fiber hodoscopes (trajectory) used in the balloon-borne instruments. The more compact readout allowed by the SSDs and silicon photomultiplier (SiPM) modules TIGERISS will use on the Cherenkov detectors instead of photomultiplier tubes (PMTs) lets us build the largest possible instrument within the allowed payload envelope. An expanded view of a TIGERISS Cherenkov detector in Figure 4c shows that the Cherenkov-light radiators, in this case acrylic, will be at the top of the detector boxes to improve light collection over the bottom placement used in the balloon-borne instruments.

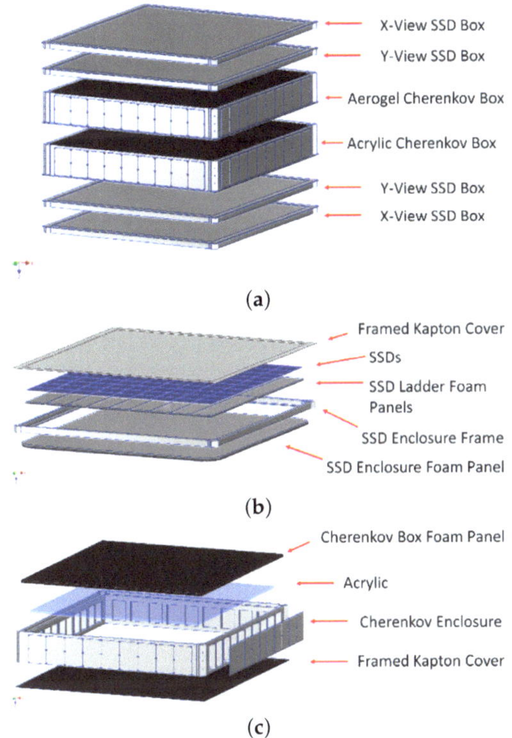

Figure 4. (**a**) Expanded view of the standard TIGERISS payload technical model. (**b**) SSD expanded view. (**c**) Acrylic Cherenkov detector expanded view.

3.2. Payload Model Development

There are similarities and major differences in the design requirements for balloon and space payloads. SuperTIGER was designed to operate in the very low atmospheric pressure at stratospheric altitudes, as well as to deal with major shocks in excess of 10 g experienced when the parachute opens following termination and on landing. TIGERISS will need to operate in hard vacuum, will experience shocks during launch, and will undergo acoustic and vibration loads that SuperTIGER did not. Analysis of TIGERISS detector component and payload models for launch environment conditions will be followed by some component model tests to address specific Technology Readiness Level (TRL) concerns, and ultimately by the full payload being put through thermal-vacuum, acoustic, and vibration tests.

All TIGERISS systems must meet TRL standards for launch and the ISS environment that exceed those of balloon payloads, and systems that are changed from SuperTIGER particularly benefit from heritage with other instruments. Silicon detectors have been used on many space missions, including ACE-CRIS [42], Light Imager for Gamma-ray

Astrophysics (AGILE) [43], Alpha Magnetic Spectrometer (AMS-02) [44], Energetic Particles: Acceleration, Composition, and Transport investigation (EPACT) on the Global Geospace Science (GGS) Wind satellite [45], Fermi-Large Area Telescope (Fermi-LAT) [46], Payload for Antimatter Matter Exploration and Light-nuclei Astrophysics (PAMELA) [47], Parker Solar Probe [48], and the Solar Terrestrial Relations Observatory (STEREO) [49]. TIGERISS will use daisy-chained detector ladders that are particularly similar to those used in AMS-02 [44] and Fermi-LAT [46]. TIGERISS SiPM components are similar to those used on two CubeSat missions, Ionospheric Neutron Content Analyzer (INCA) [50] and BurstCube [51], using carrier and summing electronics for SiPM arrays developed for APT [52] and the Antarctic Demonstrator for the Advanced Particle-astrophysics Telescope (ADAPT), Solar Neutron TRACking (SONTRAC) [53], and the High-Energy Light Isotope eXperiment (HELIX) [54]. TIGERISS will use a data acquisition (DAQ) system based on field-programmable gate arrays (FPGAs) based on that flown on the HyperAngular Rainbow Polarimeter (HARP) CubeSat [55] and in development for the HARP2 instrument on the Plankton, Aerosol, Clouds, ocean Ecosystem (PACE) mission [56].

The Japan Aerospace Exploration Agency (JAXA) Japanese Experiment Module (JEM) "Kibo" Exposed Facility Unit 10 (EFU10) location originally proposed for TIGERISS is now expected to be occupied by ECOsystem Spaceborne Thermal Radiometer Experiment on Space Station (ECOSTRESS) [57] when TIGERISS is planned to launch to the ISS in June 2026, and we were directed to investigate all possible ISS external payload accommodation sites. Until August 13, 2023, these included JEM-EFU6 and JEM-EFU7, as well as the European Space Agency (ESA) Columbus Laboratory external payload Starboard Overhead X-Direction (SOX) location. We have been notified by the ISS Program Office that the Global Ecosystem Dynamics Investigation (GEDI) payload [58] is planned until the end of the ISS for JEM-EFU6. None of the zenith-facing NASA EXpedite the PRocessing of Experiments to the Space Station (ExPRESS) Logistics Carrier (ELC) locations are expected to be available for TIGERISS. Detailed payload technical models for the SOX (Figure 5a) and JEM-EF (Figure 5b) locations are under development, including a standard JEM-EF model configuration and one 0.2 m wider for JEM-EFU7 that would require a JAXA waiver. Table 1 gives instrument dimensions and geometry factors for these models and the one used in the proposal.

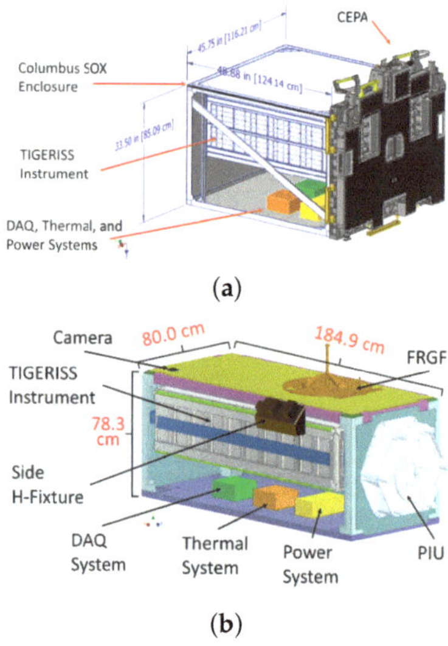

Figure 5. (a) Columbus SOX TIGERISS payload technical model. (b) JEM-EF standard TIGERISS payload technical model.

Table 1. TIGERISS instrument dimensions and geometry factors.

ISS Attachment	Length	Width	Height	Area	Geometry Factor
JEM-EF proposal	1.67 m	0.67 m	0.40 m	1.12 m^2	1.66 m^{2}sr
Columbus SOX	1.00 m	0.90 m	0.42 m	0.90 m^2	1.28 m^{2}sr
JEM-EF standard	1.50 m	0.60 m	0.42 m	0.90 m^2	1.19 m^{2}sr
JEM-EF wide	1.50 m	0.80 m	0.42 m	1.20 m^2	1.83 m^{2}sr

3.3. Thermal Analysis

The thermal environment on the ISS is significantly different than for stratospheric-balloon payloads. SuperTIGER was able to maintain all detector and electronics systems within acceptable temperature ranges with the use of insulation and thermostat-triggered heaters on the most sensitive electronics. It also used a rotator system to point the solar array toward the sun, which introduced a fixed thermal gradient from the hot to cold sides. The widely varying solar illumination and Earth albedo conditions TIGERISS will experience require both active heating and radiator heat dissipation.

TIGERISS thermal analysis efforts have been carrying both Columbus SOX and JEM-EF payload configurations. With the elimination of the JEM-EFU6 location with an active coolant loop, just the JEM-EFU7 and Columbus SOX locations remain, which only have passive thermal control and heaters. Integrated ISS thermal modeling for a range of orbital conditions has been performed, with a focus on hot and cold cases to assess radiator sizing and heater power budget needs. Figure 6a shows the TIGERISS SOX mechanical model, including thermal radiators mounted to Columbus Laboratory, and Figure 6b shows the payload as part of the Integrated ISS thermal model. The launch and orbital cases where limited power is available for survival heaters, as well as the up to seven hours without power during installation, are also being studied. Current modeling finds that expected thermal conditions will be within TIGERISS component tolerances and that heater power and radiator space needs are safely within limits. As with SuperTIGER, TIGERISS will correct for time-varying detector gain responses from changing temperatures by normalizing detector signals using $_{26}$Fe and/or other of the more abundant cosmic ray nuclei species.

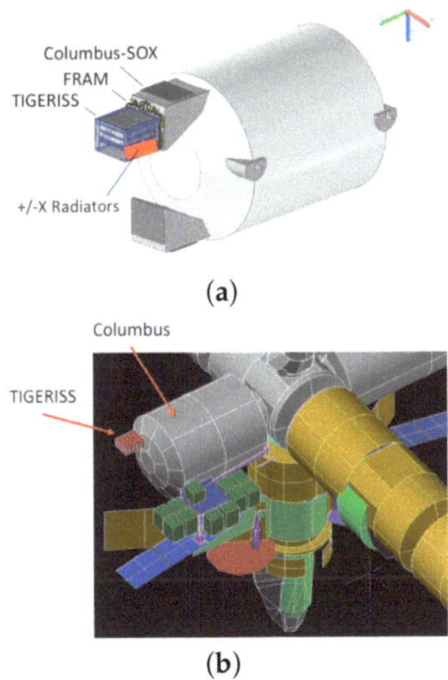

Figure 6. (**a**) Columbus SOX TIGERISS payload technical model showing radiators. (**b**) Columbus SOX TIGERISS payload thermal model.

3.4. Predicted TIGERISS Measurements

Predictions for TIGERISS event statistics incorporate cosmic ray spectra and corrections for geomagnetic screening, instrument thresholds, and interactions in the instrument based on a method originally developed for the CALET [59]. For elements from $_5$B to $_{32}$Ge, energy spectra have been measured by the ACE-CRIS at the L1 Lagrange Point [60]. For UHGCR elements for which energy spectra have not been measured, the $_{26}$Fe spectrum is scaled using SuperTIGER relative abundances for elements through $_{40}$Zr [20]. The predictions between $_{40}$Zr and $_{60}$Nd are based on the assumed 20% odd/80% even splitting of charge pairs measured by HEAO-3-HNE [14], which agree reasonably with the Super-TIGER measurements [2], and abundances of elements in charge groups above $_{60}$Nd are scaled by SS abundances [39]. The level of solar modulation does not have a strong impact on the TIGERISS UHGCR measurements due to significant geomagnetic screening in the ISS 51.6° inclination orbit.

3.4.1. Statistics from One Year

TIGERISS GCR statistics for ISS observations have been generated for the new instrument models under study [6]. Figure 7a gives predicted one-year TIGERISS measurements for the proposed JEM-EF model (pink), Columbus SOX model (black), current JEM-EF standard model (green), and JEM-EF wide model (blue) configurations [6] compared with those from the first SuperTIGER flight (red) [1–4]. The expected TIGERISS one-year statistics are comparable to or better than those for SuperTIGER where their sensitive ranges overlap.

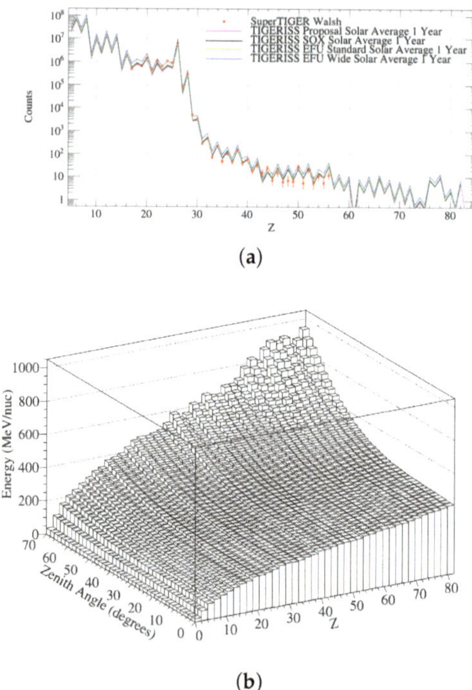

(**a**)

(**b**)

Figure 7. (**a**) Predicted abundances measured by TIGERISS after one year of operation [6] compared to those measured by SuperTIGER over its first 55-day long-duration balloon flight [1–4]. (**b**) Incident threshold energy (MeV/amu) required to trigger TIGERISS as a function of Z and zenith angle (θ) [6].

Table 1 shows that only the wide JEM-EF model has a larger geometry factor than the proposed TIGERISS instrument, but Figure 7a shows that all of the new models are expected to outperform it. Addressing subsystem interface requirements to constrain the mechanical model design envelopes for needed electronics, cabling, and thermal systems resulted in the standard JEM-EF instrument configuration in the proposal being downsized by 17 cm in length and 7 cm in width, as shown in Table 1. The superior performance of the newer models is due to the calculations used in the proposal only accepting events above a conservative energy threshold [61]. The current calculations [6] use the angle-dependent threshold energies derived for each element from Geant4 simulations, shown in Figure 7b. These results show that TIGERISS instrument models with higher confidence of design after the first year of development can deliver the scientific results promised in the proposal.

3.4.2. Statistics from Extended Observations

The ISS is now planned to operate through 2030, and if TIGERISS delivers as planned, its operations may be extended through the end of the ISS. Expected TIGERISS statistics from three years of observations under average solar modulation are shown in Figure 8 for the same payload configurations shown in Figure 7a. The increased UHGCR statistics from extended TIGERISS operations will resolve most even and many odd-Z elements, including the important $_{76}$Os, $_{78}$Pt, and $_{82}$Pb abundances, with greater statistical significance.

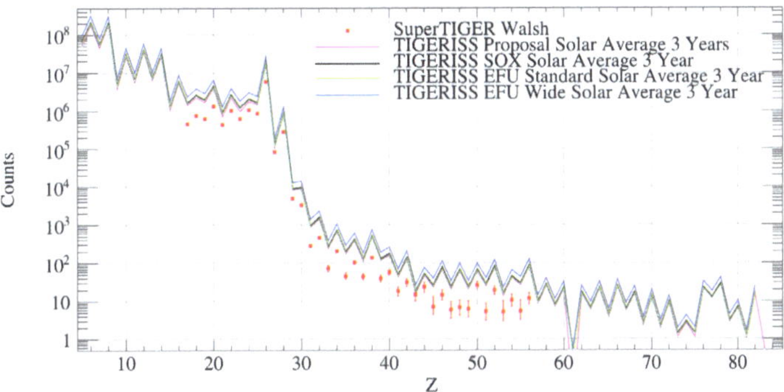

Figure 8. Predicted abundances measured by TIGERISS after three years of operation [5,6], compared to those measured by SuperTIGER over its first 55-day long-duration balloon flight [1–4].

4. Conclusions

The stratospheric balloon-borne SuperTIGER instrument has made the best single-element resolution UHGCR measurements to date through $_{56}$Ba; the TIGERISS instrument, with a planned 2026 launch, will extend these to $_{82}$Pb with superior resolution. Switching from scintillator detectors to SSDs for position and charge measurement will provide better charge resolution and linearity for TIGERISS, allowing it to measure all GCRs from $_5$B to $_{82}$Pb with a single instrument. SuperTIGER results have shown that there is something missing from the OB Association GCRS model, and TIGERISS will probe for other GCRS signatures and test GCR acceleration models through $_{82}$Pb. With the one year of observations possible under the five-year performance period of the Astrophysics Pioneers Program, TIGERISS will test SuperTIGER measurements with different systematics. If these measurements agree, they will effectively double the UHGCR single-element resolution statistics through $_{56}$Ba. Regardless, TIGERISS will provide the first single-element resolution UHGCR measurements from $_{56}$Ba to $_{82}$Pb, measuring further up the periodic table the relative contributions of r- and s-process neutron capture sources to the GCRs.

Author Contributions: All authors with † supported SuperTIGER, and all authors with ‡ have supported the conceptualization and design of TIGERISS. Conceptualization, L.P.W., J.V.M., B.F.R., W.V.Z., W.R.B., R.F.B., R.G.B., T.J.B., D.L.B., J.H.B., N.W.C., S.C., P.F.D., M.H.I., J.F.K., L.L., M.P.M., R.A.M., J.G.M., J.W.M., S.A.I.M., G.A.d.N., S.N., N.E.O., I.M.P., K.S., M.S., S.S., H.A.T., N.E.W., D.W.; methodology, L.P.W., B.F.R., W.V.Z., W.R.B., S.C., M.H.I., M.K., J.F.K., A.W.L., W.L., M.P.M., J.G.M.,

J.W.M., S.A.I.M., R.P.M., S.N., N.E.O., I.M.P., K.S., M.S., S.S., H.A.T., N.E.W., D.W.; software, L.P.W., J.V.M., B.F.R., W.V.Z., Y.A., R.F.B., R.G.B., N.W.C., J.F.K., A.W.L., W.L., S.A.I.M., R.P.M., S.N., M.A.O., N.E.O., I.M.P., K.S., M.S., H.A.T., N.E.W., D.W.; validation, L.P.W., J.V.M., R.M.C., B.F.R., W.V.Z., Q.A., R.F.B., R.G.B., D.L.B., P.F.D., M.K., W.L., L.L., M.P.M., J.G.M., S.A.I.M., R.P.M., S.N., M.A.O., N.E.O., I.M.P., M.S., S.S., H.A.T., N.E.W., D.W., A.T.W.; formal analysis, B.F.R., W.V.Z., N.W.C., S.C., M.K., J.F.K., A.W.L., W.L., M.P.M., R.A.M., S.A.I.M., R.P.M., S.N., N.E.O., K.S., M.S., S.S., H.A.T., N.E.W.; investigation, L.P.W., J.V.M., R.M.C., B.F.R., W.V.Z., Q.A., Y.A., W.R.B., R.F.B., R.G.B., T.J.B., D.L.B., J.H.B., N.W.C., S.C., P.F.D., M.H.I., M.K., J.F.K., A.W.L., W.L., L.L., M.P.M., R.A.M., J.G.M., J.W.M., S.A.I.M., R.P.M., S.N., M.A.O., N.E.O., I.M.P., K.S., M.S., S.S., H.A.T., N.E.W., J.E.W., D.W., A.T.W.; resources, J.V.M., R.M.C., B.F.R., W.R.B., J.H.B., N.W.C., S.C., J.F.K., L.L., R.A.M., J.G.M., J.W.M., S.A.I.M., G.A.d.N., S.N., S.S.; data curation, W.V.Z., N.W.C., J.F.K., M.A.O.; writing—original draft preparation, B.F.R., W.V.Z.; writing—review and editing, L.P.W., J.V.M., R.M.C., B.F.R., Q.A., Y.A., W.R.B., R.F.B., R.G.B., T.J.B., D.L.B., J.H.B., N.W.C., S.C., P.F.D., M.H.I., M.K., J.F.K., A.W.L., W.L., L.L., M.P.M., R.A.M., J.G.M., J.W.M., S.A.I.M., R.P.M., G.A.d.N., S.N., M.A.O., N.E.O., I.M.P., K.S., M.S., S.S., H.A.T., N.E.W., J.E.W., D.W., A.T.W.; visualization, L.P.W., B.F.R., W.V.Z., Y.A., W.R.B., D.L.B., N.W.C., S.C., J.F.K., W.L., M.P.M., S.A.I.M., R.P.M., S.N., N.E.O., K.S., M.S., H.A.T., N.E.W., J.E.W., D.W.; supervision, J.V.M., B.F.R., W.V.Z., W.R.B., R.F.B., R.G.B., T.J.B., J.H.B., N.W.C., S.C., P.F.D., M.H.I., M.K., J.F.K., L.L., R.A.M., J.W.M., R.P.M., G.A.d.N., S.N., M.S., S.S., N.E.W., J.E.W.; project administration, B.F.R., W.V.Z., W.R.B., N.W.C., S.C., J.F.K., L.L., R.A.M., J.W.M., G.A.d.N., S.N., M.S., S.S., J.E.W.; funding acquisition, J.V.M., B.F.R., W.V.Z., W.R.B., R.G.B., T.J.B., N.W.C., S.C., P.F.D., M.H.I., J.F.K., R.A.M., J.W.M., G.A.d.N., S.N., M.S., N.E.W. All authors have read and agreed to the published version of the manuscript.

Funding: SuperTIGER has been funded by NASA under grant numbers NNX09AC17G, NNX14AB25G, NNX15AC23G, and 80NSSC20K0405. TIGERISS is funded by NASA under cooperative agreement number 80NSSC22M0299.

Data Availability Statement: No new data were created or analyzed in this study. Data sharing is not applicable to this article.

Acknowledgments: SuperTIGER was supported by Cecil J. Waddington at the University of Minnesota until he passed on October 1, 2020. SuperTIGER has also been supported by M. E. Wiedenbeck. SuperTIGER and TIGERISS efforts at Washington University in St. Louis have also been supported by the McDonnell Center for the Space Sciences and the Peggy and Steve Fossett Foundation.

Conflicts of Interest: The authors declare no conflicts of interest.

Abbreviations

The following abbreviations are used in this manuscript:

MDPI	Multidisciplinary Digital Publishing Institute
ACE	Advanced Composition Explorer
ADAPT	Antarctic Demonstrator for the Advanced Particle-astrophysics Telescope
AGILE	Light Imager for Gamma-ray Astrophysics
AMS	Alpha Magnetic Spectrometer
APT	Advanced Particle-astrophysics Telescope
BAS	Balloon Air Sampler
BNSM	binary neutron star merger
Caltech	California Institute of Technology
CALET	CALorimetric Electron Telescope
COVID-19	coronavirus disease 2019
CR	cosmic ray
CRIS	Cosmic Ray Isotope Spectrometer
DAMPE	Dark Matter Particle Explorer
DAQ	data acquisition
EAS	extensive air shower
ECOSTRESS	ECOsystem Spaceborne Thermal Radiometer Experiment on Space Station

EFU	Exposed Facility Unit
ELC	ExPRESS Logistics Carrier
E-MIST	Exposing Microorganisms in the Stratosphere
EPACT	Energetic Particles: Acceleration, Composition, and Transport investigation
ESA	European Space Agency
ExPRESS	EXpedite the PRocessing of Experiments to the Space Station
FPGA	field-programmable gate array
GCR	galactic cosmic rays
GEDI	Global Ecosystem Dynamics Investigation
GGS	Global Geospace Science
HARP	HyperAngular Rainbow Polarimeter
HEAO	High-Energy Astronomy Observatory
HELIX	High-Energy Light Isotope eXperiment
HNE	Heavy Nuclei Experiment
INCA	Ionospheric Neutron Content Analyzer
ISM	interstellar media
ISS	International Space Station
JAXA	Japan Aerospace Exploration Agency
JEM	Japanese Experiment Module
LAT	Large-Area Telescope
LDB	Long-Duration Balloon
LDEF	Long-Duration Exposure Facility
NASA	National Aeronautics and Space Administration
NKU	Northern Kentucky University
NSM	Neutron Star Merger
PACE	Plankton, Aerosol, Clouds, ocean Ecosystem
PAMELA	Payload for Antimatter Matter Exploration and Light-nuclei Astrophysics
PMC-Turbo	Polar Mesospheric Cloud Turbulence
PMT	photomultiplier tube
PSU	Pennsylvania State University
SiPM	silicon photomultiplier
SN	supernova
SNe	supernovae
SONTRAC	Solar Neutron TRACking
SOX	Starboard Overhead X-Direction
SS	Solar System
SSD	silicon strip detector
STEREO	Solar Terrestrial Relations Observatory
SuperTIGER	Super Trans-Iron Galactic Element Recorder
TIGER	Trans-Iron Galactic Element Recorder
TIGERISS	Trans-Iron Galactic Element Recorder for the International Space Station
TRL	Technology Readiness Level
UHCRE	Ultra-Heavy Cosmic Ray Experiment
UHECR	ultra-high energy cosmic ray
UHGCR	ultra-heavy galactic cosmic ray
UMBC	University of Maryland Baltimore County
WUSTL	Washington University in St. Louis

References

1. Walsh, N.E. SuperTIGER Elemental Abundances for the Charge Range $41 \leq Z \leq 56$. Ph.D. Thesis, Washington University, St. Louis, MI, USA, 2020.
2. Walsh, N.E.; Akaike, Y.; Binns, W.; Bose, R.; Brandt, T.; Braun, D.; Cannady, N.; Dowkontt, P.; Hams, T.; Israel, M.; et al. SuperTIGER Abundances of Galactic Cosmic Rays for the Atomic Number (Z) Interval 30 to 56. In Proceedings of the 37th International Cosmic Ray Conference (ICRC2021), Berlin, Germany, 12–23 July 2021; Volume 395. [CrossRef]
3. Walsh, N.E.; Akaike, Y.; Binns, W.R.; Bose, R.G.; Brandt, T.J.; Braun, D.L.; Cannady, N.W.; Dowkontt, P.F.; Hams, T.; Israel, M.H.; et al. SuperTIGER instrument abundances of galactic cosmic rays for the charge interval $41 \leqslant Z \leqslant 56$. *Adv. Space Res.* **2022**, *70*, 2666–2673. [CrossRef]

4.   Walsh, N.E. SuperTIGER Abundances of Galactic Cosmic Rays for the Atomic Number (Z) Interval 40 to 56. In Proceedings of the 38th International Cosmic Ray Conference (ICRC2023), Nagoya, Japan, 26 July–3 August 2023; Volume 444, p. 053. [CrossRef]

5.   Rauch, B.F.; Zober, W.V.; Borda, R.F.; Bose, R.G.; Braun, D.L.; Buckley, J.; Calderon, J.; Cannady, N.W.; Caputo, R.; Coutu, S.; et al. The Trans-Iron Galactic Element Recorder for the International Space Station (TIGERISS). In Proceedings of the 38th International Cosmic Ray Conference (ICRC2023), Nagoya, Japan, 26 July–3 August 2023; Volume 444, p. 171. [CrossRef]

6.   Rauch, B.F.; Zober, W.V.; Borda, R.F.; Bose, R.G.; Braun, D.L.; Buckley, J.; Calderon, J.; Cannady, N.W.; Caputo, R.; Coutu, S.; et al. Modeling Expected TIGERISS Observations. In Proceedings of the 38th International Cosmic Ray Conference (ICRC2023), Nagoya, Japan, 26 July–3 August 2023; Volume 444, p. 172. [CrossRef]

7.   Abbott, B.P.; Abbott, R.; Abbott, T.D.; Acernese, F.; Ackley, K.; Adams, C.; Adams, T.; Addesso, P.; Adhikari, R.X.; Adya, V.B.; et al. GW170817: Observation of Gravitational Waves from a Binary Neutron Star Inspiral. *Phys. Rev. Lett.* **2017**, *119*, 161101. [CrossRef] [PubMed]

8.   Abbott, B.P.; Abbott, R.; Abbott, T.D.; Acernese, F.; Ackley, K.; Adams, C.; Adams, T.; Addesso, P.; Adhikari, R.X.; Adya, V.B.; et al. Multi-messenger Observations of a Binary Neutron Star Merger. *Astrophys. J. Lett.* **2017**, *848*, L12. [CrossRef]

9.   Zober, W.V.; Rauch, B.F.; Borda, R.F.; Bose, R.G.; Braun, D.L.; Buckley, J.; Calderon, J.; Cannady, N.W.; Caputo, R.; Coutu, S.; et al. Science Objectives and Goals of the TIGERISS mission. In Proceedings of the 38th International Cosmic Ray Conference (ICRC2023), Nagoya, Japan, 26 July–3 August 2023; Volume 444, p. 144. [CrossRef]

10.   Marrocchesi, P. CALET: A calorimeter-based orbital observatory for High Energy Astroparticle Physics. *Nucl. Instrum. Methods Phys. Res. Sect. A* **2012**, *692*, 240–245. [CrossRef]

11.   Sun, H.; Alemanno, F.; Altomare, C.; An, Q.; Azzarello, P.; Barbato, F.C.T.; Bernardini, P.; Bi, X.J.; Cagnoli, I.; Cai, M.S.; et al. Measurement of Heavy Nulei beyond Iron in Cosmic Rays with the DAMPE Experiment. In Proceedings of the 38th International Cosmic Ray Conference (ICRC2023), Nagoya, Japan, 26 July–3 August 2023; Volume 444, p. 174. [CrossRef]

12.   Israel, M.H.; Lave, K.A.; Wiedenbeck, M.E.; Binns, W.R.; Christian, E.R.; Cummings, A.C.; Davis, A.J.; de Nolfo, G.A.; Leske, R.A.; Mewaldt, R.A.; et al. Elemental Composition at the Cosmic-Ray Source Derived from the ACE-CRIS Instrument. I. $_6$C to $_{28}$Ni. *Astrophys. J.* **2018**, *865*, 69. [CrossRef]

13.   Binns, W.R.; Wiedenbeck, M.E.; von Rosenvinge, T.T.; Israel, M.H.; Christian, E.R.; Cummings, A.C.; de Nolfo, G.A.; Leske, R.A.; Mewaldt, R.A.; Stone, E.C. The Isotopic Abundances of Galactic Cosmic Rays with Atomic Number $29 \leq Z \leq 38$. *Astrophys. J.* **2022**, *936*, 13. [CrossRef]

14.   Binns, W.R.; Garrard, T.L.; Gibner, P.S.; Israel, M.H.; Kertzman, M.P.; Klarmann, J.; Newport, B.J.; Stone, E.C.; Waddington, C.J. Abundances of Ultraheavy Elements in the Cosmic Radiation: Results from HEAO 3. *Astrophys. J.* **1989**, *346*, 997. [CrossRef]

15.   Fowler, P.H.; Walker, R.N.F.; Masheder, M.R.W.; Moses, R.T.; Worley, A.; Gay, A.M. Ariel 6 Measurements of the Fluxes of Ultra-heavy Cosmic Rays. *Astrophys. J.* **1987**, *314*, 739. [CrossRef]

16.   Price, P.B.; Lowder, D.M.; Westphal, A.J.; Wilkes, R.D.; Brennen, R.A.; Afanasyev, V.G.; Akimov, V.V.; Rodin, V.G.; Baryshinikov, G.K.; Gorshkov, L.A.; et al. Trek-a Cosmic-Ray Experiment on the Russian Space Station MIR. *Astrophys. Space Sci.* **1992**, *197*, 121–143. [CrossRef]

17.   Westphal, A.J.; Price, P.B.; Weaver, B.A.; Afanasiev, V.G. Evidence against stellar chromospheric origin of Galactic cosmic rays. *Nature* **1998**, *396*, 50–52. [CrossRef]

18.   Donnelly, J.; Thompson, A.; O'Sullivan, D.; Daly, J.; Drury, L.; Domingo, V.; Wenzel, K.P. Actinide and Ultra-Heavy Abundances in the Local Galactic Cosmic Rays: An Analysis of the Results from the LDEF Ultra-Heavy Cosmic-Ray Experiment. *Astrophys. J.* **2012**, *747*, 40. [CrossRef]

19.   Murphy, R.P. Identifying the Origin of Galactic Cosmic Rays with the SuperTIGER Instrument. Ph.D. Thesis, Washington University, St. Louis, MI, USA, 2015.

20.   Murphy, R.P.; Sasaki, M.; Binns, W.R.; Brandt, T.J.; Hams, T.; Israel, M.H.; Labrador, A.W.; Link, J.T.; Mewaldt, R.A.; Mitchell, J.W.; et al. Galactic Cosmic Ray Origins and OB Associations: Evidence from SuperTIGER Observations of Elements $_{26}$Fe through $_{40}$Zr. *Astrophys. J.* **2016**, *831*, 148. [CrossRef]

21.   Rauch, B.F.; Walsh, N.E.; Zober, W.V. SuperTIGER Ultra-Heavy Galactic Cosmic Ray Atmospheric Propagation Corrections and Uncertainty Analysis. In Proceedings of the 37th International Cosmic Ray Conference (ICRC2021), Berlin, Germany, 12–23 July 2021; Volume 395, p. 089. [CrossRef]

22.   Rauch, B.F.; Bose, R.G.; West, A.T.; Lisalda, L.; Abarr, Q.; Akaike, Y.; Binns, W.; Brandt, T.; Braun, D.L.; Dowkontt, P.; et al. SuperTIGER-2 2018 Flight Payload Recovery and Preliminary Instrument Assessment. In Proceedings of the 36th International Cosmic Ray Conference (ICRC2019), Madison, WI, USA, 24 July–1 August 2019; Volume 358, p. 131. [CrossRef]

23.   Link, J.T. Measurements of Ultra-Heavy Galactic Cosmic Rays with the TIGER Instrument. Ph.D. Thesis, Washington University, St. Louis, MI, USA, 2003.

24.   Link, J.T.; Barbier, L.M.; Binns, W.R.; Christian, E.R.; Cummings, J.R.; de Nolfo, G.A.; Geier, S.; Israel, M.H.; Mewaldt, R.A.; Mitchell, J.W.; et al. Measurements of the Ultra-Heavy Galactic Cosmic-Ray Abundances between Z = 30 and Z = 40 with the TIGER Instrument. In Proceedings of the 28th International Cosmic Ray Conference, Tsukuba, Japan, 31 July–7 August 2003.

25.   Rauch, B.F. Measurement of the Relative Abundances of the Ultra-Heavy Galactic Cosmic Rays ($30 \leq Z \leq 40$) with the Trans-Iron Galactic Element Recorder (TIGER) Instrument. Ph.D. Thesis, Washington University, St. Louis, MI, USA, 2008.

26. Rauch, B.F.; Link, J.T.; Lodders, K.; Israel, M.H.; Barbier, L.M.; Binns, W.R.; Christian, E.R.; Cummings, J.R.; de Nolfo, G.A.; Geier, S.; et al. Cosmic Ray origin in OB Associations and Preferential Acceleration of Refractory Elements: Evidence from Abundances of Elements $_{26}$Fe through $_{34}$Se. *Astrophys. J.* **2009**, *697*, 2083–2088. [CrossRef]
27. Lawrence, D.J. The trans-iron galactic element recorder: A detector that will measure the elemental abundances of the ultra-heavy cosmic rays. Ph.D. Thesis, Washington University, Saint Louis, MI, USA, 1996.
28. Sposato, S.H. The 1997 balloon flight of the trans-iron galactic element recorder. Ph.D. Thesis, Washington University, St. Louis, MI, USA, 1999.
29. Lawrence, D.J.; Barbier, L.M.; Beatty, J.J.; Binns, W.R.; Christian, E.R.; Crary, D.J.; Ficenec, D.J.; Hink, P.L.; Klarmann, J.; Krombel, K.E.; et al. Large-area scintillating-fiber time-of-flight/hodoscope detectors for particle astrophysics experiments. *Nucl. Instrum. Methods Phys. Res. A* **1999**, *420*, 402–415. [CrossRef]
30. Hughes, Z.; Buckley, J.; Bergström, L.; Binns, W.; Buhler, J.; Chen, W.; Cherry, M.; Funk, S.; Hooper, D.; Mitchell, J.; et al. Report of 2019 APTlite balloon flight. In Proceedings of the 236th American Astronomical Society Meeting, online, 1–3 June 2020; Volume 236, p. 142.04.
31. Hughes, Z.D. Toward an Understanding of High-Mass Gamma-Ray Binaries: An Investigation Using Current Observatories and the Development of a Future GeV Instrument. Ph.D. Thesis, Washington University, St. Louis, MI, USA, 2021.
32. Meshik, A.; Kehm, K.; Pravdivtseva, O.; Rauch, B. Measurements of Atmospheric Noble Gases in Antarctica Captured by Autonomous Balloon Sampling. In Proceedings of the AGU Fall Meeting Abstracts, Chicago, IL, USA, 12–16 December 2022; Volume 2022, pp. P54A–02.
33. Smith, D.J.; Thakrar, P.J.; Bharrat, A.E.; Dokos, A.G.; Kinney, T.L.; James, L.M.; Lane, M.A.; Khodadad, C.L.; Maguire, F.; Maloney, P.R.; et al. A Balloon-Based Payload for Exposing Microorganisms in the Stratosphere (E-MIST). *Gravit. Space Res.* **2022**, *2*, 70–80. [CrossRef]
34. Williams, B.P.; Kjellstrand, B.; Jones, G.; Reimuller, J.D.; Fritts, D.C.; Miller, A.; Geach, C.; Limon, M.; Hanany, S.; Kaifler, B.; et al. The PMC-Turbo Balloon Mission to Study Gravity Waves and Turbulence through High-Resolution Imaging of Polar Mesospheric Clouds. In Proceedings of the AGU Fall Meeting Abstracts, New Orleans, LA, USA, 11–15 December 2017; Volume 2017, pp. SA24A–08.
35. Binns, W.R.; Bose, R.G.; Braun, D.L.; Brandt, T.J.; Daniels, W.M.; Dowkontt, P.F.; Fitzsimmons, S.P.; Hahne, D.J.; Hams, T.; Israel, M.H.; et al. The SUPERTIGER Instrument: Measurement of Elemental Abundances of Ultra-Heavy Galactic Cosmic Rays. *Astrophys. J.* **2014**, *788*, 18. [CrossRef]
36. Sanuki, T.; Motoki, M.; Matsumoto, H.; Seo, E.S.; Wang, J.Z.; Abe, K.; Anraku, K.; Asaoka, Y.; Fujikawa, M.; Imori, M.; et al. Precise Measurement of Cosmic-Ray Proton and Helium Spectra with the BESS Spectrometer. *Astrophys. J.* **2000**, *545*, 1135–1142. [CrossRef]
37. Aguilar, M.; Alcaraz, J.; Allaby, J.; Alpat, B.; Ambrosi, G.; Anderhub, H.; Ao, L.; Arefiev, A.; Arruda, L.; Azzarello, P.; et al. Isotopic Composition of Light Nuclei in Cosmic Rays: Results from AMS-01. *Astrophys. J.* **2011**, *736*, 105, [CrossRef]
38. Engelmann, J.J.; Ferrando, P.; Soutoul, A.; Goret, P.; Juliusson, E.; Koch-Miramond, L.; Lund, N.; Masse, P.; Peters, B.; Petrou, N.; et al. Charge composition and energy spectra of cosmic-ray nuclei for elements from Be to Ni - Results from HEAO-3-C2. *Astron. Astrophys.* **1990**, *233*, 96–111.
39. Lodders, K. Solar System Abundances and Condensation Temperatures of the Elements. *Astrophys. J.* **2003**, *591*, 1220–1247. [CrossRef]
40. Woosley, S.E.; Heger, A. Nucleosynthesis and remnants in massive stars of solar metallicity. *Phys. Rep.* **2007**, *442*, 269–283. [CrossRef]
41. Lingenfelter, R.E. The Origin of Cosmic Rays: How Their Composition Defines Their Sources and Sites and the Processes of Their Mixing, Injection, and Acceleration. *Astrophys. J. Suppl. Ser.* **2019**, *245*, 30. [CrossRef]
42. Stone, E.C.; Cohen, C.M.S.; Cook, W.R.; Cummings, A.C.; Gauld, B.; Kecman, B.; Leske, R.A.; Mewaldt, R.A.; Thayer, M.R.; Dougherty, B.L.; et al. The Cosmic-Ray Isotope Spectrometer for the Advanced Composition Explorer. *Space Sci. Rev.* **1998**, *86*, 285–356. [CrossRef]
43. Prest, M.; Barbiellini, G.; Bordignon, G.; Fedel, G.; Liello, F.; Longo, F.; Pontoni, C.; Vallazza, E. The AGILE silicon tracker: An innovative γ-ray instrument for space. *Nucl. Instrum. Methods Phys. Res. A* **2003**, *501*, 280–287. [CrossRef]
44. Rapin, D.; AMS-Tracker Collaboration. The AMS-02 silicon tracker: First year on ISS in space. *Nucl. Instrum. Methods Phys. Res. A* **2013**, *718*, 524–526. [CrossRef]
45. von Rosenvinge, T.T.; Barbier, L.M.; Karsch, J.; Liberman, R.; Madden, M.P.; Nolan, T.; Reames, D.V.; Ryan, L.; Singh, S.; Trexel, H.; et al. The Energetic Particles: Acceleration, Composition, and Transport (EPACT) investigation on the WIND spacecraft. *Space Sci. Rev.* **1995**, *71*, 155–206. [CrossRef]
46. Atwood, W.B.; Abdo, A.A.; Ackermann, M.; Althouse, W.; Anderson, B.; Axelsson, M.; Baldini, L.; Ballet, J.; Band, D.L.; Barbiellini, G.; et al. The Large Area Telescope on the Fermi Gamma-Ray Space Telescope Mission. *Astrophys. J.* **2009**, *697*, 1071–1102. [CrossRef]
47. Straulino, S.; Adriani, O.; Bonechi, L.; Bongi, M.; Castellini, G.; D'Alessandro, R.; Gabbanini, A.; Grandi, M.; Papini, P.; Ricciarini, S.; et al. The PAMELA silicon tracker. *Nucl. Instrum. Methods Phys. Res. A* **2004**, *530*, 168–172. [CrossRef]

48. Wiedenbeck, M.E.; Burnham, J.A.; Cohen, C.M.S.; Cook, W.R.; Crabill, R.M.; Cummings, A.C.; Davis, A.J.; Kecman, B.; Labrador, A.W.; Leske, R.A.; et al. Thin silicon solid-state detectors for energetic particle measurements-Development, characterization, and application on NASA's Parker Solar Probe mission. *Astron. Astrophys.* **2021**, *650*, A27. [CrossRef]
49. Mewaldt, R.A.; Cohen, C.M.S.; Cook, W.R.; Cummings, A.C.; Davis, A.J.; Geier, S.; Kecman, B.; Klemic, J.; Labrador, A.W.; Leske, R.A.; et al. The Low-Energy Telescope (LET) and SEP Central Electronics for the STEREO Mission. *Space Sci. Rev.* **2008**, *136*, 285–362. [CrossRef]
50. Mitchell, J.G.; Bruno, A. Performance Characteristics of the Ionospheric Neutron Content Analyzer (INCA). In Proceedings of the 36th International Cosmic Ray Conference (ICRC2019), Madison, WI, USA, 24 July–1 August 2019; Volume 36, p. 53. [CrossRef]
51. Perkins, J.S.; Brewer, I.; Briggs, M.S.; Bruno, A.; Burns, E.; Caputo, R.; Cenko, B.; Cucchiara, A.; De Nolfo, G.; Dumonthier, J.; et al. BurstCube: A CubeSat for gravitational wave counterparts. *Proc. SPIE* **2020**, *11444*, [CrossRef]
52. Buckley, J.; Alnussirat, S.; Altomare, C.; Bose, R.G.; Braun, D.L.; Buckley, J.H.; Buhler, J.; Burns, E.; Chamberlain, R.D.; Chen, W.; et al. The Advanced Particle-astrophysics Telescope (APT) Project Status. In Proceedings of the 37th International Cosmic Ray Conference (ICRC2021), Berlin, Germany, 12–23 July 2021; Volume 395, p. 655. [CrossRef]
53. de Nolfo, G.A.; Mitchell, J.G.; Suarez, G.; Ryan, J.M.; Bruno, A.; Dumonthier, J.; Legere, J.; Messner, R.; Tatoli, T.; Williams, L. Next-generation SOlar Neutron TRACking (SONTRAC) instrument. *Nucl. Instrum. Methods Phys. Res. A* **2023**, *1054*, 168352. [CrossRef]
54. Jeon, H. The Design and Status of the HELIX Ring Imaging Cherenkov Detector and Hodoscope Systems. In Proceedings of the 38th International Cosmic Ray Conference (ICRC2023), Nagoya, Japan, 26 July–3 August 2023; Volume 444, p. 121. [CrossRef]
55. Martins, J.V.; Fernandez-Borda, R.; McBride, B.; Remer, L.; Barbosa, H.M.J. The Harp Hype Ran Gular Imaging Polarimeter and the Need for Small Satellite Payloads with High Science Payoff for Earth Science Remote Sensing. In Proceedings of the 2018 IEEE International Geoscience and Remote Sensing Symposium (IGARSS 2018), Valencia, Spain, 22–27 July 2018; pp. 6304–6307. [CrossRef]
56. Remer, L.A.; Knobelspiesse, K.; Zhai, P.W.; Xu, F.; Kalashnikova, O.V.; Chowdhary, J.; Hasekamp, O.; Dubovik, O.; Wu, L.; Ahmad, Z.; et al. Retrieving Aerosol Characteristics From the PACE Mission, Part 2: Multi-Angle and Polarimetry. *Front. Environ. Sci.* **2019**, *7*. [CrossRef]
57. Hulley, G.; Hook, S.; Fisher, J.; Lee, C. ECOSTRESS, A NASA Earth-Ventures Instrument for studying links between the water cycle and plant health over the diurnal cycle. In Proceedings of the 2017 IEEE International Geoscience and Remote Sensing Symposium (IGARSS), Fort Worth, TX, USA, 23–28 July 2017; pp. 5494–5496. [CrossRef]
58. Dubayah, R.; Blair, J.B.; Goetz, S.; Fatoyinbo, L.; Hansen, M.; Healey, S.; Hofton, M.; Hurtt, G.; Kellner, J.; Luthcke, S.; et al. The Global Ecosystem Dynamics Investigation: High-resolution laser ranging of the Earth's forests and topography. *Sci. Remote Sens.* **2020**, *1*, 100002. [CrossRef]
59. B. F. Rauch for the CALET Collaboration. Predicted CALET measurements of ultra-heavy cosmic ray relative abundances. *Adv. Space Res.* **2014**, *53*, 1444–1450. [CrossRef]
60. George, J.S.; Lave, K.A.; Wiedenbeck, M.E.; Binns, W.R.; Cummings, A.C.; Davis, A.J.; de Nolfo, G.A.; Hink, P.L.; Israel, M.H.; Leske, R.A.; et al. Elemental Composition and Energy Spectra of Galactic Cosmic Rays During Solar Cycle 23. *Astrophys. J.* **2009**, *698*, 1666–1681. [CrossRef]
61. Rauch, B.F.; Walsh, N.E.; Zober, W.V. Determination of Expected TIGERISS Observations. In Proceedings of the 37th International Cosmic Ray Conference (ICRC2021), Berlin, Germany, 12–23 July 2021; Volume 395, p. 088. [CrossRef]

 *instruments*

Article

Antideuteron Identification in Space with Helium Calorimeter

Francesco Nozzoli [1,2,*], **Irina Rashevskaya** [1], **Leonardo Ricci** [2], **Francesco Rossi** [1,2], **Piero Spinnato** [1], **Enrico Verroi** [1], **Paolo Zuccon** [1,2] **and Gregorio Giovanazzi** [2]

1 TIFPA-INFN, Via Sommarive 14, I-38123 Trento, Italy; francesco.rossi-9@unitn.it (F.R.); piero.spinnato@tifpa.infn.it (P.S.)
2 Physics Department, University of Trento, Via Sommarive 14, I-38123 Trento, Italy; leonardo.ricci@unitn.it (L.R.); gregorio.giovanazzi@unitn.it (G.G.)
* Correspondence: francesco.nozzoli@unitn.it

Abstract: The search for low-energy antideuterons in cosmic rays allows the addressing of fundamental physics problems testing for the presence of primordial antimatter and the nature of Dark Matter. The PHeSCAMI (Pressurized Helium Scintillating Calorimeter for AntiMatter Identification) project aims to exploit the long-living metastable states of the helium target for the identification of low-energy antideuterons in cosmic rays. A space-based pressurized helium calorimeter would provide a characteristic identification signature based on the coincident detection of a prompt scintillation signal emitted by the antideuteron energy loss during the slowing-down phase in the gas, and the ($\approx$μs) delayed scintillation signal provided by the charged pions produced in the subsequent annihilation. The performance of a high-pressure (200-bar) helium scintillator prototype, tested in the INFN-TIFPA laboratory, will be summarized.

Keywords: antimatter; dark matter; annihilation; helium scintillator; metastable states

Citation: Nozzoli, F.; Rashevskaya, I.; Ricci, L.; Rossi, F.; Spinnato, P.; Verroi, E.; Zuccon, P.; Giovanazzi, G. Antideuteron Identification in Space with Helium Calorimeter. *Instruments* **2024**, *8*, 3. https://doi.org/10.3390/instruments8010003

Academic Editor: Antonio Ereditato

Received: 15 October 2023
Revised: 27 December 2023
Accepted: 4 January 2024
Published: 6 January 2024

1. Introduction

The presence of low-energy antideuterons $\bar{d}$ in cosmic rays is considered to be a golden channel for the identification of Dark-Matter annihilations in the galaxy. The expected astrophysical background due to secondary antinuclei produced by high-energy protons colliding with the interstellar medium is kinematically suppressed for kinetic energies below a few GeV/n. Thus, the search for a rare component of low-energy antinuclei in cosmic rays allows testing for the presence of primordial antimatter and the nature of Dark Matter [1,2].

The AMS02 magnetic spectrometer is currently the most sensitive experiment for antinuclei search in cosmic rays. However, AMS02 cannot efficiently explore the sub-GeV region with the mass reconstruction based on the particle time of flight. The current $\bar{d}$ search of AMS02 in the [2–3.8] GeV/n region with the powerful identification technique based on the ring imaging Cherenkov detector provides $\approx$7 candidates in the $\bar{d}$ mass region, where a few events are expected due to $\bar{p}$ background [3]. Regarding antihelium, conversely, the AMS02 spectrometer provided a tantalizing hint for an unexpected presence of antihelium in cosmic rays; $\approx$10 events are reported in the rigidity region from −40 GV to −15 GV with a mass compatible with antihelium. A careful study of all the hypothetical systematics on the evaluation of the expected, negligible background for these events is still ongoing [4].

The GAPS balloon experiment develops a different signature with respect to existing and past magnetic spectrometers, where stopping antinuclei will form an exotic atom whose characteristic X-rays should be detected to identify the antiparticle mass [5,6]. The first flight of the GAPS balloon is scheduled for the austral summer of 2024–2025. The GAPS experiment will explore $\bar{d}$ in the kinetic energy region 100–250 MeV/n. To pursue the peculiar X-ray signature, the GAPS collaboration developed 2.5 mm thick Si(Li) detectors with 1 keV resolution but also with a very large dynamic range (10 keV–100 MeV). The 1440 Si(Li) sensors of the GAPS tracker will be cooled to $\approx -40\,°C$ thanks to a large oscillating heat pipe cooling system. A challenging task of the GAPS experiment is to reduce the very large ($\approx$MHz) particle rate down to a (still quite large) >50 kHz trigger rate [5]. A ToF-based trigger system should be able to identify and reject "on the fly" most of the incoming p and He nuclei. This is one of the main difficulties also for the trigger strategy of future large space-based spectrometers like ALADInO [7] or AMS100 [8].

The innovative detection approach for $\bar{d}$, developed for the PHeSCAMI (Pressurized Helium Scintillating Calorimeter for AntiMatter Identification) project, will allow a relatively simple trigger strategy and provide an additional identification signature for $\bar{d}$ in helium gas in a room temperature detector.

2. Metastable States in Helium

The typical lifetime for stopping antinuclei in matter is of the order of picoseconds. However, since 1991, the existence of long-living ($\approx\mu$s) metastable states for stopping $\bar{p}$ in helium targets has been measured [9–11]. These metastable states in helium have also been measured for other heavy negative particles such as negative pions and kaons [12,13]. The theoretical description of the effect [11,14–17] predicts that the metastable state lifetimes increase as the reduced mass squared of the exotic atom [18]. Thus, a slightly larger delay is expected for $\bar{d}$ captured in helium as compared to the $\bar{p}$ case. The antiprotonic-helium metastable states are well understood, and their existence is already exploited for other fundamental physics measurements like the antiproton-to-electron-mass ratio at the CERN Antiproton Decelerator [19].

The phenomenology for the formation of metastable states in helium can be summarized following the scheme of Figure 1.

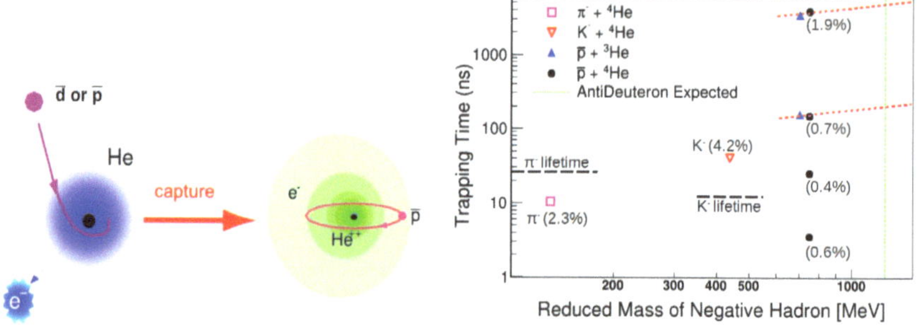

Figure 1. [**Left Panel**] Slow $\bar{p}$ or $\bar{d}$ (but also π^- and K^-) can be captured by He and trapped in (μs living) metastable states. [**Right Panel**] Summary of measured trapping probability and trapping time for different mass of negative hadrons [12,18].

An exotic metastable atom can be produced when $\bar{p}$ or $\bar{d}$ are stopping near an ordinary helium atom. In this case, a capture of the antiparticle from the helium nucleus is possible, and the atom spontaneously removes one of its two electrons. The antiparticle typically is captured in a state with a large principal and angular momentum quantum numbers ($n\sim 38$ for $\bar{p}$). Because of the large mass ratio, the orbits of the antinuclei are smaller with respect to the typical size of the electron orbits. However, the annihilation is suppressed by the relatively large principal and angular momentum quantum numbers that imply a small superposition of nucleus-antinucleus wavefunctions. The annihilation probability increases when the bound system de-excites towards the fundamental level. For the He target, the remaining electron cannot provide the (fast) Auger de-excitation, i.e., the main de-excitation process for the $Z > 2$ antiprotonic atoms. On the other hand, the relatively large size of the orbit of the single remaining electron also suppresses the Stark collisional de-excitation of the inner antiparticle (that is the main de-excitation process for the Protonium, $\bar{p} - p$, naked system). Thus, for antiprotonic-He, the (slow) radiative channel is the main remaining de-excitation process. This metastability is a unique (and well-measured) feature for the He target that is not expected/observed for other target nuclei [10]. The captured antiproton/antideuteron can thus orbit the He nucleus for a few microseconds before annihilating, providing a few charged pion tracks. This process is expected to happen in the case of a few percent of the captures, and this characteristic delayed annihilation signal in He is a distinctive signature to identify the antimatter nature of the stopping particle that can be used to detect antideuterons in space.

Measurements for liquid and gas helium targets [9–11] have shown that about 3.5% of the antiproton annihilations are delayed in several decay components from a few ns to a few μs. A simplified model with two decay components λ_{fast} and λ_{slow} can be adopted to roughly describe the time distribution of delayed annihilations:

$$n(t) = A\left(F\,\lambda_{fast}e^{-\lambda_{fast}t} + (1-F)\lambda_{slow}e^{-\lambda_{slow}t}\right) \tag{1}$$

where $A \approx 3.5\%$ is the fraction of delayed annihilation and F is the fraction of the *"fast"* annihilation component.

In Figure 2, experimental measurements of these parameters at different temperatures and pressures for antiprotonic-He are shown as a function of the helium density. The measurements for highly pressurized helium gas are scarce. However, in principle, handling helium gas at a pressure of 400 bar at room temperature is feasible, and the expected gas density in that condition is just half of the density of liquid helium. For 400 bar helium gas, we expect that roughly half of the delayed annihilations belong to the fast, $\tau_{fast} = 1/\lambda_{fast} = 250 \pm 70$ ns or to the slow, $\tau_{slow} = 1/\lambda_{slow} = 3.2 \pm 0.1$ μs components. Despite the poor knowledge of these parameters for helium gas at 400 bar, we can evaluate that $(63 \pm 4)\%$ of delayed annihilations occurs in a time window from 50 ns to 2 μs for antiprotonic-He. Knowing that the delayed annihilation time is proportional to the squared reduced mass of the exotic atom [18], we can evaluate that $\approx 50\%$ of $\bar{d}$ annihilations should occur in the same time window for a 400-bar helium target.

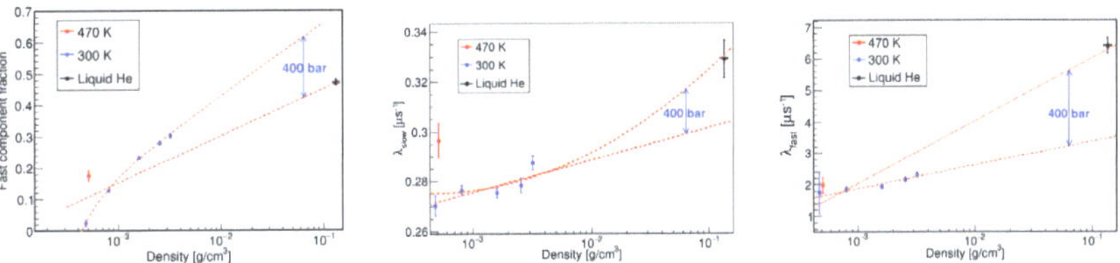

Figure 2. Measurements of antiprotonic-He delayed decay parameters for helium gas at different pressure or temperature [10] and for liquid helium [9]. The dashed lines depict two hypothetical models to extrapolate the decay parameters and uncertainties for He at 400 bar.

3. Antideuteron Identification with Helium Calorimeters

Helium gas is a fast UV scintillator, with a light yield similar to other fast plastic/liquid scintillators and capable of $\sim$ns timing performance [20,21].

Pressurized helium gas scintillators are currently adopted in fast neutron detection [21,22]; however, the gas in these detectors is typically stored in a thick and heavy steel vessel. Therefore, commercially available pressurized helium gas detectors are not suitable for investigating low-energy $\bar{d}$ that would stop in the thick vessel material. The idea of the PHeSCAMI project is to design a large helium calorimeter (HeCal) using a composite overwrapped pressure vessel (COPV) that would provide a small grammage of the walls, allowing for the detection of $\bar{d}$ with kinetic energy down to ≈ 50 MeV/n. A COPV is a pressure-containing vessel, typically composed of a metallic liner, a composite overwrap, and two bosses at the edges. COPVs are commonly manufactured by winding resin-impregnated high tensile strength fiber tape directly onto a cylindrical or spherical metallic liner. The inner liner contains the gas and limits permeation through the tank wall, while the outer fiber overwrap absorbs the stresses generated by the high-pressure gas within. COPVs have been developed for spaceflight due to their high strength and low weight as compared to metallic gas cylinders. They are also used in the automotive industry for hydrogen or compressed natural gas storage. ArianeGroup has developed a space-qualified COPV for helium: working pressure is 400 bar, volume is 300 L, dry mass is 80 kg, and the average vessel grammage is $\approx$3.5 g/cm^2 [23]. A smaller space-qualified HeHPV was developed in the ESA-ARTES program: working pressure is 310 bar, volume is 40 L, dry mass is 8.5 kg, and average grammage is $\approx$1.5 g/cm^2 [24]. To measure the UV scintillation light emitted by the helium stored in the COPV a possible strategy is depicted in Figure 3a): one of the two bosses of the COPV tank can be equipped with Wavelength Shifting fibers (WLS) that are able to convert the UV scintillation light into visible light and to guide the photons to an external Photomultiplier Tube (PMT).

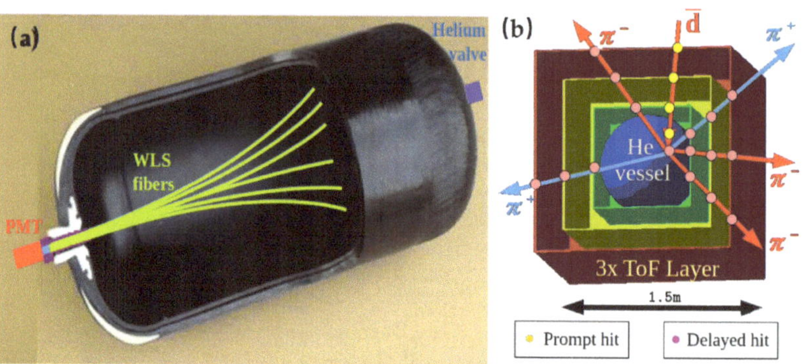

Figure 3. (**a**) Example of a possible light readout system for the COPV. (**b**) The design for a possible PHeSCAMI demonstrator prototype: 300 L calorimeter filled with 400 bar He (HeCal) surrounded by three plastic scintillator layers (ToF).

A possible configuration for a detector prototype of the PHeSCAMI technique is depicted in Figure 3. The response of a similar detector to $\bar{d}$, $\bar{p}$ and to the main cosmic ray components (p, He, C, e$^-$) has been simulated with the Geant4 11.1.3 package and FTFP-BERT physics list [25]. The inner part is a $\sim$20 kg scintillating helium calorimeter (HeCal) where the 400-bar gas is filling the ArianeGroup space-qualified COPV [23].

The HeCal is surrounded by three layers, made by 4 mm thick plastic scintillator bars, providing velocity measurement (β) by Time of Flight (ToF) and charge measurement (Z) by ionization energy loss measurement (dE/dX). It is assumed that with current technology, such a ToF detector is capable of measuring β with 5% resolution and deposited energy with 10% resolution. A time resolution of 1 ns and energy resolution of 10% have been assumed in the simulation for the HeCal detector. These assumptions are supported by preliminary measurements on an HeCal prototype (see next sections). Considering the energy loss in the ToF detector and the vessel, a minimum kinetic energy of $\sim$60 MeV/n is necessary for $\bar{d}$ to reach the He target. On the other hand, $\bar{d}$ with kinetic energy larger than 140 MeV/n would typically cross the 400 bar He active region without stopping inside.

This defines the 60–140 MeV/n energy window of sensitivity for $\bar{d}$ by this detector configuration. Figure 3 also shows the typical event topology for a stopping $\bar{d}$ within the He gas. The antiparticle initially produces three prompt hits (yellow) in the ToF detector and one prompt energy deposit (S1) in the HeCal; these prompt hits occur within 10 ns. This is the typical signal produced by any ionizing particle stopping in the vessel. Then, only for $\bar{p}$ and $\bar{d}$, the antiparticle can be captured in the He metastable states and after a time delay going from several tens of ns to a few μs the annihilation occurs (pink delayed hits in Figure 3). Typical $\pi^\pm$ multiplicity is 3.0 ± 0.2 for each anti-nucleon annihilation at rest [26]; therefore, twice the number of delayed out-going tracks is expected for $\bar{d}$ regarding $\bar{p}$. For the same reason, the delayed signal (S2) in the HeCal for $\bar{d}$ is also expected to have a double amplitude regarding the delayed signal for $\bar{p}$. The characteristic temporal structure of S1/S2 signals as measured by HeCal for d̄ is shown in Figure 4. The time gap from S1 (prompt) to S2 (delayed) is related to the metastability of He capture and is statistically distributed with $\tau_c \sim O\,(1\,\mu s)$. The S1 signal is related to the energy loss in the scintillating He; the amplitude measures the residual particle kinetic energy after the energy losses due to the ToF detector and vessel crossing.

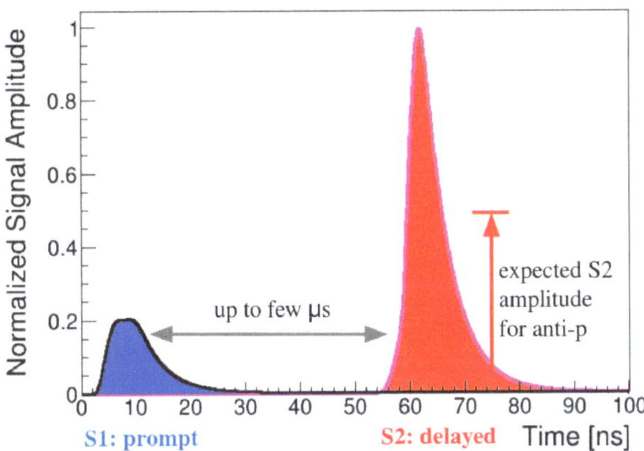

Figure 4. Typical HeCal timing signature for d̄ expected by the simulation of the PHeSCAMI detector, the scintillation components of helium have been considered (see next sections).

One advantage of the PHeSCAMI approach is that it relies on the relatively simple trigger condition, which is not purely ToF-based, as in the case of the GAPS experiment. Most cosmic rays (90%) are relativistic protons [27]. Thus, they would deposit ~10 MeV crossing the HeCal diameter, and this energy is much lower than the energy deposited by stopping d̄ (60–140 MeV/n). Moreover, crossing helium nuclei (~10% of cosmic rays [27]) would be identified by six "prompt" hits in the ToF detector with ~4 MIP (4 × the energy deposited by a Minimum Ionizing Particle). Therefore, a "start trigger" condition can be defined as:

$[HeCal_{Energy} > 25\,\text{MeV}]\,AND\,[5 > \#ToF_{E>2MIP} > 1]$. The "start trigger" would reject most of the crossing protons and a large fraction of helium nuclei, opening a 50 ns–2 µs time gate where delayed annihilation signals are searched for. The delayed annihilation signal provides a relatively large amount of energy deposited in HeCal (due to d̄ annihilation), and several hits in the ToF detector are expected due to produced charged pions. Thus, a "stop trigger" condition, enabling the data acquisition and the event storage on disk, can be defined as:

$[HeCal_{Energy} > 25\,\text{MeV}]\,AND\,[\#ToF > 4]\,AND\,[\#ToF_{E>2MIP} < 4]\,AND\,[\Delta T < 2\,\mu s]$. This provides a strong suppression of the random coincidences due to ordinary cosmic rays casually detected within the 2 µs time gate. A precise identification of d̄ and p̄ is possible in the offline event analysis. In particular, Figure 5 shows the spectrometric separation power of the "prompt" part of the event.

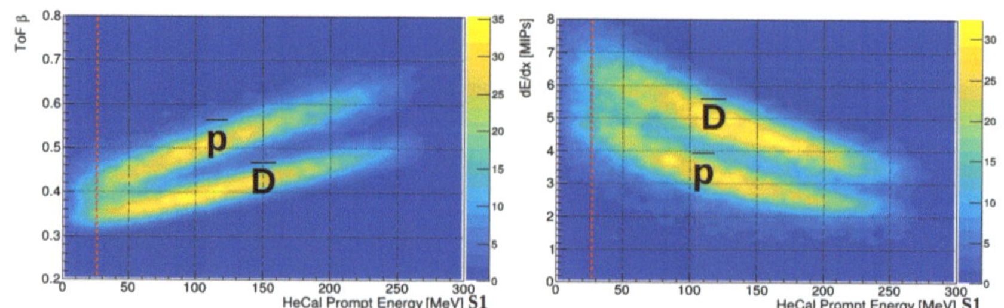

Figure 5. The comparison of simulated particle kinetic energy expected in HeCal (S1) with the velocity β expected in the ToF detector can identify the slower $\bar{\text{d}}$ from the faster $\bar{\text{p}}$ (**left**). Similarly, the energy deposited in the ToF detector layers is larger for the slower $\bar{\text{d}}$ and smaller for the faster $\bar{\text{p}}$ (**right**). Dashed red lines are the thresholds defined in the "start trigger".

On the other hand, for the PHeSCAMI technique, also the "delay" HeCal signal, S2, and the reconstruction of the number of charged pions produced in the annihilation (ToF activity classifier [28]) allow a good separation of $\bar{\text{p}}$ from $\bar{\text{d}}$, as shown in Figure 6.

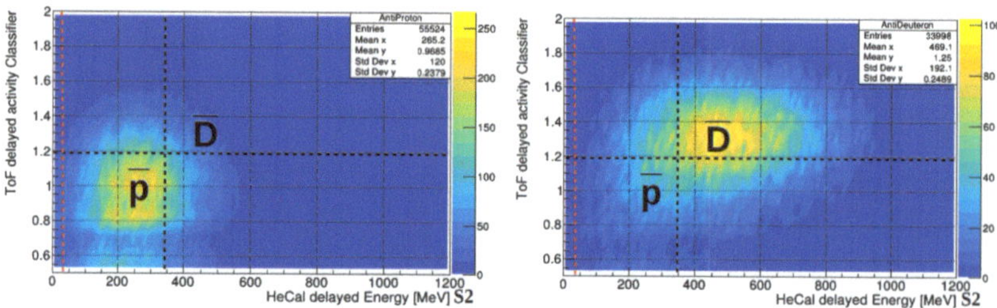

Figure 6. Comparison of simulated delayed events. The $\bar{\text{d}}$ annihilation provides twice the number of charged pions as compared to $\bar{\text{p}}$ annihilation. This implies an identification based on HeCal "delayed" energy, S2, and the number of ToF detector delayed hits, here combined to ToF detector delayed hit energy in an overall ToF activity classifier [28]. Vertical dashed red lines are the S2 energy thresholds defined in the "stop trigger" condition.

Combining prompt and delayed event information, the PHeSCAMI technique can identify a single $\bar{\text{d}}$ over 1000 background $\bar{\text{p}}$ in the 60–140 MeV/n range. Considering the expected $\bar{\text{p}}$ flux, this technique would be able to test the presence of $\bar{\text{d}}$ in cosmic rays down to a flux of $2\text{–}3 \times 10^{-6}$ $(\text{m}^2\text{s sr GeV/n})^{-1}$ with less than 1 $\bar{\text{p}}$ as background.

4. Test of HeCal Performance with Arktis B-470 Detector

The timing and energy resolutions of the HeCal detector are key parameters for the PHeSCAMI project. Some measurements on a prototype based on the fast neutron detector B-470 Arktis Radiation Detectors [21,22,29] have been conducted at INFN-TIFPA laboratories to test the response of pressurized helium gas as a scintillator.

The Arktis B-470 detector consists of a 5mm thick stainless steel cylindrical vessel filled with (209 bar) pressurized He gas and two Hamamatsu-R580 photomultiplier tubes (PMTs, $\varnothing \approx 38$ mm, Q.E. $\approx 27\%$) installed at the two ends of the vessel (see Figure 7). The inner wall of the vessel is lined with a wavelength shifter to convert ultraviolet He scintillation to the optical wavelengths for the PMTs.

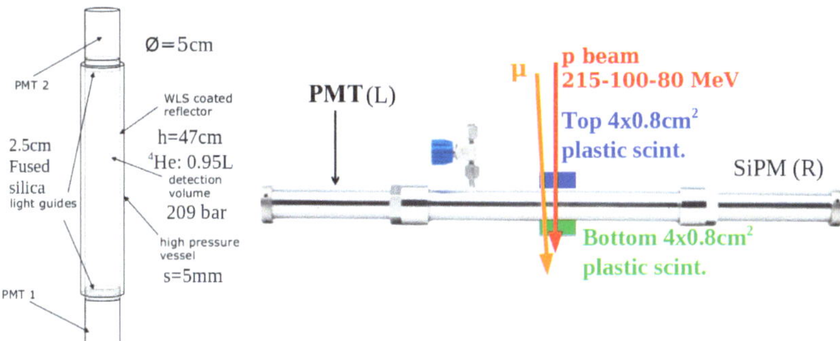

Figure 7. The Arktis B-470 detector used as a preliminary test of HeCal performance.

Performances of this detector for fast neutron identification are extensively studied; however, for the aim of the PHeSCAMI project, a characterization of the response of He scintillation to charged (crossing/stopping) particles is necessary. To allow the detection of the charged particles in the calorimeter avoiding the passage of the particle through the PMT, one PMT of the Arktis B-470 detector was replaced with an array of $8 \times$ Silicon PhotoMultipliers (SensL MicroFJ-60035 6×6 mm^2 Fill Factor 65%); see left panel of Figure 8. The SiPM circular array is shielded by 20 cm of iron, and a central hole, $\varnothing \approx 1$ cm, allows the particles to enter the helium target, crossing only the ($\approx$2.5 cm thick) fused silica optical window.

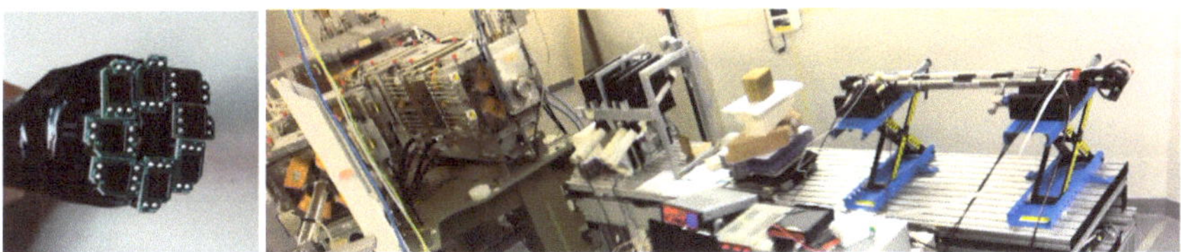

Figure 8. [**Left Panel**] An array of 8xSiPM replaces one PMT of the Arktis B-470 detector. [**Right Panel**] The detector prototype during the test at the Trento Proton Therapy facility.

4.1. Muon Calibration

A preliminary test with muons, μ, from cosmic rays was performed. The B-470 detector was operated in coincidence with two ($40 \times 8 \times 4$ mm^3) plastic scintillators, placed in a telescopic configuration, to detect crossing μ. The data were acquired with a LeCroy HDO9104-MS oscilloscope by sampling the detector waveforms at 20 Gs/s. The "minimum" energy deposition in the helium volume was obtained for μ transversely crossing the detector diameter (depositing $\approx$0.26 MeV in 200 bar helium). Conversely, the maximum energy deposition was obtained for μ crossing the whole detector (vertically placed, 250 μ detected in 4 months, depositing ~3 MeV in 200 bar helium). The muon calibration is analyzed along with the proton calibration to measure the detector performance described in the following sections (Figures 9–11).

4.2. Proton Calibration

A test on the proton beam line in the experimental room of the Trento Proton Therapy Facility (Italy) [30] was pursued to characterize the B-470 detector response to protons in the energy range 70–230 MeV that is the same range of interest for $\bar{\mathrm{d}}$ detection in CR.

The detector was tested with transversely crossing protons (Right Panel of Figure 7). In particular, for beam energy of 215 MeV, 100 MeV, and 80 MeV, we expect an energy deposit in He of 0.54 MeV, 0.93 MeV, and 1.1 MeV, respectively. Moreover, the B-470 was tested with protons entering longitudinally in the detector through the hollow SiPM array and the fused silica optical window (left panel of Figure 12). In this latter case, the energy measured by the helium scintillator follows the typical behavior expected for the Bragg peak. (Right Plot of Figure 12). Two plastic scintillators (4 mm thickness) are placed in front of the detector to provide the DAQ trigger, T_0, and behind the detector, to provide a crossing/veto trigger, T_2. Waveforms were acquired with a CAEN DT5742B 5 GS/s digitizer based on the DRS4 chip.

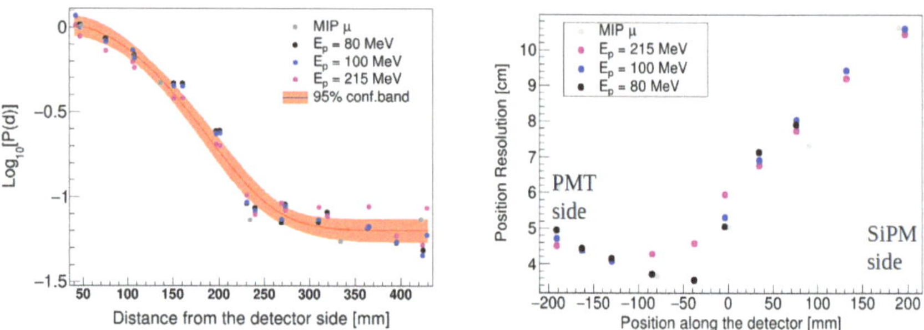

Figure 9. [**Left Panel**] Position-dependent light collection efficiency measured by testing the Arktis B-470 with Protons and μtransversely crossing the detector. [**Right panel**] Position resolution inferred from the measured signal asymmetry, the position resolution for particles passing near the SiPM side is worst due to the smaller light collection.

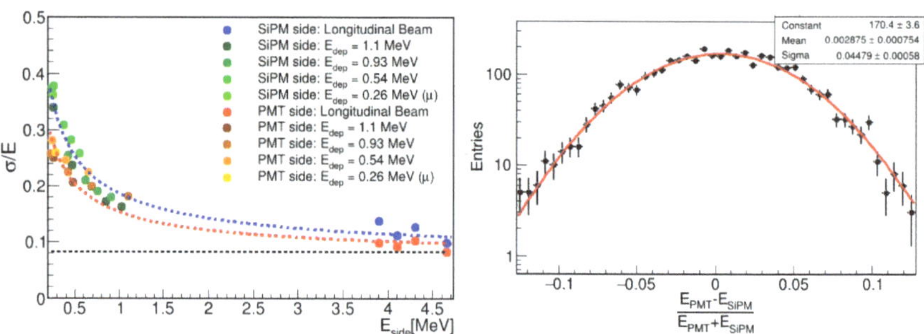

Figure 10. [**Left Panel**] Energy resolution for each side of the Arktis B-470 detector measured by transversely crossing μand p and by longitudinal proton runs. [**Right Panel**] Energy resolution inferred with 112 MeV longitudinal protons; 16 MeV were deposited in helium, and the relative difference in measured signal amplitude is ≈4.5%.

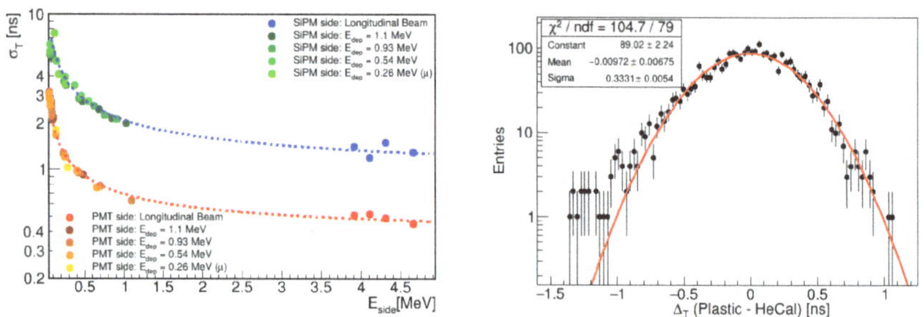

Figure 11. [**Left Panel**] Time resolution for each side of the Arktis B-470 detector measured by transversely crossing μ and p and by longitudinal proton runs. [**Right Panel**] A time resolution better than 330 ps is achieved with the 112 MeV longitudinal proton beam.

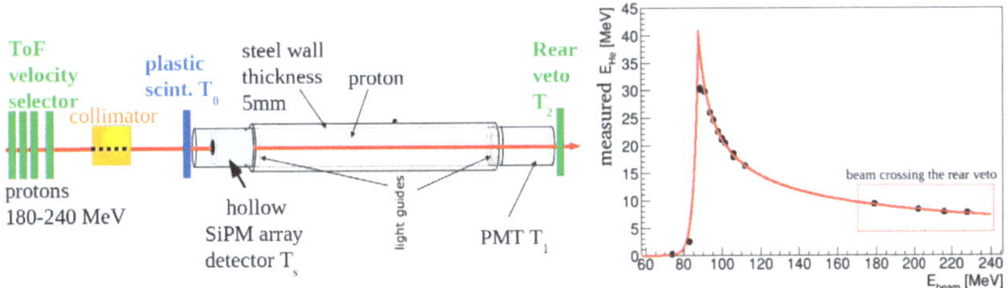

Figure 12. [**Left Panel**] Set-up adopted for the proton beam test of the HeCal performance. [**Right Panel**] Measured energy deposition of protons in helium, evidence of the Bragg peak.

4.3. Arktis B-470 Light Collection Efficiency

The measurement of the Arktis B-470 detector response at different transversal positions allows the testing of the effect of position-dependent light collection efficiency. The signal amplitude variation for particles crossing different positions along the tube is shown in Figure 9. Measurements as a function of the distance from the SiPM side and from the PMT side agree and are superimposed in the left plot of Figure 9.

From the measured signal amplitudes, the probability of photon detection at each side can be modeled as: $P(d) = P_0 e^{-(d/d_0)^2} + P_\infty$ where: d is the distance from the photon detector (SiPM or PMT), $d_0 = 123 \pm 2$ mm is a photon absorption length scale and $P_\infty = 1 - P_0 = 0.063 \pm 0.003$. The measured d_0 value can be attributed mainly to the peculiar B-470 detector geometry. By inverting the efficiency relationship, it is possible to infer the position of the crossing particle with a resolution of 5–10 cm by considering the asymmetry of the signals collected on both sides. Position resolution for particles passing near the SiPM (right) side is worse than the one for particles passing near the PMT (left) side due to the smaller detection surface of the 8xSiPM array (1.9 cm^2 vs. 3 cm^2 considering the 65% SiPM Fill Factor and the 27% PMT Quantum Efficiency).

4.4. Arktis B-470 Energy Resolution

The energy resolution of each side of the B-470 detector was investigated by considering the measured position-dependent collection efficiency (see left panel of Figure 10).

In particular, the relative energy resolution for each side is shown as a function of the photon collection efficiency corrected Energy: $E_{side} = E_{dep}P(d)$. For the four runs of longitudinally crossing protons (tagged by the rear veto), a rough approximation, $P(d) \approx 0.5$, was adopted. A simplified resolution model dominated by Poisson statistics was adopted to fit the energy resolutions measured for each side: $\sigma/E = (E_{side} n_{ph})^{-1/2} \oplus K$ (red and blue dotted line in the left panel of Figure 10). Both sides provide an asymptotic energy resolution of $K \approx 8\%$. Moreover, the numbers of collected photoelectrons evaluated at the side (P(d) = 1) are $n_{ph}^{SiPM} = 34.5 \pm 1.6$ ph.e/MeV and $n_{ph}^{PMT} = 56.4 \pm 4.1$ ph.e/MeV for SiPM and PMT side, respectively. The ratio $n_{ph}^{PMT}/n_{ph}^{SiPM} = 1.63 \pm 0.14$ is compatible with the expected ($\approx$1.6) collection efficiency ratio among the two sides (knowing the photodetection surfaces, SiPM Fill Factor and PMT Q.E.).

Finally, in the Right Panel of Figure 10, the asymmetry distribution of the Energy measured by both detector sides for longitudinal protons with beam Energy of 112 MeV is shown. In this case, a deposited energy of 16 MeV is measured in the helium scintillator. The obtained width of the relative energy difference, $\approx$4.5%, is reasonably compatible with $K/\sqrt{2}$ as expected from the simplified resolution model.

In summary, there are a lot of unknowns and uncertainties when relating our measurements based on the Arktis B-470 detector to the expected performance of the PHeSCAMI detector. These include the nature of helium gas scintillation (wavelength, photon yield, pressure dependence, etc.), the wavelength shifter used in the Arktis B-470, and the differences in the geometry. Despite these, we feel that our measurements based on the Arktis

B-470 detector show that it is plausible to achieve the assumed hypothesis that the HeCal detector would be able to measure energy depositions larger than 10 MeV with an energy resolution better than 10%. It is our future project to study the unknowns and uncertainties mentioned above.

4.5. Arktis B-470 Time Resolution

The time resolution of each side of the B-470 detector was investigated by measuring the time difference between the helium scintillation and the plastic scintillator signals. The time resolution of the plastic scintillator is negligible in this difference. It is observed that the measured time resolution improves for larger signals collected by the detector side.

The red and blue dotted lines in the left panel of Figure 11 are the fit of a simplified model of time resolution dominated by the Poisson statistics: $\sigma_T = \sigma_1 (E_{side} n_{ph})^{-1/2} \oplus \sigma_0$. In this model, σ_1 can be interpreted as the effective time uncertainty of the single photoelectron, while σ_0 is the asymptotic time resolution expected for large signals. The SiPM side provides the worst time resolution: $\sigma_1^{SiPM} = 11.0 \pm 0.5$ ns and $\sigma_0^{SiPM} = 0.9 \pm 0.2$ ns while for the PMT channel: $\sigma_1^{PMT} = 4.4 \pm 0.2$ ns and $\sigma_0^{PMT} = 0.39 \pm 0.03$ ns. The better timing performance of PMT with respect to the SiPM array is related to the larger photon detection efficiency and the relatively fast single photoelectron signal shape. The asymptotic value of time resolution obtained for the PMT channel suggests that, for this preliminary measurement, the ultimate time resolution is dominated by the typical non-uniform sampling time step, varying from cell to cell, of the CAEN DT5742B 5GS/s digitizer based on the DRS4 chip. As a summary, the time resolution tested with the Arktis B-470 detector confirms the capability of a pressurized helium calorimeter to detect the >50 ns delayed annihilation, i.e., the signature for antinuclei of the PHeSCAMI technique.

5. Scintillation Components of Helium at 200 Bar

Both the slow and fast scintillation components of helium were investigated by sampling the Arktis B-470 signal waveforms at 20 Gs/s with a LeCroy HDO9104-MS oscilloscope. In Figure 13, the measured scintillation signal, obtained as the average of many different scintillation pulses, is shown.

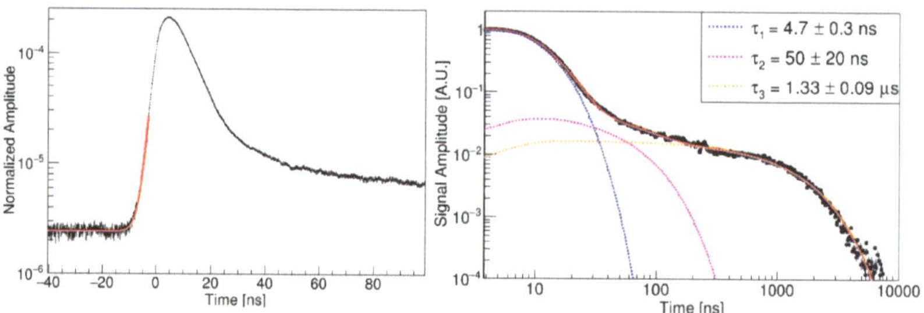

Figure 13. [**Left Plot**] Scintillation signal for the Arktis B-470, the rise time is $\tau_{rise} = 1.60 \pm 0.05$ ns. [**Right Plot**] The scintillation decay time of helium can be described by three components.

The signal rise time was inferred by fitting the first part of the sampled waveform with the function: $A(t) = B + [A(0) - B] e^{t/\tau_{rise}}$. The measured $\tau_{rise} = 1.60 \pm 0.05$ ns confirms that pressurized helium is a suitable scintillator for fast calorimetry (see left plot of Figure 13). The scintillation decay of helium was described with a three-component model:

$$A(t) = \frac{N_1}{\tau_1} e^{-t/\tau_1} + \frac{N_2}{\tau_2} e^{-t/\tau_2} + \frac{N_3}{\tau_3} e^{-t/\tau_3} \qquad (2)$$

where $\tau_1 = 4.7 \pm 0.3$ ns, $\tau_2 = 50 \pm 20$ ns and $\tau_3 = 1.33 \pm 0.09$ µs are the fast, intermediate and slow scintillation decay times, respectively. The measured relative amplitudes are: $N_3/N_1 = (70 \pm 15)\%$ and $N_2/N_1 = (18 \pm 7)\%$. The quoted errors for τ_i and N_i are dominated by the systematic uncertainty related to the possibility of multiple intermediate components. The presence of fast and slow scintillation components in helium was well known, and the ratio N_3/N_1 was found to be much larger for nuclear recoils. For this reason, helium scintillators are adopted to identify fast neutrons from gamma rays thanks to the PSD technique [20–22,29]. In this work, we identify a hint for a (so far undetected) small intermediate component ($\tau_2 = 50 \pm 20$ ns), and we measured a fast decay component ($\tau_1 = 4.7 \pm 0.3$ ns). These measurements confirm that pressurized helium is quite a fast scintillator. In particular, it is known that scintillation decay time in helium is dependent on pressure and gas impurities. In Figure 14, the lifetime of the fast decay component, $\tau_1 = 4.7$ ns, measured at 200 bar is compared with scintillation decay times for helium measured at lower pressure in [31].

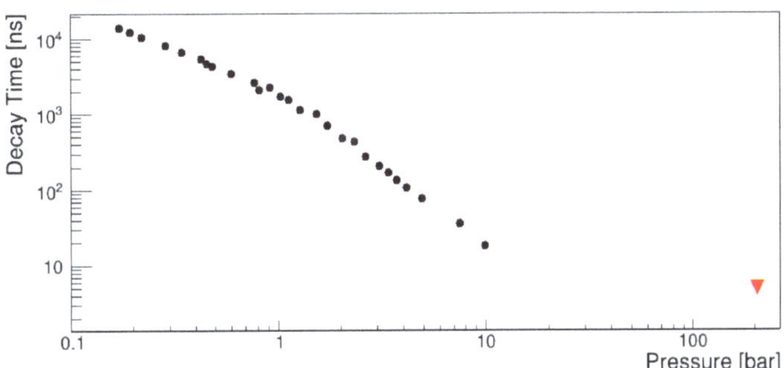

Figure 14. The scintillation decay time of helium as a function of pressure [31]. The red triangle is the upper limit inferred from the lifetime of the fast decay component, $\tau_1 = 4.7$ ns, observed in Arktis B-470.

Since our system does not directly measure the UV scintillation light from helium, but only the visible light emitted by the wavelength shifter adopted in the Arktis B-470 detector, both the measured values of the intermediate and fast components could be related to this WLS emission stimulated by a faster UV helium scintillation. Therefore, the measured value of $\tau_1 = 4.7$ ns should be interpreted as an upper limit for the helium scintillation fast decay time at 200 bar.

6. Conclusions

The signature for antideuteron identification in cosmic rays, offered by the PHeSCAMI (Pressurized Helium Scintillating Calorimeter for AntiMatter Identification) project, has been summarized. Preliminary measurements on the Arktis B-470 prototype at INFN-TIFPA laboratory have confirmed the capability of pressurized helium calorimeters to provide suitable energy and time resolutions. A helium scintillating calorimeter, based on a commercial COPV vessel, able to test the PHeSCAMI signature, is under development at INFN-TIFPA.

Author Contributions: Conceptualization, F.N. and P.Z.; methodology, F.N.; software, F.N. and F.R.; validation, F.N., F.R. and P.Z.; formal analysis, F.N. and F.R.; investigation, F.N. and P.Z.; resources, F.N., I.R., L.R., F.R., P.S., E.V., P.Z. and G.G.; data curation, F.N., F.R. and P.Z.; writing—original draft preparation, F.N.; writing—review and editing, F.N., I.R., L.R., F.R., P.S., P.Z. and G.G.; visualization, F.N.; supervision, F.N. and P.Z.; project administration, F.N. and P.Z.; funding acquisition, F.N., I.R., L.R., P.S., E.V. and P.Z. All authors have read and agreed to the published version of the manuscript.

Funding: This work has been funded by INFN (Grant73/2018) "ADHD-Anti-Deuteron Helium Detector" and by the Italian Ministerial grant PRIN-2022 n.2022LLCPMH "PHeSCAMI-Pressurized Helium Scintillating Calorimeter for AntiMatter Identification", CUP E53D23002100006.

Institutional Review Board Statement: Not applicable.

Informed Consent Statement: Not applicable.

Data Availability Statement: The data presented in this study are available on request from the corresponding author.

Conflicts of Interest: The authors declare no conflict of interest.

Abbreviations

The following abbreviations are used in this manuscript:

PHeSCAMI	Pressurized Helium Scintillating Calorimeter for AntiMatter Identification
HeCal	Helium Calorimeter
ToF	Time of Flight
UV	Ultraviolet
COPV	Composite Overwrapped Pressure Vessel
PMT	Photomultiplier Tube
SiPM	Silicon Photomultiplier
Q.E.	Quantum Efficiency
WLS	Wavelength Shifter

References

1. Donato, F.; Fornengo, N.; Maurin, D. Antideuteron fluxes from dark matter annihilation in diffusion models. *Phys. Rev. D* **2008**, *78*, 043506. [CrossRef]
2. Doetinchem, P.V.; Perez, K.; Aramaki, T.; Baker, S.; Barwick, S.; Bird, R.; Boezio, M.; Boggs, S.; Cui, M.; Datta, A.; et al. Cosmic-ray antinuclei as messengers of new physics: Status and outlook for the new decade. *J. Cosmol. Astropart. Phys.* **2020**, *2020*, 035. [CrossRef] [PubMed]
3. Lu, S. *Antideuterons in AMS*. MIAPbP Workshop: Antinuclei in the Universe? Munich. 2023. Available online: https://www.munich-iapbp.de/antinuclei-2022/schedule (accessed on 3 January 2024).
4. Kounine, A. The Latest Results from the Alpha Magnetic Spectrometer. *PoS* **2020**, *364*, 028. [CrossRef]
5. Munini, R.; Lenni, A. The identification of the cosmic-ray light nuclei with the GAPS experiment. *PoS* **2023**, *444*, 179. [CrossRef]
6. Feldman, S.N.; Aramaki, T.; Boezio, M.; Boggs, S.E.; Bonvicini, V.; Bridges, G.; Campana, D.; Craig, W.W.; von Doetinchem, P.; Everson, E.; et al. The GAPS Time-of-Flight Detector. *PoS* **2023**, *444*, 120. [CrossRef]
7. Battiston, R.; Bertucci, B.; Adriani, O.; Ambrosi, G.; Baudouy, B.; Blasi, P.; Boezio, M.; Campana, D.; Derome, L.; De Mitri, I.; et al. High precision particle astrophysics as a new window on the universe with an Antimatter Large Acceptance Detector In Orbit (ALADInO). *Exper. Astron.* **2021**, *51*, 1299–1330; Erratum in: *Exper. Astron.* **2021**, *51*, 1331–1332. [CrossRef]
8. Kirn, T. The AMS-100 experiment: The next generation magnetic spectrometer in space. *Nucl. Instrum. Meth. A* **2022**, *1040*, 167215. [CrossRef]
9. Iwasaki, M.; Nakamura, S.N.; Shigaki, K.; Shimizu, Y.; Tamura, H.; Ishikawa, T.; Hayano, R.S.; Takada, E.; Widmann, E.; Outa, H.; et al. Discovery of antiproton trapping by long-lived metastable states in liquid helium. *Phys. Rev. Lett.* **1991**, *67*, 1246–1249. [CrossRef]
10. Widmann, E.; Sugai, I.; Yamazaki, T.; Hayano, R.S.; Iwasaki, M.; Nakamura, S.N.; Tamura, H.; Ito, T.M.; Kawachi, A.; Nishida, N.; et al. Effects of impurity atoms and molecules on the lifetime of antiprotonic helium atoms. *Phys. Rev. A* **1996**, *53*, 3129–3139. [CrossRef]
11. Widmann, E.; Sugai, I.; Yamazaki, T.; Hayano, R.S.; Iwasaki, M.; Nakamura, S.N.; Tamura, H.; Ito, T.M.; Kawachi, A.; Nishida, N.; et al. Phase and density dependence of the delayed annihilation of metastable antiprotonic helium atoms in gas, liquid, and solid helium. *Phys. Rev. A* **1995**, *51*, 2870–2880. [CrossRef]
12. Nakamura, S.N.; Iwasaki, M.; Outa, H.; Hayano, R.S.; Watanabe, Y.; Nagae, T.; Yamazaki, T.; Tada, H.; Numao, T.; Kuno, Y.; et al. Negative-pion trapping by a metastable state in liquid helium. *Phys. Rev. A* **1992**, *45*, 6202–6208. [CrossRef] [PubMed]
13. Yamazaki, T.; Aoki, M.; Iwasaki, M.; Hayano, R.S.; Ishikawa, T.; Outa, H.; Takada, E.; Tamura, H.; Sakaguchi, A. Trapping of negative kaons by metastable states during the atomic cascade in liquid helium. *Phys. Rev. Lett.* **1989**, *63*, 1590–1592. [CrossRef] [PubMed]
14. Condo, G. On the absorption of negative pions by liquid helium. *Phys. Lett.* **1964**, *9*, 65–66. [CrossRef]
15. Russell, J.E. Metastable States of $\alpha\pi^- e^-$, $\alpha K^- e^-$, and $\alpha \bar{p} e^-$ Atoms. *Phys. Rev. Lett.* **1969**, *23*, 63–64. [CrossRef]
16. Russell, J.E. Interactions of an $\alpha K^- e^-$ Atom with a He Atom. *Phys. Rev.* **1969**, *188*, 187–197. [CrossRef]
17. Wright, D.E.; Russell, J.E. Energies of Highly Excited Heliumlike Exotic Atoms. *Phys. Rev. A* **1972**, *6*, 2488–2492. [CrossRef]
18. Ketzer, B.; Hartmann, F.J.; Daniel, H.; von Egidy, T.; Niestroj, A.; Schmid, S.; Schmid, W.; Yamazaki, T.; Sugai, I.; Nakayoshi, K.; et al. Isotope effects on delayed annihilation time spectra of antiprotonic helium atoms in a low-temperature gas. *Phys. Rev. A* **1996**, *53*, 2108–2117. [CrossRef] [PubMed]
19. Hori, M.; SóTéR, A.; Barna, D.; Dax, A.; Hayano, R.; Friedreich, S.; Bertalan, J.; Pask, T.; Widmann, E.; Horváth, D.; et al. Two-photon laser spectroscopy of antiprotonic helium and the antiproton-electron mass ratio. *Nature* **2011**, *475*, 484–488. [CrossRef]

20. Huffman, R.E.; Larrabee, J.C.; Chambers, D. New Excitation Unit for Rare Gas Continua in the Vacuum Ultraviolet. *Appl. Opt.* **1965**, *4*, 1145–1150. [CrossRef]
21. Zhu, T.; Liang, Y.; Rolison, L.; Barker, C.; Lewis, J.; Gokhale, S.; Chandra, R.; Kiff, S.; Chung, H.; Ray, H.; et al. Improved fission neutron energy discrimination with 4He detectors through pulse filtering. *Nucl. Instrum. Methods Phys. Res. Sect. Accel. Spectrometers Detect. Assoc. Equip.* **2017**, *848*, 137–143. [CrossRef]
22. Jebali, R.; Scherzinger, J.; Annand, J.; Chandra, R.; Davatz, G.; Fissum, K.; Friederich, H.; Gendotti, U.; Hall-Wilton, R.; Håkansson, E.; et al. A first comparison of the responses of a 4He-based fast-neutron detector and a NE-213 liquid-scintillator reference detector. *Nucl. Instrum. Methods Phys. Res. Sect. Accel. Spectrometers Detect. Assoc. Equip.* **2015**, *794*, 102–108. [CrossRef]
23. Benedic, F.; Leard, J.; Lefloch, C. Helium High Pressure Tanks at EADS Space Transportation New Technology with Thermoplastic Liner. Doc ID: ADA445482: EADS Report. 2005. Available online: https://www.ariane.group/wp-content/uploads/2019/01/11_300LHE_MPCV.pdf (accessed on 3 January 2024).
24. HeHPV (Helium High-Pressure Vessel)-Qualification of a Helium High Pressure Tank. Available online: https://artes.esa.int/projects/hehpv-helium-highpressure-vessel (accessed on 3 January 2024).
25. Agostinelli, S.; Allison, J.; Amako, K.; Apostolakis, J.; Araujo, H.; Arce, P.; Asai, M.; Axen, D.; Banerjee, S.; Barrand, G.; et al. Geant4—A simulation toolkit. *Nucl. Instrum. Methods Phys. Res. Sect. Accel. Spectrometers Detect. Assoc. Equip.* **2003**, *506*, 250–303. [CrossRef]
26. Amsler, C. Proton-antiproton annihilation and meson spectroscopy with the Crystal Barrel. *Rev. Mod. Phys.* **1998**, *70*, 1293–1339. [CrossRef]
27. Workman, R.L.; et al. Review of Particle Physics. *Prog. Theor. Exp. Phys.* **2022**, *2022*, 083C01. [CrossRef]
28. Therhaag, J.; Team, T.C.D. TMVA-Toolkit for multivariate data analysis. *AIP Conf. Proc.* **2012**, *1504*, 1013–1016. [CrossRef]
29. Arktis Radiation Detectors Limited, 8045 Zurich, Switzerland. Available online: https://www.arktis-detectors.com/ (accessed on 3 January 2024).
30. Tommasino, F.; Rovituso, M.; Fabiano, S.; Piffer, S.; Manea, C.; Lorentini, S.; Lanzone, S.; Wang, Z.; Pasini, M.; Burger, W.; et al. Proton beam characterization in the experimental room of the Trento Proton Therapy facility. *Nucl. Instrum. Methods Phys. Res. Sect. Accel. Spectrometers Detect. Assoc. Equip.* **2017**, *869*, 15–20. [CrossRef]
31. Guazzoni, P.; Pignanelli, M. Pressure dependence of radiative processes in helium. *Phys. Lett. A* **1968**, *28*, 432–433. [CrossRef]

Article

Real-Time Monitoring of Solar Energetic Particles Using the Alpha Magnetic Spectrometer on the International Space Station

Andrea Serpolla [1,*]**, Matteo Duranti** [2]**, Valerio Formato** [3] **and Alberto Oliva** [4]

[1] Département de Physique Nucléaire et Corpusculaire, Université de Genève, 1205 Geneva, Switzerland
[2] Istituto Nazionale Fisica Nucleare, Sezione di Perugia, 06100 Perugia, Italy; matteo.duranti@pg.infn.it
[3] Istituto Nazionale Fisica Nucleare, Sezione di Roma Tor Vergata, 00133 Rome, Italy; valerio.formato@roma2.infn.it
[4] Istituto Nazionale Fisica Nucleare, Sezione di Bologna, 40126 Bologna, Italy; alberto.oliva@bo.infn.it
* Correspondence: andrea.serpolla@unige.ch

Abstract: The International Space Station (ISS) orbits at an average altitude of 400 km, in the Low Earth Orbit (LEO) and is regularly occupied by astronauts. The material of the Station, the residual atmosphere and the geomagnetic field offer a partial protection against the cosmic radiation to the crew and the equipment. The solar activity can cause sporadic bursts of particles with energies between $\sim$10 keV and several GeVs called Solar Energetic Particles (SEPs). SEP emissions can last for hours or even days and can represent an actual risk for ISS occupants and equipment. The Alpha Magnetic Spectrometer (AMS) was installed on the ISS in 2011 and is expected to take data until the decommissioning of the Station itself. The instrument detects cosmic rays continuously and can also be used to monitor SEPs in real-time. A detection algorithm developed for the monitoring measures temporary increases in the trigger rates of AMS, using McIlwain's L-parameter to characterize different conditions of the data-taking environment. A real-time monitor for SEPs has been realized reading data from the AMS Monitoring Interface (AMI) database and processing them using the custom algorithm that was developed.

Keywords: cosmic rays; solar energetic particles; space weather

Citation: Serpolla, A.; Duranti, M.; Formato, V.; Oliva, A. Real-Time Monitoring of Solar Energetic Particles Using the Alpha Magnetic Spectrometer on the International Space Station. *Instruments* **2023**, *7*, 38. https://doi.org/10.3390/instruments7040038

Academic Editor: Antonio Ereditato

Received: 16 October 2023
Revised: 27 October 2023
Accepted: 30 October 2023
Published: 31 October 2023

1. Introduction

The International Space Station (ISS) orbits at an average altitude of about 400 km, in the Low-Earth Orbit (LEO) and is exposed to a flux of Cosmic Rays (CRs), i.e., energetic particles that wander around in the universe and which can enter into our Solar System. Equipment and crew members inside the ISS are partially shielded from the cosmic radiation by the surrounding material of the station itself, the residual atmosphere and the geomagnetic field.

CRs are mainly injected by galactic and extra-galactic sources [1]; however, the solar activity itself produces a flux of energetic particles, which affects the CR spectrum observed from Earth. The Sun ejects a flow of plasma into the interplanetary space, the Solar Wind (SW), which influences the CR energetic spectrum. In addition, violent phenomena taking place on the surface of the Sun (e.g. reconnection of magnetic field lines; Coronal Mass Ejections (CMEs)) can eject bursts of energetic particles, Solar Energetic Particles (SEPs), into space, with energies ranging from $\sim$10 keV and several GeVs. SEP events can last for hours or even days and are primarily composed by protons [2,3].

The ISS is not completely shielded from cosmic radiation and, during extra-vehicular activities, the exposure is high. Therefore, the radiation hazard that astronauts are subjected to has to be monitored.

The Alpha Magnetic Spectrometer 02 (AMS-02) is installed on one of the side arms of the ISS and measures the CR flux continuously [4]. The instrument is composed of different detectors to measure the characteristics of passing particles: a Silicon Tracker, to measure the rigidity and particle charge sign; a Time-of-Flight (ToF), to measure the velocity, the moving direction and the particle charge magnitude Z; a Transition Radiation Detector (TRD), to identify and separate $e^{\pm}$ from p and nuclei; a Ring Imaging Cherenkov (RICH), to measure the velocity and Z; and an electromagnetic calorimeter (ECAL), to measure the energy. Due to its equipment, AMS-02 is also suitable for detecting SEPs. In particular, ongoing SEP events can be detected by looking for sudden increments of the trigger rates [5]. The low latency of the trigger system allows for building an effective real-time monitor for SEP fluxes affecting the ISS, which could also be used to alert the astronauts onboard.

2. Materials and Methods

Intense SEP fluxes increase AMS trigger rates above the usual averages. This feature can be used to detect and monitor ongoing SEP events [5].

The AMS Monitoring Interface (AMI) contains all the data needed to monitor SEPs [6]. The AMI is made of InfluxDB databases to store the data and Grafana instances to visualize them. The monitor described in this manuscript adopts an architecture similar to the one used for the AMI. First, a series of fillers processes AMI data, utilizing a custom algorithm developed for this work [7]. Then, the fillers results are written in an InfluxDB database, connected in turn to a Grafana instance, which acts as an interface for SEP monitoring [8].

2.1. Primary Data

The AMI database is written by a feeder that recurrently processes the raw data flow of AMS directly on board the ISS. Data writing occurs approximately every minute when the connection to the ground is available. When the connection is lost, scientific data are buffered until it is re-established, usually within 20–30 min.

The delays in the writing operations and the interruptions of the data flow from the ISS can decrease the number of entries found in real-time in the AMI database. To prevent data losses in the SEP monitor, the AMI database is queried multiple times, first in real-time and then with different delays, at intervals of 30 s, 1 min, 2 min, 5 min, and 1 h.

The analysis algorithm uses AMS trigger rates and live-time to search for increases in the instrument activity that could be due to an SEP event. In addition, ISS positions and flight angles are also used to calculate the geomagnetic field intensity, the zenith angle and the McIlwain's L-parameter [9], useful for characterizing the impact of the geomagnetic field on the data. To avoid the use of outdated information, records with positions or flight angles updated respectively more than 10 s or 1 min before are discarded.

2.2. AMS-02 Trigger Rates

AMS-02 uses two types of triggers: fast triggers (FTs) and level-1 (LV1) triggers [10]. FTs are evaluated in $\sim$40 ns, while LV1 triggers can take up to 1 μs and are evaluated only in the case of a positive FT. Fast trigger is in turn the logical OR of other three signals:

- Fast Trigger Charged (FTC), which evaluates the possible presence of events with charged particles;
- Fast Trigger big-Z (FTZ), which evaluates the possible presence of events with highly-charged particles;
- Fast Trigger ECAL (FTE), which evaluates the possible presence of events with electromagnetic interactions in the ECAL.

On the other hand, the LV1 trigger can compose up to 8 sub-level-1 (subLV1) triggers; currently only 7 are defined and used:

0   unbiased trigger for events with charged particles;
1   trigger for events with single charged particles;
2   trigger for events with normal ions;
3   trigger for events with slow ions;

4   trigger for events with electrons;
5   trigger for events with photons;
6   unbiased trigger for events with electromagnetic interactions.

The fast and LV1 triggers make use of signals from the ToF, the ACC and the ECAL.

The SEP monitor takes into consideration all the trigger rates, except for subLV1-6, which has an extremely low rate.

The status of the trigger is monitored in AMS by scalers placed in the LVL1 board that count how many times each trigger condition has been fired during the previous second. At the same time, the board measures the experiment live-time, i.e., the fraction of the previous second during which the detector was available for data acquisition.

Because of the dead-time of the instrument, measured trigger rates λ_{meas} need to be adjusted to their true values λ, dividing them by the live-time η:

$$\lambda = \lambda_{\mathrm{meas}}/\eta. \tag{1}$$

In our study we measure trigger rates in 1-min time interval. The rates corrected by the live-time are of the order of 100–1000 Hz.

A special case occurs for trigger rates that are generally low, i.e., $\lesssim 10\,\mathrm{Hz}$. For those, the rate distribution can be discrete and simply dividing by the live-time would be inadequate to get the true rate. Because of that, in this work all the measured trigger rates are corrected by the live-time, except for FTZ and subLV1-3. In the Supplemental Materials the live-time and the trigger rates distributions are shown in detail.

2.3. Impact of the Geomagnetic Field

The nominal level of a trigger rate varies along the ISS orbits. The particle flux observed on satellites in LEO is strongly influenced by the geomagnetic environment, which permits only particles with a rigidity greater than the local cutoff. Cutoff rigidity can be determined by tracing particles of different energies in the magnetosphere and depends on the satellite location and particle direction. Figure 1 illustrates how AMS activity varies in function of McIlwain's L-parameter, which is a convenient parameter to describe the cutoff rigidity for a particle entering from the zenithal direction, as a function of the satellite position in the geomagnetic field [9].

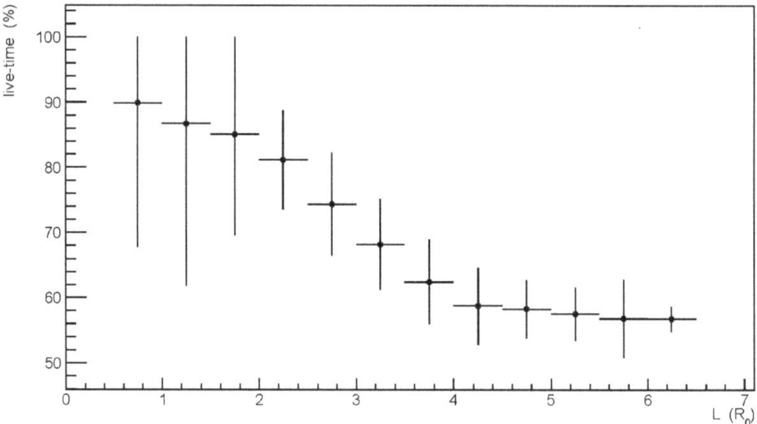

Figure 1. Average live-time of AMS-02 in January 2020 as a function of the L-value of the position of detection. Lower live-times show a higher activity of the instrument, because of a greater flux of particles crossing it. Live-time decreases with increasing L-values, indicating a dependence between the two quantities.

The relationship between the L-parameter and the vertical cutoff rigidity R_{VC} is shown by the equation

$$L = \sqrt{k/R_{VC}} \, , \tag{2}$$

with $k \simeq 16.2\,\mathrm{GV} \cdot R_0^2$ and R_0 the Earth radius [11]. Figure 2 shows a map of the L-values calculated for 1 month of ISS orbit data.

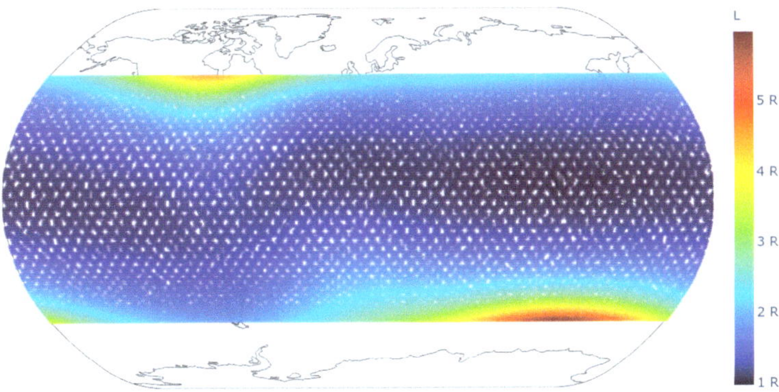

Figure 2. Map of the L-values calculated on ISS orbit data of January, 2020. The values are expressed in Earth radii units and are averaged in cells $1°$ (latitude) $\times$ $1°$ (longitude), with an error $\lesssim 4\%$. ISS orbit projection on Earth's surface spans a latitude range approximately from $-52°$ to $+52°$.

The ISS orbits project onto the Earth's surface within a latitude range spanning from approximately $-52°$ to $+52°$. Within this range, the motion of the ISS does not entirely cover the Earth's surface, as evident from the lack of data points in the central region of the map depicted in Figure 2. Furthermore, Figure 2 reveals that the majority of the path traversed by the ISS is associated with low L-values, typically less than $3\,R_0$. However, as the ISS approaches the poles, the L-parameter can reach higher values, peaking at around $6\,R_0$. Higher L-values correspond to lower vertical cutoff rigidities, expanding the potential for observing SEP events with lower energy levels.

For this work, trigger rate distributions are grouped in fixed bins of L from 0 to $6.5\,R_0$, $0.5\,R_0$ wide. Figure 3 shows the statistics that characterizes the defined L bins.

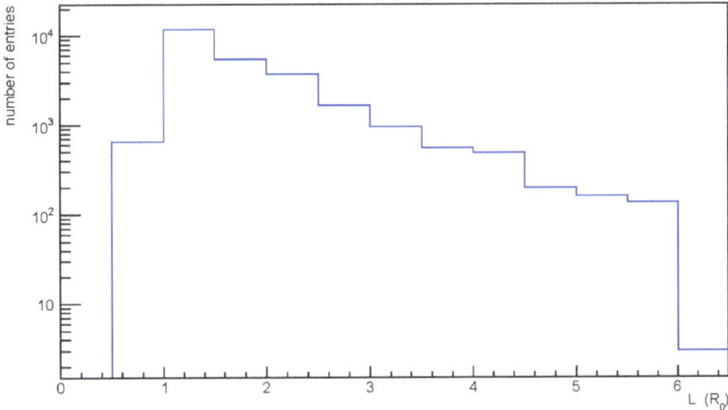

Figure 3. Distribution of the L-values calculated for AMI data of a 3-day period, i.e., 1–3 January 2020. L bins span between 0.5 and $6.5\,R_0$ and are $0.5\,R_0$ wide. Every bin collects more than 100 entries, except for the last bin, characterized in general by a much lower statistics (i.e., 10 entries per month).

Due to the non-perfect dipolar shape of the geomagnetic field, the field lines come closer to the Earth's surface in correspondence of the South Atlantic Anomaly (SAA), a region where particles trapped in the Van Allen belts cross the satellite orbit. This feature causes an observed higher flux, with a variability dependent on how the SAA is crossed by the ISS. Those increments are not related to a possible SEP event, therefore these regions are excluded from further analysis imposing a lower threshold of 25,000 nT on the geomagnetic field intensity. The intensities are obtained from the GPS positions, using the 13th generation of the International Geomagnetic Reference Field (IGRF) [12].

Another situation where AMS could experience a sudden change in the number of detected particles is when the ISS flight orientation is changed significantly, e.g., during the docking of a spacecraft. In those cases, AMS might assume a great inclination with respect to the zenith axis. ISS flight angles, i.e., yaw, pitch and roll can be used to calculate the zenith angle of AMS. To exclude increments of the trigger rates related to fluxes of particles trapped along the geomagnetic field lines, data collected with a zenith angle greater than $15°$ are discarded.

2.4. Detection of SEPs

An SEP event can increase AMS trigger rates over their usual levels observed in a specific geomagnetic region. To quantify the increments, a score s is defined and assigned to each data entry:

$$s = \int_{\bar{\lambda}}^{+\infty} f(\lambda; \mu, [\sigma]) \mathrm{d}\lambda \,, \tag{3}$$

with $\bar{\lambda}$ representing the observed trigger rate, f the PDF that models its distribution, and μ, σ respectively the average and the standard deviation of rates from the preceding 3 days; σ is actually an optional parameter, used only when f is a continuous function. Regarding the time window used for assessing μ and σ, the ISS orbit around the Earth takes $\sim$90 min and its precession $\sim$16 orbits; therefore, the window spans $\sim$3 ISS motion cycles. The trigger rates corrected by live-time are modeled by a Gaussian distribution, while the lower-rate ones, not corrected by the live-time (i.e., FTZ and subLV1-3), are modeled by a Poisson distribution.

Time-series of μ and σ show a step trend that can lead to apparent increments in the score calculus, when L is about to change its bin. In order to remove this effect, μ and σ are smoothed using the linear interpolation between the values obtained from the two closest bins.

The score defined in Equation (3) is eventually used to detect the presence of an SEP event. Specifically, increments in the trigger rates resulting from intense SEPs would lead to low score values, tending to 0. Because of its definition, the score value can also be seen as the p-value of the observed increment to be outcome of just a statistical fluctuation. This perspective is valuable for establishing a cut for the scores to avoid biasing the reference distributions defining the trigger rates nominal levels. Indeed, data with a score smaller than 10^{-6} are excluded from entering the 3-days time window, eliminating $\sim$2 entries per year due to statistical fluctuations.

3. Results

The real-time SEP monitor resulting from the work described in this manuscript is published at [8]. AMI data are analyzed in real-time and again after 30 s, 1, 2, 5 min, and 1 h. Real-time fillers are kept running on a separate deployment platform and the results are stored in an InfluxDB database to which the Grafana interface is connected.

3.1. Background Rejection

Looking at the score time series resulting from the data analysis, many low scores compatible with an SEP event can be found, even when no other experiments confirmed the occurrence of an event. To reduce the background, the same cuts applied on the geomagnetic field intensity, the ISS zenith angle, and the ages of position and flight orientation information for the reference trigger rate definition can be used. Specifically, data collected with a geomagnetic field intensity less than 25,000 nT, a zenith angle greater than $15°$, an age older than 10 s for the ISS position data, and older than 1 min for the ISS flight orientation, are excluded from SEP events search. Figure 4 shows the effectiveness of those cuts on the score time-series obtained during a confirmed event [13].

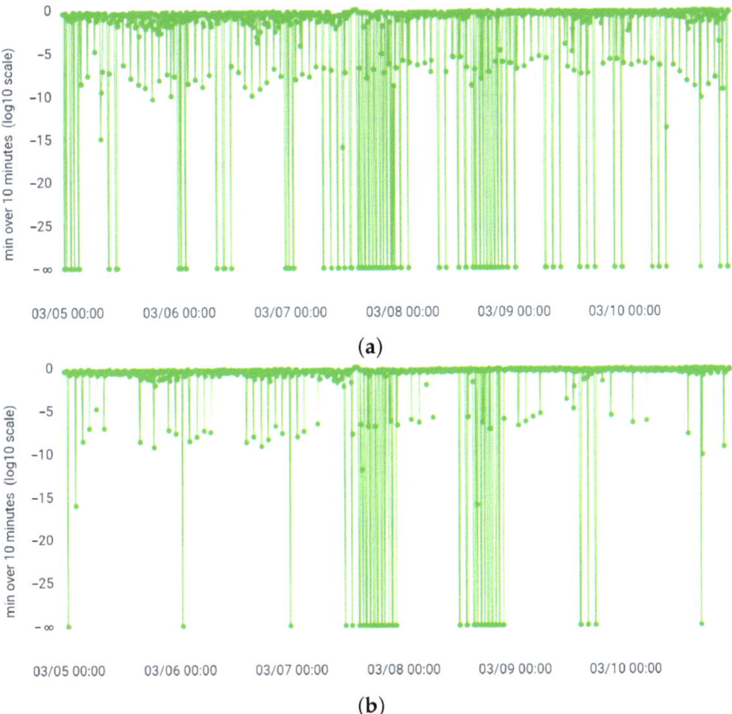

Figure 4. Score time-series for the FT rate (corrected by the live-time) in the period 5–11 March 2012, during a confirmed SEP event. (**a**) Top panel shows the time-series with all the data. (**b**) Bottom panel shows the same time-series, with the background rejection cuts applied.

In addition to the previous standard cuts, the additional condition $L \geq 1.5\,R_0$ results in being able to reject an important fraction of noise for the LV1 trigger rate. The effectiveness of this cut is most likely related to the errors introduced by the rate modeling in the low-L bins and is further treated in the Discussion section. Figure 5 compares the LV1 trigger scores obtained by applying only the standard cuts, with those obtained with the additional L cut.

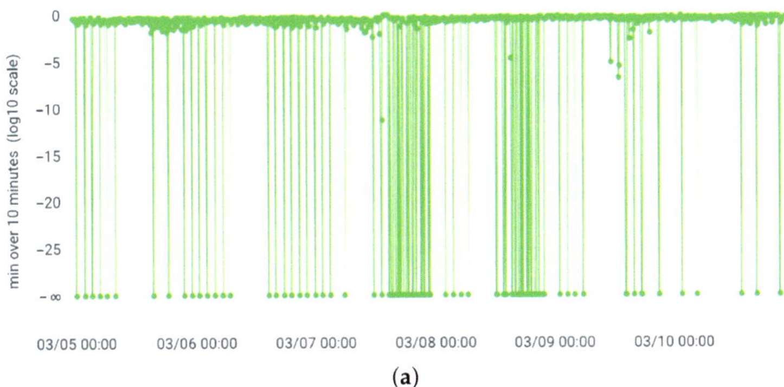

Figure 5. *Cont.*

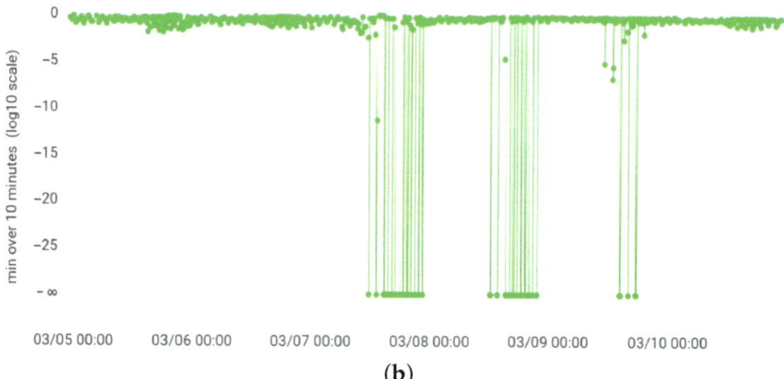

(b)

Figure 5. Score time-series for the LV1 trigger rate (corrected by the live-time) in the period March 5–11, 2012, during a confirmed SEP event. (**a**) Top panel shows the time-series with the standard cuts for background rejection applied. (**b**) Bottom panel shows the same time-series, with the additional requirement $L \geq 1.5\, R_0$.

3.2. Efficiency of the Real-Time Monitoring

Due to the delays of the AMI data downlink described in the previous section, the monitoring in real-time can experience some data losses. Figure 6 shows the efficiencies measured for data queries performed with different delays.

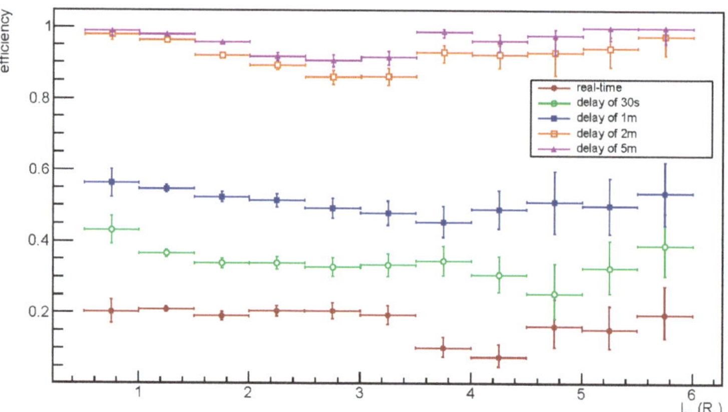

Figure 6. Efficiency of data requests to the AMI database. Efficiencies were measured querying data in real-time and with different delays for 10 days, between 10–20 May 2023. The number of entries obtained after 1 h are used as totals for the efficiencies measurement. Overall, the efficiency increases with the delay and in real-time $\epsilon \lesssim 20\%$, after 30 s $\sim$20–30%, after 1 min $\sim$40–60% and after more than 2 min $\gtrsim 80\%$.

4. Discussion

The monitor described in this manuscript uses an architecture similar to the one used for the AMI, foreseeing a later implementation in the AMI itself. However, Figure 6 shows how relying on AMI data negatively affects the real-time monitoring, mining the possibility of launching live alerts for SEP events. This issue could be solved by implementing the analysis directly onboard the ISS. In that case, data would be immediately available and the access to the primary data flow of AMS would also provide access to a broader information. In addition, performing the analysis in space would also give the opportunity to provide an alert system directly on the ISS.

As Figure 3 shows, most of the L bins used for the analysis collect $\gtrsim 100$ entries in a 3-day time window, except the bin $[6.0, 6.5)\, R_0$. For $L \geq 6\, R_0$, $\sim$10 entries per month of data can be found. The use of a larger time window to collect more statistics would decrease the possibility of performing a real-time analysis efficiently and the use of a larger bin would lower the sensibility in regions where the shielding of the geomagnetic field is minimum. The score values used to highlight SEP events benefits from the use of the information of multiple bins at the same time, therefore the last bin, even if less representative, can still provide a foothold for the weighted averages used in the score calculus.

Another issue in the developed analysis regards the discrepancy between the modeling PDFs used and the actual distributions, which are shown in the Supplementary Materials. Inadequacies in data characterization also introduce some noise that notably affects the LV1 trigger, as indicated by the additional L cut for background rejection. The efficacy of this cut can be understood by examining the detailed trigger rate distributions in the Supplemental Material. Low-L bins exhibit statistics that deviate from Gaussian and Poisson distributions, with a higher prevalence of low rates. A higher proportion of low values results in a generally lower μ, thus leading to a greater frequency of low scores. Triggers other than LV1, as exemplified in Figure 4, are less affected by this problem. In contrast, the LV1 score time-series displays a significant amount of noise when applying only the standard cuts, as seen in Figure 5. Because of these problems, the work presented in this manuscript can be extended by exploring supplemental parameters, beyond L, which could enhance the characterization of the reference distributions.

The background rejection presented here is a fast and preliminary way to lower the possibility of detecting false SEP events. However, the cuts applied by this work are not completely effective and excesses not related to any confirmed SEP event can still be found in the monitor [8].

As Figures 4b and 5b show, SEPs do not cause a single continuous increment in the trigger rates, but rather a series of increments. This is due to the variation of the geomagnetic field intensity during ISS motion. This dynamic could be modeled and a specific algorithm developed to predict the excesses following the initial one and the end of the event itself.

Another feature visible in Figures 4b and 5b is the presence of null scores in the time-series. The occurrence of scores precisely equal to 0 is due to the very high increments SEPs can cause with respect to the nominal rates. Null scores do not affect the possibility of launching SEP alerts and makes choosing a threshold for the alert trigger much easier.

In the end, AMS collects much more information than the one its triggers provide. The instrument detectors can offer additional data about the observed particles, which could improve the SEP identification and the background rejection.

5. Conclusions

The work presented in this manuscript is the first implementation of a real-time monitor for SEP flows near the ISS. Ref. [5] showed how it is possible to detect SEP events using the trigger rates of AMS-02. This monitoring system makes use of that information to allow for the detection of SEP events in real time. However, in order to provide reliable alerts, the delays in the primary data availability need to be reduced and the background rejection has to be improved.

An SEP monitoring system directly onboard AMS on the ISS represents the best solution for a real-time system, avoiding all problems of latency and buffering that are experienced with the on-ground monitoring. Eventually, the use of more information, coming not only from the trigger system but also from all the other AMS detectors, would improve the SEP detection, reducing the number of false positives and improving the efficiency.

Supplementary Materials: The following supporting information can be downloaded at: https://www.mdpi.com/article/10.3390/instruments7040038/s1, Figure S1: Distributions of AMS-02 live-time of January, 2020. Each panel (**a–l**) shows the distribution obtained for a specific L bin; the bins span between $0.5\,R_0$ and $6.5\,R_0$ and are $0.5\,R_0$ wide. Figure S2: Distributions of AMS-02 trigger rate FT (corrected by the live-time) of January, 2020. Each panel (**a–l**) shows the distribution obtained for a specific L bin; the bins span between $0.5\,R_0$ and $6.5\,R_0$ and are $0.5\,R_0$ wide. Figure S3:

Distributions of AMS-02 trigger rate FTC (corrected by the live-time) of January, 2020. Each panel (**a–l**) shows the distribution obtained for a specific L bin; the bins span between 0.5 R_0 and 6.5 R_0 and are 0.5 R_0 wide. Figure S4: Distributions of AMS-02 trigger rate FTZ (raw) of January, 2020. Each panel (**a–l**) shows the distribution obtained for a specific L bin; the bins span between 0.5 R_0 and 6.5 R_0 and are 0.5 R_0 wide. Figure S5: Distributions of AMS-02 trigger rate FTE (corrected by the live-time) of January, 2020. Each panel (**a–l**) shows the distribution obtained for a specific L bin; the bins span between 0.5 R_0 and 6.5 R_0 and are 0.5 R_0 wide. Figure S6: Distributions of AMS-02 trigger rate LV1 (corrected by the live-time) of January, 2020. Each panel (**a–l**) shows the distribution obtained for a specific L bin; the bins span between 0.5 R_0 and 6.5 R_0 and are 0.5 R_0 wide. Figure S7: Distributions of AMS-02 trigger rate subLV1-0 (corrected by the live-time) of January, 2020. Each panel (**a–l**) shows the distribution obtained for a specific L bin; the bins span between 0.5 R_0 and 6.5 R_0 and are 0.5 R_0 wide. Figure S8: Distributions of AMS-02 trigger rate subLV1-1 (corrected by the live-time) of January, 2020. Each panel (**a–l**) shows the distribution obtained for a specific L bin; the bins span between 0.5 R_0 and 6.5 R_0 and are 0.5 R_0 wide. Figure S9: Distributions of AMS-02 trigger rate subLV1-2 (corrected by the live-time) of January, 2020. Each panel (**a–l**) shows the distribution obtained for a specific L bin; the bins span between 0.5 R_0 and 6.5 R_0 and are 0.5 R_0 wide. Figure S10: Distributions of AMS-02 trigger rate subLV1-3 (raw) of January, 2020. Each panel (**a–l**) shows the distribution obtained for a specific L bin; the bins span between 0.5 R_0 and 6.5 R_0 and are 0.5 R_0 wide. Figure S11: Distributions of AMS-02 trigger rate subLV1-4 (corrected by the live-time) of January, 2020. Each panel (**a–l**) shows the distribution obtained for a specific L bin; the bins span between 0.5 R_0 and 6.5 R_0 and are 0.5 R_0 wide. Figure S12: Distributions of AMS-02 trigger rate subLV1-5 (corrected by the live-time) of January, 2020. Each panel (**a–l**) shows the distribution obtained for a specific L bin; the bins span between 0.5 R_0 and 6.5 R_0 and are 0.5 R_0 wide.

Author Contributions: Conceptualization, A.S. and A.O.; methodology, A.S. and A.O.; software, A.S.; formal analysis, A.S.; resources, V.F. and A.O.; data curation, A.S.; writing—original draft preparation, A.S., M.D. and A.O.; writing—review and editing, A.S., M.D. and A.O.; visualization, A.S.; supervision, M.D. and A.O.; project administration, A.O.; funding acquisition, A.S., M.D. and A.O. All authors have read and agreed to the published version of the manuscript.

Funding: This research was funded by Commissione Scientifica Nazionale 2 (CSN2)—Istituto Nazionale Fisica Nucleare (INFN), study grant number 24156/2022.

Data Availability Statement: The code used in this work, which is in development and continuously updated, is publicly available at [7]. Historical and real-time data processed using the developed code are publicly accessible at [8].

Conflicts of Interest: The authors declare no conflict of interest. The funders had no role in the design of the study; in the collection, analyses, or interpretation of data; in the writing of the manuscript; or in the decision to publish the results.

Abbreviations

The following abbreviations are used in this manuscript:

AMI	AMS Monitoring Interface
AMS	Alpha Magnetic Spectrometer
CME	Coronal Mass Ejection
CR	Cosmic Rays
CSN2	Commissione Scientifica Nazionale 2
ECAL	electromagnetic calorimeter
FT	fast trigger
FTC	fast trigger charged
FTE	fast trigger ECAL
FTZ	fast trigger big-Z
INFN	Istituto Nazionale di Fisica Nucleare
IGRF	International Geomagnetic Reference Field
ISS	International Space Station

LEO	Low-Earth Orbit
LV1	level 1
RICH	Ring Imaging Cherenkov
SAA	South Atlantic Anomaly
SEP	Solar Energetic Particle
TRD	Transition Radiation Detector
ToF	Time-of-Flight

References

1. Longair, M. *High Energy Astrophysics*, 3rd ed.; Cambridge University Press: Cambridge, UK, 2011.
2. Raouafi, N.E.; Patsourakos, S.; Pariat, E.; Young, P.R.; Sterling, A.C.; Savcheva, A.; Shimojo, M.; Moreno-Insertis, F.; DeVore, C.R.; Archontis, V.; et al. Solar Coronal Jets: Observations, Theory, and Modeling. *Space Sci. Rev.* **2016**, *201*, 1–53.
3. Kahler, S.W.; Sheeley, N.R., Jr.; Howard, R.A.; Koomen, M.J.; Michels, D.J.; McGuire, R.V.; Von Rosenvinge, T.T.; Reames, D.V. Associations between coronal mass ejections and solar energetic proton events. *J. Geophys. Res.* **1984**, *89*, 9683–9693. [CrossRef]
4. AMS-02. Available online: https://ams02.space/ (accessed on 15 October 2023).
5. Faldi, F.; Bertucci, B.; Tomassetti, N.; Vagelli, V. Real-time monitoring of solar energetic particles outside the ISS with the AMS-02 instrument. *Rend. Lincei. Sci. Fis. Nat.* **2023** *34*, 339–345. [CrossRef]
6. Hashmani, R.; Konyushikhin, M.; Shan, B.; Cai, X.; Demirköz, M.B. New monitoring interface for the AMS experiment. *NIM-A* **2023**, *1046*, 167704. [CrossRef]
7. AMI SEP Repository. Available online: https://gitlab.cern.ch/aserpoll/ami-sep (accessed on 15 October 2023).
8. AMI SEP Grafana App. Available online: https://ami-sep-monitor.app.cern.ch (accessed on 15 October 2023).
9. McIlwain, C. Coordinates for mapping the distribution of magnetically trapped particles. *J. Geophys. Res.* **1961**, *66*, 3681–3691. [CrossRef]
10. Lin, C. Trigger Logic Design Specification, 2005. Internal Document of AMS-02 Collaboration (n. AMS-JT-JLV1-LOGIC-R02c). Available online: https://ams.cern.ch/AMS/DAQsoft/trigger_logic_v02c.pdf (accessed on 15 October 2023).
11. Shea, M.; Smart, D.; Gentile, L. Estimating cosmic ray vertical cutoff rigidities as a function of the McIlwain L-parameter for different epochs of the geomagnetic field. *Phys. Earth Planet. Inter.* **1987**, *48*, 200–205. [CrossRef]
12. Alken, P.; Thébault, E.; Beggan, C.D.; Amit, H.; Aubert, J.; Baerenzung, J.; Bondar, T.N.; Brown, W.J.; Califf, S.; Chambodut, A.; et al. International Geomagnetic Reference Field: The thirteenth generation. *Earth Planets Space* **2021**, *73*, 49.
13. Major SEP Events. Available online: https://cdaw.gsfc.nasa.gov/CME_list/sepe/ (accessed on 15 October 2023).

 *instruments*

Article

Charge Resolution Study on AMS-02 Silicon Layer-0 Prototype

Alessio Ubaldi [1,2] **and Maura Graziani** [1,2,*]

[1] Dipartimento di Fisica, Università degli Studi di Perugia, 06132 Perugia, Italy; alessio.ubaldi@studenti.unipg.it or alessio.ubaldi@pg.infn.it

[2] INFN Sezione di Perugia, Via Alessandro Pascoli, 23c, 06123 Perugia, Italy

[*] Correspondence:maura.graziani@unipg.it or maura.graziani@pg.infn.it

Abstract: The work presented in this paper represents a preliminary study on the performance of the new Silicon tracker layer, Layer 0 (L0), that will be installed on top of the Alpha Magnetic Spectrometer (AMS-02), at the end of 2024. AMS-02 is a cosmic ray (CR) detector that has been operating on the International Space Station (ISS) since May 2011. Thanks to its nine-layer Silicon tracker, this apparatus can perform high-energy CR measurements with an unprecedented level of statistics and precision. However, high-Z ($Z \geq 15$) CR nuclei statistics is strongly affected by fragmentation along the detector: with the installation of the new Silicon layer, it will be possible to achieve new unique high-energy (TeV region) measurements of those nuclei along with increased statistics for all nuclei up to Zinc. To achieve this, a Silicon ladder prototype, which will be part of the final Silicon layer, was exposed to an ion test beam at the super-proton synchrotron (SPS) of CERN to characterize its charge resolution and the readout electronics. Preliminary results have shown a charge resolution of 10 % for nuclei up to Z = 7.

Keywords: AMS-02 Layer 0 upgrade; silicon micro-strip detector; nuclei; charge resolution; ADC; cosmic ray

Citation: Ubaldi, A.; Graziani, M. Charge Resolution Study on AMS-02 Silicon Layer-0 Prototype. *Instruments* **2023**, *7*, 45. https://doi.org/10.3390/instruments7040045

Academic Editor: Antonio Ereditato

Received: 9 November 2023
Revised: 15 November 2023
Accepted: 17 November 2023
Published: 24 November 2023

1. Introduction

By CRs, we mean various species of energetic particles, charged or not, coming from space with galactic and extra-galactic origin. After the discovery of radioactivity (1896, A. H. Becquerel), it was observed that the rate of discharge of an electroscope increased considerably when it approached radioactive sources. Between 1901 and 1903, numerous researchers noticed that electroscopes discharged even when shielded, deducing that highly penetrating radiation contributed to the spontaneous discharge. The evidence of CRs' extraterrestrial origin is mainly due to the Austrian–American physicist Victor Franz Hess and the Italian physicist Domenico Pacini in the early Twentieth Century. Hess discovered an increase in radiation intensity with altitude in 1912 [1] and was awarded the Nobel Prize in 1936 for that. As well as having established the foundation of particle physics, CRs' discovery and study have provided important contributions to understanding the physical processes underlying the astrophysics phenomenon and have allowed obtaining a closer to complete and more-detailed comprehension of the fundamental mechanisms of particle physics.

CRs are divided into primary ones, which are produced by astrophysical sources, and secondary ones, which are produced by the interactions of the primaries with the interstellar medium. At the top of Earth's atmosphere, the CR radiation is composed of ∼90% of protons, ∼8% of Helium nuclei, ∼1% higher-charge nuclei, and ∼1 % of electrons, positrons, and antiprotons. Most CRs arriving at Earth's surface are constituted by muons, which are a by-product of particle showers formed in the atmosphere by galactic CRs, starting from a single energetic particle. The study of CRs allows one to investigate a wide range of phenomena such as: the production, acceleration, and propagation of the latter. Currently, CRs and accelerator particle physics represent two complementary studies with the aim of solving the current physics mysteries such as the presence of dark matter or the absence of primordial anti-matter in our universe.

CRs' spectrum (number of particles per energy unit, time unit, surface unit, and solid angle) is well described by a power law of the energy, with a power index of ~-2.7 for primary nuclei up to 10^{15} eV. The most-common way to describe the spectrum is by particles per rigidity R: the rigidity R, measured in volts, is defined as R = cp/q, where p and q are, respectively, the momentum and charge of the particle. Particles with different charges and masses have the same dynamics in a magnetic field if they have the same rigidity R.

The AMS-02 experiment is capable of performing precise and continuous measurements of CRs, providing a large amount of statistics and data since its installation on the ISS in May 2011. The apparatus is composed of different subdetectors to measure the characteristics of traversing particles. The core of the instrument is formed by a Silicon tracker composed of nine layers of Silicon micro-strip sensors. A permanent magnet surrounds six layers, forming the spectrometer (inner tracker), which is able to measure the charge sign of a traversing particle. The Transition Radiation Detector (TRD), located at the top, identifies and separates leptons ($e^{\pm}$) from hadrons (p and nuclei). Time-of-flight (ToF) systems determine the direction and velocity of incoming particles and measure their charge. Anti-coincidence counters (ACCs), surrounding the tracker in the magnet bore, reject particles entering sideways. The ring imaging Cherenkov counter (RICH) provides a high-precision measurement of the velocity. The electromagnetic calorimeter (ECAL) is a three-dimensional calorimeter of 17 radiation lengths, which provides energy measurements of positrons and electrons.

AMS-02 has the unique capability of distinguishing matter from anti-matter, thanks to its capability of measuring the charge sign from the track deflection within its magnetic field. No other experiment currently taking data has a similar capability, nor is it foreseen to have one in the near future. In January 2020, AMS-02 was serviced with the installation of a new cooling system, the Upgraded Tracker Thermal Pump System (UTTPS). In the new configuration, the AMS is supposed to take data for the whole life of the ISS, which is currently extended to 2030.

The latest report from the AMS collaboration [2] has highlighted an unprecedented observation: primary CRs have at least two distinct classes of rigidity dependence (Ne, Mg, Si and He, C, O). Moreover, it has been observed that the rigidity dependencies of primary and secondary CR fluxes (Li, Be, B) are distinctly different. These results together with ongoing measurements of heavier elements in CRs will enable determining how many classes of rigidity dependence exist in both primary and secondary CRs and provide important information for the development of the theoretical models.

Measuring both nuclei charge and the sign of the charge, with high precision, is a fundamental requirement to acquire a significant amount of data and supply important information about CR fluxes. In order to do that, an upgrade (Layer 0 upgrade) will be installed on top of the AMS-02 experiment. The AMS-02 Layer 0 upgrade consists of two planes of Silicon micro-strip sensors, both composed by 36 electromechanical units called a "ladder". The upgrade will provide an increase by a factor of three of the acceptance in many analysis channels, along with two new measurements of charge. Elements from Z = 15 to Z = 30 have limited statistics: the upgrade will enable performing complete and accurate measurements of the spectra of the elements up to Zn, where data from the AMS and spectrometry in the TeV region are statistically poor. It will also provide the foundation for a comprehensive theory of CRs. Moreover, the study of secondary CRs with Z > 14 will contribute a complete and unique understanding of the CRs' propagation charge dependence, which is of widespread interest in physics.

In order to achieve these goals, a complete and accurate characterization of the performances of the Silicon sensorsthat will be installed on the apparatus is of fundamental importance. This preliminary work focused on the study of one of those ladders that will be mounted on the Layer 0 planes: in particular, after a description of the components present in the detector, the process of analysis will be reviewed, starting from the calibration of the

Silicon sensors and the electronics, going through the corrections applied to the signal and, finally, arriving at the evaluation of the actual charge resolution of the ladder.

2. Materials and Methods

The Layer 0 Silicon ladder prototype is a fundamental electromechanical unit composed of 10 Silicon sensors and an electronics front-end (LEF) board, which allows the measurements of the charge and position of a passing particle. The main characteristics of the Silicon sensorsused are reported in Table 1.

Table 1. Main characteristics of the ladder prototype and the Silicon sensors used.

Parameter	Rating	Unit
Device type	Single-sided AC readout	-
Silicon type	n-type Phosphorus-doped	-
Crystal orientation	$< 100 >$	-
Thickness	320 ± 15	μm
Front-side metal	AL	-
Back-side metal	AL	-
Chip size	$113{,}000 \pm 20 \ \times 80{,}000 \pm 20$	μm
Active area	$111{,}588.75 \times 78{,}840$	μm
Number of strips	4096	ch
Strip pitch	27.25	μm
Number of readout strips	1024	-
Readout strip pitch	109	μm
Strip width	10	μm
Readout AL width	12	μm
Readout PAD size	56×300	μm

Sixteen application-specific integrated circuits (ASICs) located on the LEF, named IDE1140 or VA, read out 64 Silicon micro-strips each. Each ASIC includes an array of 64 spectrometric channels, an analog multiplexer (MUX), the registers, and the logic elements. An individual spectrometric channel contains a charge-sensitive preamplifier (PA), a shaping amplifier (Shaper), and a sample-and-hold unit. The sample-and-hold units are triggered by a common external signal (HOLD), which is generated by a field-programmable gate array (FPGA) after receiving an external trigger signal. While the HOLD signal is high, the FPGA sends 64 clock pulses to the MUX, providing the sequential readout of the signal values held in the sample-and-hold units. Then, the picked up values are amplified by an internal differential amplifier (DA) and a two- stage separate amplifier (Amp). Finally, all signals are digitized by analog-to-digital converters (ADCs). The scheme in Figure 1 shows all the described components of the VA.

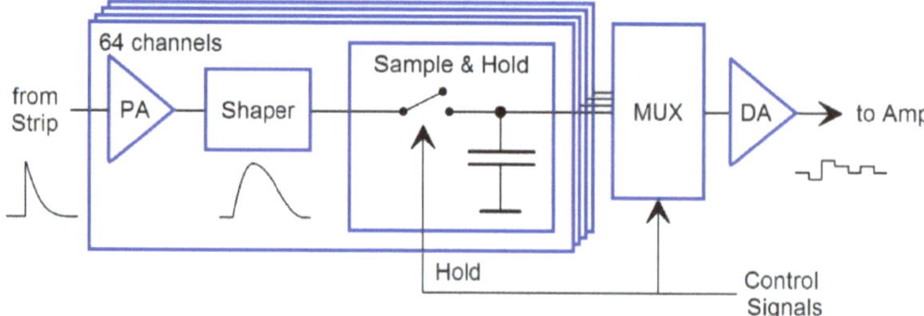

Figure 1. Structural schematic of the IDE1140 [3] demonstrating the signal shape in the critical points of the chip. Abbreviations used here are explained in the text.

An ion beam test was performed in November 2022 at the super-proton synchrotron of CERN: A 40 mm Beryllium target was hit by a primary beam of Pb (379 GV/c), which produced ions by fragmentation. The fragments were selected magnetically, in the rigidity interval of a few percent around 300 GV/c. At this scale of rigidity, every ion is considered a minimum ionizing particle (MIP).

The Bethe–Bloch formula describes the average energy loss by a particle with charge Z that traverses a target: for a fixed $\beta = v/c$ and a fixed target, that quantity only depends on the charge-squared Z^2 of the incident particle. These average ionization losses are stochastic in nature, and the Bethe–Bloch formula gives the mean value of these losses: the fluctuations around this value, in thin materials, are well described by the convolution of a Gaussian and a Landauian (LanGauss) distribution [4]. Having a beam with a population of different ions with different charges, the population distribution will be the sum of the single convolutions provided by the individual species.

2.1. Calibration and Clusterization

The ADC values of the readout strips for the i-th channel on the j-th VA preamplifier in the k-th event can be written as:

$$x_{ij}^k = p_{ij} + c_j^k + s_{ij}^k + q_{ij}^k \tag{1}$$

where p_{ij} is a constant offset pedestal (unique for each channel), c_j^k a coherent common noise component (which affects in the same way all the channels belonging to the same VA), s_{ij}^k the strip noise, and q_{ij}^k an eventual signal due to the passage of an ionizing particle in the depleted Silicon. The calibration procedure consists of the determination of the noise (σ_s) for each readout channel, recording n events in absence of incident particles ($q_{ij}^k = 0$):

$$\sigma_s = \sqrt{\frac{1}{n}\sum_{k=1}^{n}(s_{ij}^k)^2} = \sqrt{\frac{1}{n}\sum_{k=1}^{n}(x_{ij}^k - c_j^k - p_{ij})^2} \tag{2}$$

To determine the noise, it is necessary to evaluate the pedestal p and the common noise values c. The first half part of the n events taken establishes the preliminary values of the strip pedestals (p_{ij}^{RAW}):

$$p_{ij}^{RAW} = \frac{2}{n}\sum_{k=1}^{n/2} x_{ij}^k \tag{3}$$

and their standard deviations:

$$\sigma_{ped}^{RAW} = \sqrt{\frac{2}{n}\sum_{k=1}^{n/2}(x_{ij}^k - p_{ij}^{RAW})^2} \tag{4}$$

The final values of the strip pedestals are computed using the second half of the n events taken using:

$$p_{ij} = \frac{2}{n}\sum_{k=n/2}^{n} x_{ij,good}^k \tag{5}$$

where the ADC values $x_{ij,good}^k$ are the ones inside $\pm 3\sigma_{ped}^{RAW}$ with respect to p_{ij}^{RAW}. Thanks to this procedure, the too-noisy channels for a given event are excluded from the evaluation of the pedestals. The common noise is produced by the fluctuations of the power supply and other electromagnetic interferences, and it is constant for all the preamplifiers contained in the same VA. It is evaluated event by event for each VA after subtracting the pedestal, calculating the median value. This procedure defines a valid signal by applying a threshold to the signal-to-noise ratio (S/N) of the strip:

$$\frac{S}{N} = \frac{x_{ij}^k - c_j^k - p_{ij}}{\sigma_s} \tag{6}$$

After the calibration procedure, every channel contains two contributions: the strip noise and a possible value due to the crossing particle.

To correctly measure the charge of a crossing particle, it is necessary to identify all the strips that are interested in collecting all the released signal in the Silicon from that particle. This process is called clusterization. A cluster is a group formed by all the strips involved in the collection of the ionization energy loss by a particle. This process is performed by checking the S/N of every readout strip: The first strip found with this ratio above a certain threshold (n_H) is defined as the seed of the cluster. All strips adjacent to the seed are added to the cluster until their S/N ratio is above a second lower threshold (n_L). This procedure is performed for all 1024 readout strips of the Silicon ladder. An example is reported in Figure 2.

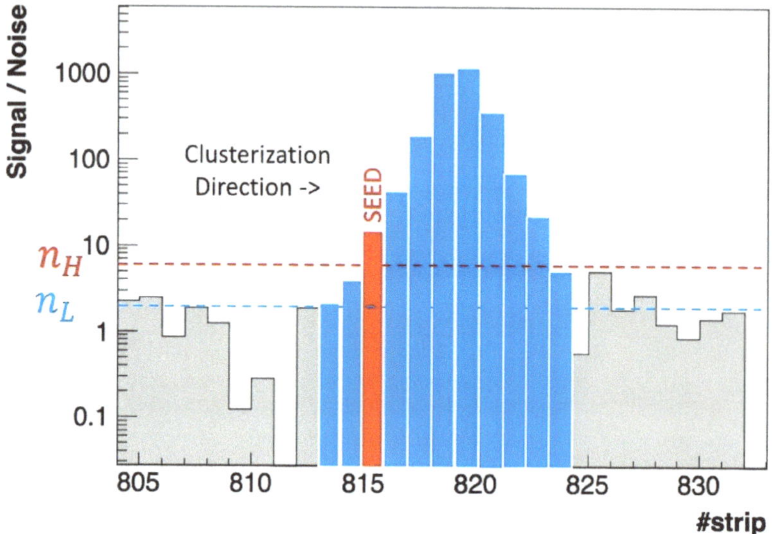

Figure 2. Signal-to-noise ratio (in logarithmic scale) as a function of the strip number for a single event. The red line indicates the higher threshold n_H, which defines the cluster seed. All strips adjacent to the seed are added to the cluster until their S/N ratio is above the blue lower threshold n_L. The cluster will be formed by all the highlighted strips.

2.2. Trigger-to-Hold Time

The time, or delay, between the arrival of the external trigger and the sampling of the signal is the so-called trigger-to-hold time: waiting for the correct amount of time between these two events is a crucial point in order to sample the peak of the shaped signal. In order to find the best value for the trigger-to-hold time, a dedicated study on CERN beam test data was performed. During the data acquisition, different runs with about the same amount of data were made with different values of the trigger-to-hold time. In total, six datasets with, respectively, 3.5 μs, 5.5 μs, 6.5 μs, 7.5 μs, 8.5 μs, and 9.5 μs of the trigger-to-hold time were analyzed. For each dataset, the distribution of the total cluster amplitude (ADC), where the amplitude is the sum of all the contributions of all the individual cluster strips, was fit using a Landauian function. The behavior of the most-probable values extrapolated from the fits as a function of the trigger-to-hold time is shown in Figure 3.

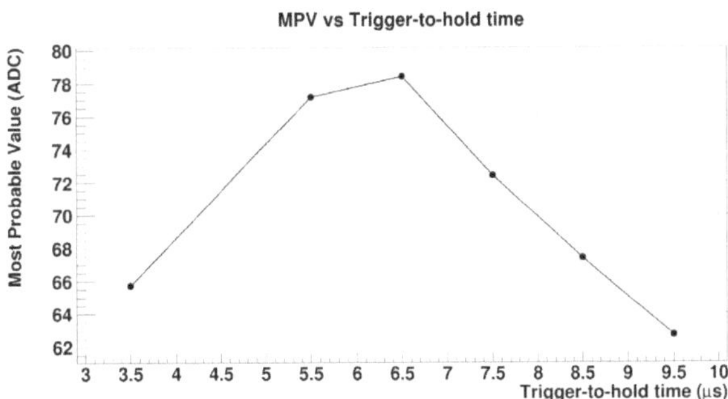

Figure 3. Most-probable values as a function of the trigger-to-hold time. When the trigger signal for sampling is coming, it is necessary to wait a certain amount of time in order to sample and hold the signal peak. The value that allows this is around 6.5 µs according to our study. As can be seen in the figure, waiting too much or too little time compared to 6.5 µs leads to a smaller sampled signal amplitude.

2.3. Eta Correction

Once all the events are clusterized, it is possible to proceed with the evaluation of the charge resolution. The dataset used for the evaluation of the charge resolution was acquired with high and low clusterization thresholds of 5.5 and 2.0, respectively, and with a trigger-to-hold time of $\sim$6 µs. Selecting the most-energetic cluster per event, i.e., the cluster with maximum amplitude, allows a noise rejection and good cluster choosing. The considered ladder has a total of 4096 Silicon micro-strips, but only one every four adjacent strips (1024) is effectively read out by the electronics: The intermediates, called floating strips, are capacitively coupled with the readout ones. All the strips are also capacitively coupled with the metalized back plane, allowing the operation of the Silicon sensorsin overdepleted mode [5]. This electrical scheme leads to an inter-strip energy loss. When collecting ionization, the floating strips share all the acquired signal with the nearest readout strips, but when doing this, part of the signal is lost due to the capacitive coupling with the back plane. As a first approximation, to quantify the inter-strips' energy loss, it is sufficient to study the signal shared between the two strips closest to the particle impact position. A more-realistic description of the capacitive charge sharing has to take into account not only the direct inter-strip capacitance with the first neighboring strips, but also indirect coupling to the second and even third readouts [6]. The inter-strips' energy loss is quantified by η, defined as follows:

$$\eta = \frac{S_1}{S_1 + S_2} \qquad \eta \in (0,1) \tag{7}$$

where S_1 and S_2 are the signals in the ADC of the two highest strips of the cluster (coinciding with the two closest to the impact position). The dependency of the total cluster amplitude on eta is shown in Figure 4: the region between the two black lines corresponds to the energy deposited by $Z = 2$ particles. Different eta values, i.e., different impact positions with respect to the two highest strips of the cluster, correspond to different ADC values for the same charge. To take into account this dependency, the ADC distribution is supposed to be parabolic in eta and constant for every amplitude:

$$f(\eta) = a\eta^2 + b\eta + c \tag{8}$$

To find the coefficients of the parabola, the $Z = 2$ sample was used. The cluster amplitude distribution was fit with a Landauian function around the maximum, for three different eta intervals: $\eta \in \{[0,0.08], [0.46,0.54], [0.92,1]\}$. The regions chosen for this purpose are shown in Figure 5, and the fits on the $Z = 2$ peak for the different regions are shown in Figure 6. The passage of the parabola was imposed on three points, each one composed of the eta values (0,0.5,1), and the most-probable values are shown in Figure 6. The eta correction is, finally, defined as:

$$\omega = \frac{c}{f(\eta)} \tag{9}$$

To get rid of the inter-strip energy loss, every ADC value was multiplied by $\omega = c/f(\eta)$, where c is the known term of the parabola and $f(\eta)$ will be the parabola value at the eta point corresponding to the ADC value that we want to correct.

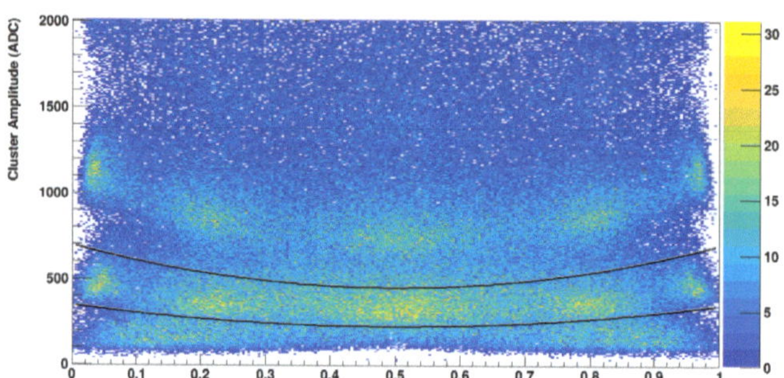

Figure 4. Cluster amplitude distribution as a function of eta. The region between the two black lines is the sample corresponding to $Z = 2$ chosen to characterize the eta dependency.

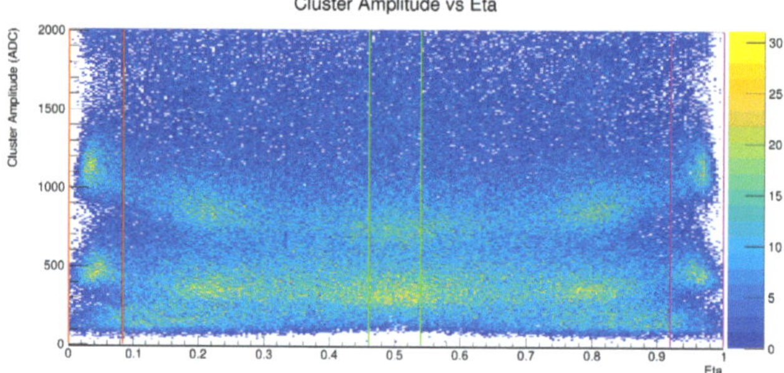

Figure 5. Cluster amplitude as a function of eta. In red, green, and purple are highlighted the three selected eta intervals.

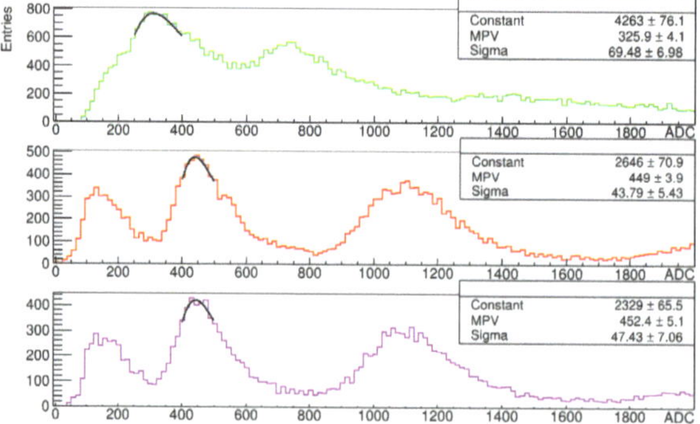

Figure 6. Distribution of the cluster amplitude for the three selected eta intervals: green (0.46 ÷ 0.54), red (0 ÷ 0.08), and purple (0.92 ÷ 1). The black lines represent the Landau fit around the maximum of the $Z = 2$ sample.

2.4. VA Equalization

Another signal correction was performed considering the VAs: ideally, one wants to observe the same response, i.e., the same ADC value, for each VA for a given Z. This did not happen, and different VAs had different response functions, which provide different ADC values for the same charges. An equalization of 9 of the 16 VAs (from Number 5 to 13) was made, considering VA Number 10 as a reference.

Figure 7 shows the corrected cluster amplitude distribution as a function of the strip number: in red is highlighted the VA number, going from 1 to 16. To equalize the VAs with respect to VA Number 10, the corrected cluster amplitude distribution inside a 64-channel range (which corresponds to a full VA) was studied. As mentioned, the distribution of a population containing different ions will be the sum of the single convolutions (between a Gaussian and a Landauian) provided by each ion. Figure 8 reports the corrected cluster amplitude distribution for VA Number 10: the red lines represent the fit performed around the peaks with the convolution between a Gaussian and a Landauian in order to estimate the most-probable values for the energy deposited by charge from Z = 2 to Z = 7. The first peak corresponds to Z = 1; despite being performed, Z = 1 was excluded from the analysis.

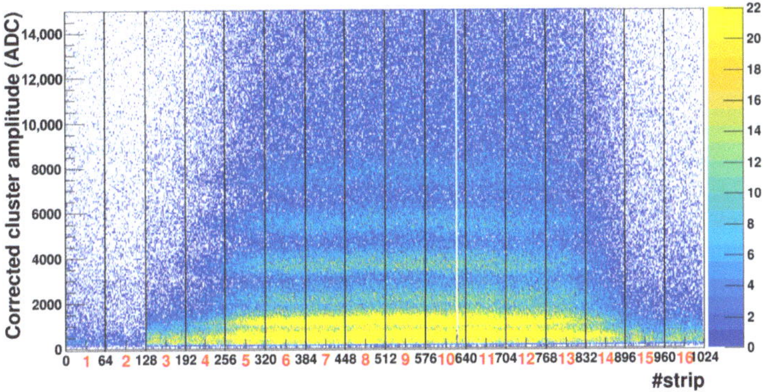

Figure 7. Corrected cluster amplitude distribution as a function of strip number. In red is reported the VA number.

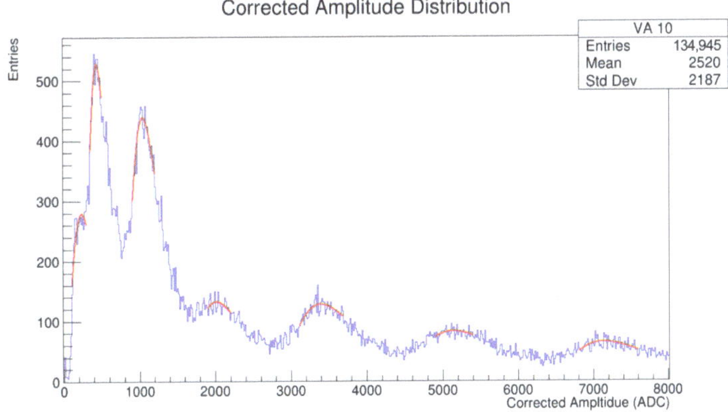

Figure 8. Corrected cluster amplitude distribution for VA Number 10. The red lines are the fit performed with the convolution of a Gaussian and a Landauian around the peaks. The first peak corresponds to Z = 1, but was excluded from the analysis.

The same procedure was applied for the remaining eight VAs. For the k-th VA (for a total of nine VAs from Number 5 to 13), the fits gave six most-probable values of MPV_i^k, with $i = 2, \dots, 7$ and $k = 5, \dots, 13$. Then, the response functions of every VA with respect to VA 10 were built by using the ratio between the most-probable values of VA 10 (MPV_i^{10}) and the most-probable values of the remaining VAs (MPV_i^k) as a function of MPV_i^k, $k \in \{5, 6, 7, 8, 9, 11, 12, 13\}$. To clarify, Figure 9 shows the ratio between VA Number 10 and VA Number 11.

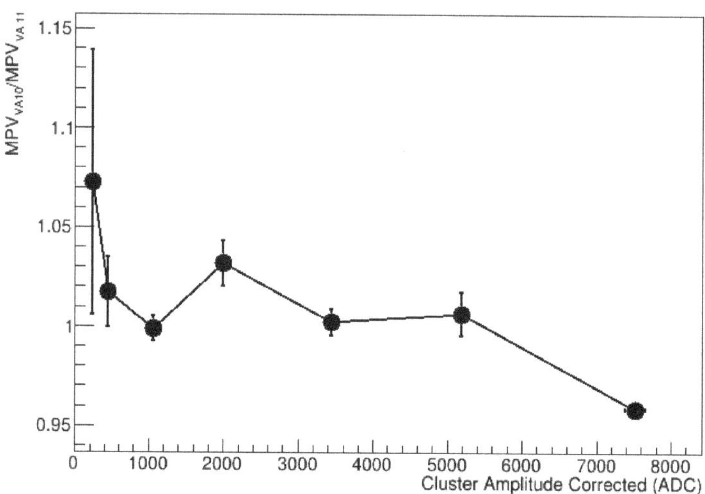

Figure 9. VA 11 equalization function (with respect to VA Number 10). On the x-axis is the corrected cluster amplitude for VA Number 11, and on the y-axis is the ratio between the most-probable values of VA Number 10 and VA Number 11. Despite being reported, Z = 1 was excluded from the analysis.

The first point corresponds to Z = 1: despite being reported, Z = 1 was excluded from the analysis because the trigger conditions were set in order to minimize the acquisition of that type of event. So, the statistics for Z = 1 is very poor and inappropriate to perform any type of statistical analysis. In reference to the same figure, the polyline that joins the points represents the function used for the equalization of VA Number 11, f^{11}. The signal measured by the k-th VA, S^k, was equalized with respect to VA Number 10 by:

$$f^k(S^k) \cdot S^k \tag{10}$$

where f^k is the equalization function for the k-th VA, obtained with the same procedure explained for k = 11.

2.5. Saturation

The analysis performed on the Silicon ladder was performed up to Z = 7, and it was not possible to acquire higher charges because of electronics saturation. This behavior is due to the dynamic range of the VA and the preamplifier. Figure 10 shows the output of the VA as a function of the input signal. The VA output is a linear function of the input signal only below a certain value, that is 172 fC. As long as the input charge is below 172 fC, the VA output is linear with the charge, but above this threshold, the VA gain decreases rapidly, leading to the same output for a large range of input charges. Incident particles generate an amount of ionization and, so, a VA input that is increasing with Z^2: the non-linear behavior of the VA for high charges limited the analysis to only those charges with $Z \leq 7$.

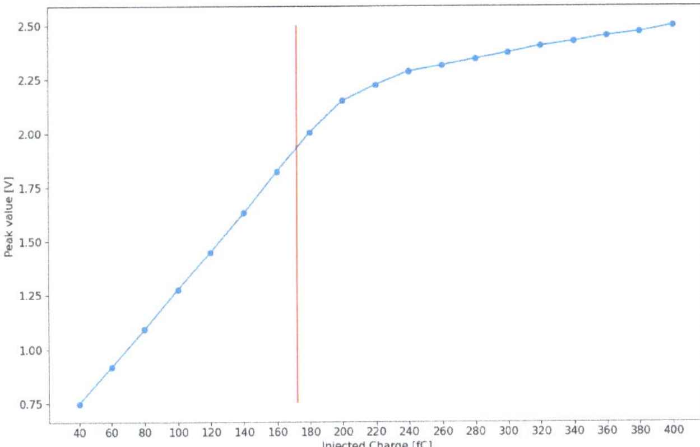

Figure 10. Voltage output of the VA as a function of the injected charge: the red line represents the declared limit of the linear range of the VA, which corresponds to 172 fC.

2.6. Charge Resolution

Figure 11 shows the distribution of the total cluster amplitude corrected by eta and equalized with respect to VA Number 10 and the six convolution functions used to fit that distribution. The applied procedure to measure the final charge resolution was the following:

- The total cluster amplitude corrected by eta and equalized with respect to VA Number 10 was fit with six different LanGauss functions;
- The parameters obtained from the fits were used to generate a Monte Carlo (MC) toy for each charge sample by the square root of a random event generated using the probability density functions (PDFs). Thanks to the Bethe–Bloch formula, the mean energy loss by a particle was proportional to Z^2, which was measured by the detector in ADC counts. In order to evaluate Z, it was necessary to study the $\sqrt{ADC}$ distribution;
- The MC toy was used to apply the central limit theorem (CLT) to estimate the charge resolution.

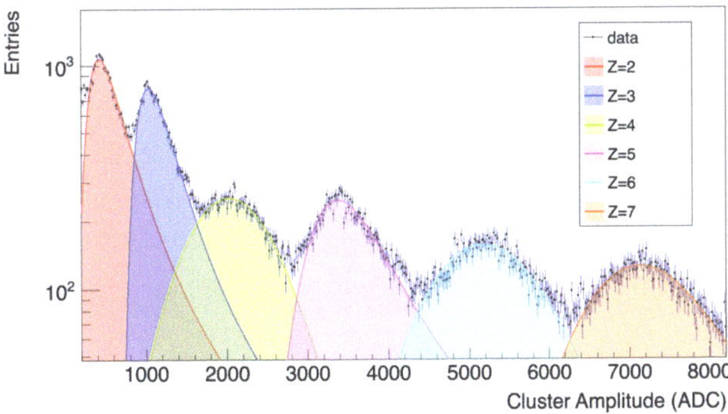

Figure 11. Distribution of the total cluster amplitude corrected by eta and equalized with respect to VA Number 10. Every peak was fit using the convolution of a Gaussian and a Landauian.

The PDFs $f_i(Z^2)$ with i = 2, ... ,7, were built. A sample of N = 1000 events for the i-th charge was generated by the square root of a random event created using $f_i(Z^2)$.

As an example, in Figure 12a is reported the $\sqrt{ADC}$ distribution for Z = 3 and its arithmetic mean generated with the MC toy. The $\sqrt{ADC}$ distribution reported in the same figure follows a PDF, $f(Z)$, with an expectation value of $\hat{z}$ and variance $(\Delta z)^2$. The resolution of the charge will be $\Delta z / \hat{z}$.

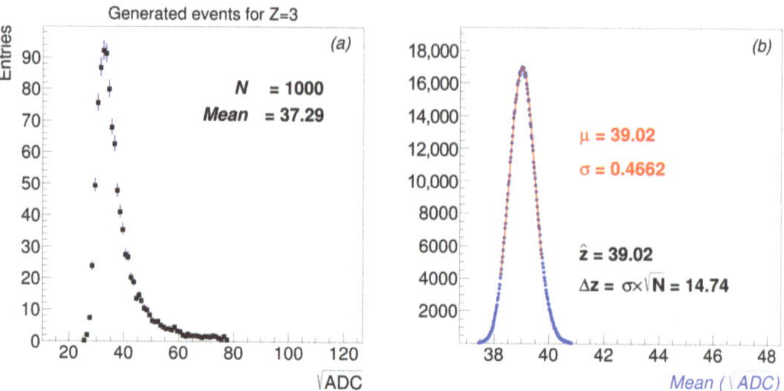

Figure 12. (**a**) Sample of N = 1000 events containing the distribution of $\sqrt{ADC}$ generated with a Monte Carlo experiment with $f_3(Z^2)$ for Z = 3. (**b**) Distribution of the mean ($\sqrt{ADC}$) for M = 10^6 Monte Carlo experiments (each one with N = 1000 events) for Z = 3. According to the central limit theorem, it is possible to evaluate $\hat{z}$ and Δz.

According to the central limit theorem (CLT), for a variable x with expectation value $E[x] = \hat{z}$ and variance $V[x] = (\Delta z)^2$, the distribution of the mean is Gaussian with mean μ and variance σ^2 linked to $\hat{z}$ and $(\Delta z)^2$ by:

$$\mu = \hat{z} \qquad \sigma = \frac{\Delta z}{\sqrt{N}}, \quad N \to \infty \tag{11}$$

Figure 12b shows the mean distribution for Z = 3 for M = 10^6 Monte Carlo experiments (each one with N = 1000 events): it is possible to evaluate $\hat{z}$ as the mean of the Gaussian distribution and Δz as $\sigma \cdot \sqrt{N}$. Figure 13 shows the distributions of the means for all the charges under study (from Z = 2 to Z = 7) and the relative Gaussian fit with mean μ and standard deviation σ.

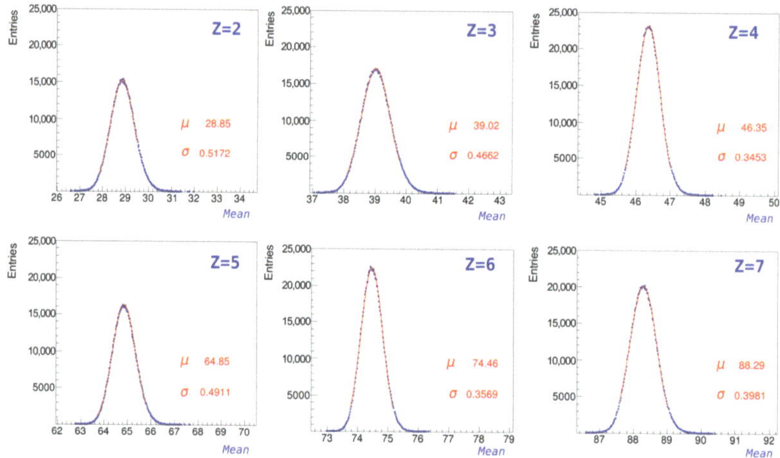

Figure 13. Distributions of the means for all the charges under study (from Z = 2 to Z = 7) and the relative Gaussian fits with mean μ and standard deviation σ.

3. Results

Charge Resolution

Figure 14 shows the preliminary results we obtained for the charge resolution of the Layer 0 prototype for a single layer (red points) and for two layers (square blue points) compared with the charge resolution of the AMS-02 inner tracker (L2 to L8; hollow orange points). The Layer 0 upgrade will be composed by two planes, and its overall charge resolution can be evaluated by the combination of two independent measurements. We evaluated the resolutions for two layers assuming that the charge resolution is the same for both: in this case, it was $1/\sqrt{2}$-times the resolution of a single layer. In Table 2 are reported the charge resolution values we evaluated for charges from Z = 2 to Z = 7 for a single layer of Layer 0 and for two layers. For comparison, the values of the inner tracker (L2 to L8) charge resolution for the same charges are also reported.

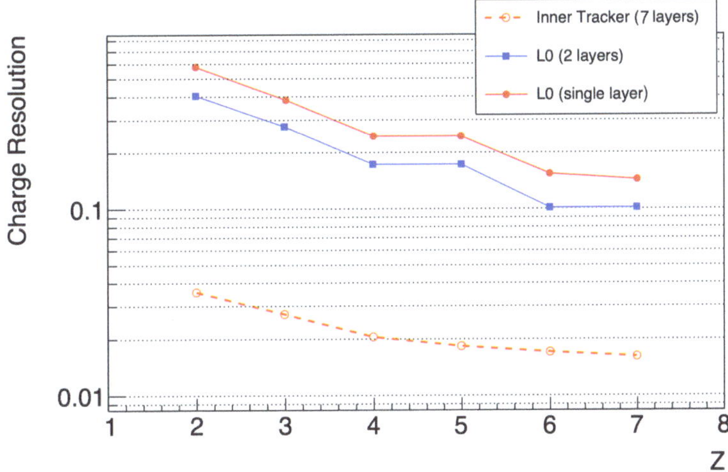

Figure 14. Preliminary charge resolution values as a function of charge Z we obtained for a single layer (red points) and for two layers (square blue points) of Layer 0 compared with the current charge resolution of the AMS-02 inner tracker, from Layer 2 (L2) to Layer 8 (L8) [7] (hollow orange points), obtained by the combination of 7 layers.

Table 2. Values of charge resolution we obtained for different Z values (first column) both for single layer (second column) and two layers (third column) of Layer 0. In the fourth column is reported the current AMS-02 inner tracker (L2 to L8) charge resolution.

Z	L0 (Single Layer)	L0 (Two Layer)	Inner Tracker (L2 to L8)
2	0.57	0.40	0.035
3	0.38	0.27	0.027
4	0.24	0.17	0.02
5	0.24	0.17	0.018
6	0.15	0.10	0.017
7	0.14	0.10	0.016

To conclude, in Figure 15 is reported the comparison between the charge resolution we evaluated for a single layer of L0 with the charge resolution for a single layer of the AMS-02 inner tracker (L2 to L8).

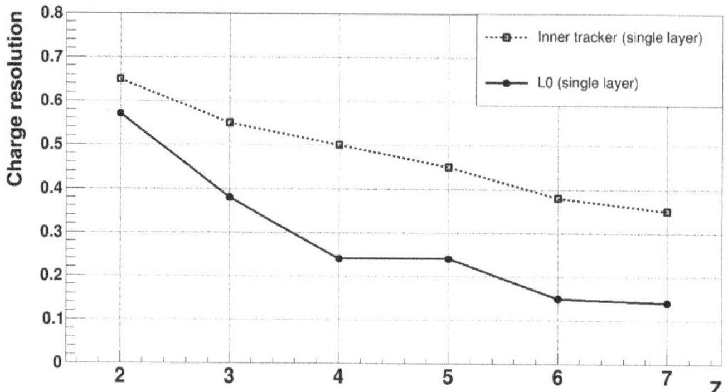

Figure 15. Charge resolution we evaluated for a single layer of Layer 0 (filled circles) compared with the charge resolution of a single layer of the AMS-02 inner tracker (L2 to L8, hollow squares) [7].

4. Discussion

The study performed on the Silicon AMS-02 Layer 0 prototype showed agreement in terms of the charge resolution with respect to the AMS-02 inner tracker. The signal collected by the Silicon sensors was analyzed after the calibration to find an algorithm that allows discriminating the signal from the noise. After the selection of the signal, an accurate characterization of the signal released by different species was performed with $2 \leq Z \leq 7$ in the Silicon sensors that will be used for the construction of L0. In the current state, the new layer on top of AMS-02 will be able to measure the charge at least up to $Z = 7$ with a resolution of 10%.

The obtained resolutions can be further improved. For example, the lack of statistics for $Z = 1$, due to trigger conditions, can be compensated by future data acquired with a suitable setup to maximize the $Z = 1$ particles' acquisition.

Furthermore, the applied correction for η can be improved considering the different dependencies for different charges.

Moreover, the saturation of the electronics (VA) limited the acquisition of high charges to $Z = 7$. Indeed, considering that the analog-to-digital converter has an input range of $0 \div 4$ V (the full scale equals $2^{14} - 1$ ADC = 16 383 ADC), this is consistent with the fact that saturation appeared at a VA output value of 2 V, which corresponds approximately to 8000 ADC. By investigating the dynamic range of the VA itself and by studying the various amplify stages, it will be possible to study even higher charges.

Moreover, other improvements would be possible considering the fits performed on the cluster distribution of Figure 11: in the current state, every contribution has been fit with a single LanGauss function described by five parameters, for a total of six different LanGauss functions. The fit could be improved by using a single function constituted by the sum of six LanGauss functions, which could possibly be more accurate.

5. Conclusions

The work presented in this manuscript is the first preliminary characterization of the performance, in terms of charge resolution, of the new Silicon sensors that will be mounted on the AMS-02 Layer 0 upgrade.

The values we evaluated for the charge resolution for charges from $Z = 2$ to $Z = 7$ for L0 differed by an order of magnitude with respect to the current overall resolution of the AMS-02 inner tracker: this was due to the fact that the latter values were obtained by the combination of seven layers.

Despite this, the values we evaluated for the charge resolution for a single layer of Layer 0 were significantly smaller (2.5-times for $Z = 7$) than the values for the single layer of the AMS-02 inner tracker, as shown in Figure 15; this is promising, although the signal correction can be further improved. This will be a starting point for a future analysis and a complete characterization of the final Layer 0 charge resolution.

Author Contributions: Conceptualization, A.U. and M.G.; methodology, A.U. and M.G.; software, A.U. and M.G.; formal analysis, A.U. and M.G.; data curation, A.U.; writing—original draft preparation, A.U.; writing—review and editing, A.U. and M.G.; visualization, A.U. and M.G.; supervision, M.G. All authors have read and agreed to the published version of the manuscript.

Funding: This research received no external funding.

Institutional Review Board Statement: Not applicable.

Informed Consent Statement: Not applicable.

Data Availability Statement: The data presented in this study can be made available upon request from the corresponding author.

Acknowledgments: The authors would like to acknowledge the continuous support of the AMS Perugia group, Alberto Oliva (INFN of Bologna) and Valerio Formato (INFN of Roma Tor Vergata). Research supported by: INFN and ASI under ASI-INFN Agreements No. 2019-19-HH.0, and ASI-University of Perugia Agreement No. 2023-10-HH.0

Conflicts of Interest: The authors declare no conflict of interest.

Abbreviations

The following abbreviations are used in this manuscript:

L0	Layer 0
AMS	Alpha Magnetic Spectrometer
ISS	International Space Station
CR	cosmic ray
eV	electron volt
TeV	teraelectron volt
SPS	super-proton synchrotron
TRD	transition radiation detector
ToF	time-of-flight
ACC	anti-coincidence counter
RICH	ring imaging Cherenkov
ECAL	electromagnetic calorimeter
UTTPS	Upgraded Tracker Thermal Pump System
LEF	L0 electronics front-end
ASIC	application-specific integrated circuit
MUX	analog multiplexer
PA	preamplifier
FPGA	field-programmable gate array
DA	differential amplifier
ADC	analog-to-digital converter
MIP	minimum ionizing particle
LanGauss	convolution of a Gaussian and a Landauian
S/N	signal-to-noise ratio
MPV	most-probable value
fC	femtocoulomb
MC	Monte Carlo
PDF	probability density function
CLT	central limit theorem
L2	Layer 2
L8	Layer 8

References

1. Hess, V.F. Über Beobachtungen der durchdringenden Strahlung bei sieben Freiballonfahrten *Phys. Z.* **1912**, *13*, 1084–1091
2. Aguilar, M.; Cavasonza, L.A.; Ambrosi, G.; Arruda, L.; Attig, N.; Barao, F.; Barrin, L.; Bartoloni, A.; Pree, S.B.; Bates, J.; et al. The Alpha Magnetic Spectrometer (AMS) on the international space station: Part II—Results from the first seven years. *Phys. Rep.* **2021**, *894*, 1–116 [CrossRef]
3. Bogdanov, A.A.; Eremin, I.V.; Chichagov, Y.V.; Eremin, V.K.; Tuboltsev, Y.V.; Verbitskaya, E.M.; Bezbah, A.A.; Fomichev, A.S.; Kiselev, O.A.; Kostyleva, D.A. Two-dimensional position-sensitive spectrometer for registration of ionizing radiation. *J. Phys. Conf. Ser.* **2019**, *1400*, 055050 [CrossRef]
4. Hancock, S. James, F.; Movchet, J.; Rancoita, P.G.; VanRossum, L. Energy loss and energy straggling of protons and pions in the momentum range 0.7 to 115 GeV/c. *Phys. Rev. A* **1983**, *28*, 615. [CrossRef]
5. Lutz, G.; Schwarz, A.S. Silicon Devices for Charged-Particle Track and Vertex Detection. *Annu. Rev. Nucl. Part. Sci.* **1995**, *45*, 295–335 [CrossRef]
6. Dabrowski, W.; Grybos, P.; Idzik, M. Study of spatial resolution and efficiency of silicon strip detectors with charge division readout. *Nucl. Instrum. Methods Phys. Res. Sect. A Accel. Spectrom. Detect. Assoc. Equip.* **1996**, *383*, 137–143 [CrossRef]
7. Jia, Y.; Yan, Q.; Choutko, V.; Liu, H.; Oliva, A. Nuclei charge measurement by the Alpha Magnetic Spectrometer silicon tracker. *Nucl. Instrum. Methods Phys. Res. Sect. A Accel. Spectrom. Detect. Assoc. Equip.* **2020**, *972*, 164169. [CrossRef]

 *instruments*

Article

Collection of Silicon Detectors Mechanical Properties from Static and Dynamic Characterization Test Campaigns

Edoardo Mancini [1,2,*], Lorenzo Mussolin [1,3], Giulia Morettini [2], Massimiliano Palmieri [2], Maria Ionica [1], Gianluigi Silvestre [1], Franck Cadoux [4], Agnese Staffa [2], Giovanni Ambrosi [1], Filippo Cianetti [2], Claudio Braccesi [2], Lucio Farnesini [1], Mirco Caprai [1], Gianluca Scolieri [1], Roberto Petrucci [2] and Luigi Torre [2]

[1] Perugia Department, National Institue of Nuclear Physics, Via A. Pascoli, 06123 Perugia, Italy; gianluigi.silvestre@pg.infn.it (G.S.); giovanni.ambrosi@pg.infn.it (G.A.)
[2] Engineering Department, University of Perugia, Via G. Duranti 93, 06124 Perugia, Italy; filippo.cianetti@unipg.it (F.C.); roberto.petrucci@unipg.it (R.P.);
[3] Physics Department, University of Perugia, Via A. Pascoli, 06123 Perugia, Italy
[4] Nuclear Physics Department, University of Geneve, 24 Rue du Général-Dufour, 1205 Geneve, Switzerland
* Correspondence: edoardo.mancini@pg.infn.it

Abstract: Physics research is constantly pursuing more efficient silicon detectors, often trying to develop complex and optimized geometries, thus leading to non-trivial engineering challenges. Although critical for this optimization, there are few silicon tile mechanical data available in the literature. In an attempt to partially fill this gap, the present work details various mechanical-related aspects of spaceborne silicon detectors. Specifically, this study concerns three experimental campaigns with different objectives: a mechanical characterization of the material constituting the detector (in terms of density, elastic, and failure properties), an analysis of the adhesive effect on the loads, and a wirebond vibrational endurance campaign performed on three different unpotted samples. By collecting and discussing the experimental results, this work aims to fulfill its purpose of providing insight into the mechanical problems associated with this specific application and procuring input data of paramount importance. For the study to be complete, the perspective taken is broader than mere silicon analysis and embraces all related aspects; i.e., the detector–structure adhesive interface and the structural integrity of wirebonds. In summary, this paper presents experimental data on the material properties of silicon detectors, the impact of the adhesive on the gluing stiffness, and unpotted wirebond vibrational endurance. At the same time, the discussion of the results furnishes an all-encompassing view of the design-associated criticalities in experiments where silicon detectors are employed.

Keywords: silicon detectors; mechanical characterization; random vibrations; shock; mechanical space qualification

Citation: Mancini, E.; Mussolin, L.; Morettini, G.; Palmieri, M.; Ionica, M.; Silvestre, G.; Cadoux, F.; Staffa, A.; Ambrosi, G.; Cianetti, F.; et al. Collection of Silicon Detectors Mechanical Properties from Static and Dynamic Characterization Test Campaigns. *Instruments* **2023**, *7*, 46. https://doi.org/10.3390/instruments 7040046

Academic Editor: Antonio Ereditato

Received: 14 October 2023
Revised: 7 November 2023
Accepted: 22 November 2023
Published: 24 November 2023

1. Introduction

At the present date, single-sided or double-sided silicon detectors (SSSD or DSSD) are widely utilized in space missions [1]; specifically, in experiments like AMS-02 [2–4] and DAMPE [5–7]. In addition, a substantial commitment has been made to the development of new and more efficient detectors, like mini-PAN [8,9], HERD [10–12], and ALADInO [13,14]. While, on the one hand, the research on the performance enhancement of the sensitive unit is of paramount importance, on the other hand, the overall experiment geometry is often as crucial for the development of improved detectors. Indeed, the performance-enhancement problem is faced at two scales: one deals with the active component of the detector, and the other with the experiment as a whole. Debating the latter, the critical parameters are hermeticity and non-active-material reduction. Within this framework, the scarcity of knowledge about detector mechanics could obstruct the maximization of such critical parameters. More extensively, the design of a silicon-based space experiment consists of developing a stiff, strong, and thermally stable structure. The design process is an iterative balance between structural material minimization (opaque to the particles), mechanical integrity (during the launcher flight), and positional stability (during on-orbit thermal cycles). The design procedure is depicted in Figure 1, starting with an iterative phase where the design is proposed, structurally analyzed, and then updated, and followed by a production phase, a testing phase, and the final launch (in the ideal scenario where tests are successful and there is no need to iterate the design after them).

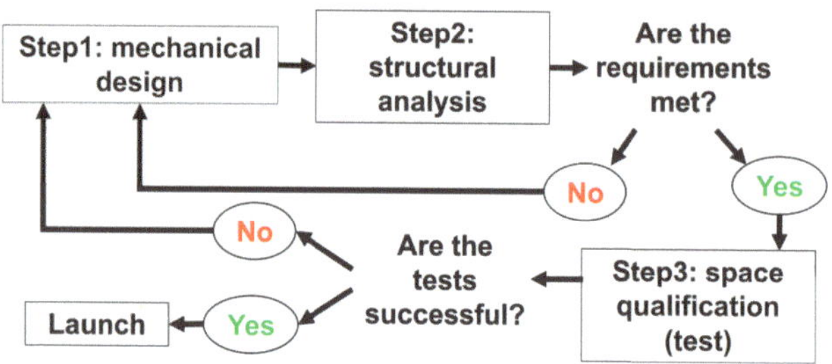

Figure 1. Standard design procedure.

The described procedure, which is standard in mechanical design studies (for more information on space structure design refer to dedicated literature [15,16]), has also been proved suitable for particle physics space experiments. Nevertheless, the unavailability of silicon-related mechanical data reduces model accuracy, and although missions are successful, they may operate with non-necessary structural mass. Indeed, in the design of the supporting structures, the contribution of the silicon to the system's mechanical properties is often neglected and only the contribution in mass is considered. Instead, considering the silicon mechanical properties by accounting for their stiffness and structural properties, or exploring a precise knowledge of silicon failure parameters, could open new and fascinating possibilities (like the one presented in [17]). Adding this concept to the design procedure leads to the new approach presented in Figure 2.

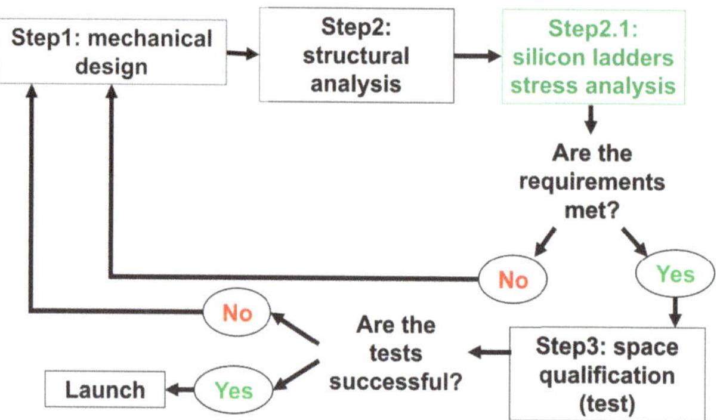

Figure 2. Procedure involving silicon detector properties.

To further clarify, structural analysis targets the provision of a priori information about an experiment's mechanical behavior during the flight and its operative life. As for any other model, a higher level of detail means more accurate results. With structural integrity being mandatory for the mission's success, the designers fill the gap between model and reality with safety margins, hence additional mass.

Conversely, introducing a proper silicon description into the model could reduce the uncertainties (ergo the margin), resulting in a mass reduction and performance increase. To build better models, two kinds of data are needed: stiffness-related and strength-related.

The first are important to account for the presence of detector assemblies, not only as additional masses but as mechanical objects. The second are to predict the detector's failure conditions more precisely. Once again, the need is for information on the whole payload and not only the sensors; the analysis should embrace the whole system installed on the structure. Consequently, to provide a comprehensive analysis, the present document does not only dwell on the characterization of silicon tile stiffness and strength (presented in Section 2) but also discusses the role of the structure–detector mechanical interface (glue) in the dynamics (in Section 3), and finally also on the wirebonds, which are crucial for the detector's functionality (Sections 4.1 and 4.2).

In conclusion, the present work is a comprehensive study of silicon detector assemblies' mechanical aspects. The work breaks this into three main parts. The first concerns a set of flexural tests performed on silicon detector tiles to gain knowledge of the stiffness and strength properties of the detective material. The second presents studies testing the impact of different glues (specifically silicon-based and epoxy-based) on the dynamic response of xSSDs. The third evaluates the vibrational tolerance of wirebonds. The wirebonds analysis is broken into two parts: a report of pull test activities, ensuring the quality of the micro bonding manufacturing process; and the proper set of high-level random vibrations introduced before.

An in-depth analysis of all the presented aspects can not be summarized in a single work, and this is not the intention. Instead, this research intends to provide a holistic view of the problem and collect the authors' experiences on the matter. This aim is attained through the provision of useful data and the discussion of possible design issues. As a whole, this paper introduces various topics paving the way for future investigations.

Silicon Detector Description

This section briefly describes silicon detector assemblies, with details relevant to the presented studies.

Silicon detectors are widely used in particle physics and represent a valid means of obtaining information in this area. Their main constituent is doped silicon enriched with a superficial metalization on both sides. In the case of SSSDs, one side shows a continuous metalization, while the other has conductive stripes in correspondence with implanted doped silicon areas. The case of the DSSD is more complex (please refer to the dedicated literature [18,19] for a comprehensive description). An SSSD is shown in Figure 3.

Figure 3. Single-sided silicon detector or tile .

The silicon dimensions are in the order of 100 mm. Therefore, to cover large areas, it is necessary to use more than one SSSD, or tile. Commonly, an active plane is created by placing sub-assemblies, called *ladders*, one next to the other. A ladder is a line of sensitive tiles electrically connected one to another. Figure 4 depicts a ladder. The silicons in the figure are 97 mm × 97 mm. A PCB is visible at the end of the silicon ladder, it carries the readout ASICs and interfaces to the off-detector electronics.

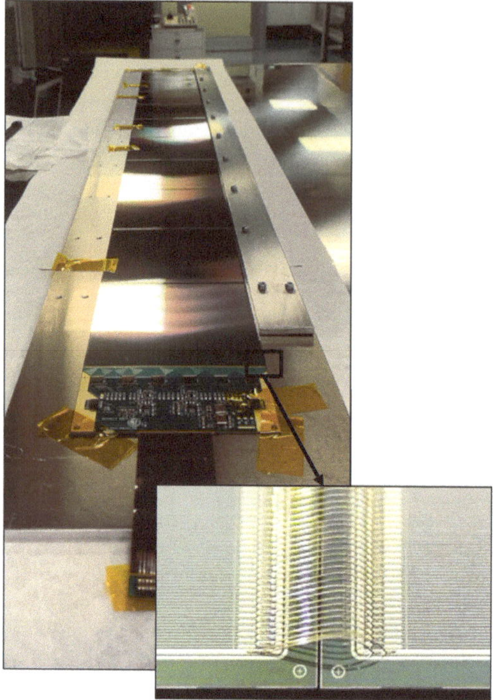

Figure 4. Ladder.

Figure 4 introduces the ladder and the wirebonds (on the left); i.e., the electric connections between adjacent tiles and between tiles and electronics. These tiny soldering junctions are how tiles are electrically connected to the read-out electronics (or *front end*). The "micro" nomenclature derives from the dimension of the wire used in the connection. The said dimension is contained by the distance between adjacent strips (each of which has to be connected to a single electronic channel), ranging from tens to hundreds of μm. The touching of wires results in measurement failure. Therefore, the size of the wire used for the connections should be comparable to the strip pitch. The diameter of the aluminum wire used is 25 μm, and a specific machine completes the attachment. Wirebonds are extremely fragile and break if touched or pulled. Thus, the silicon tiles and the front-end electronics are glued together on a substrate. The substrate exploits electrical duties by conducting the bias voltage from the electronics to the bottom of each detector. Hence, it is necessary to provide both a mechanical and electrical interface between the tiles/electronics and the substrate. In other words, the glue layer connecting the silicon and the PCB to the substrate should be at least partially conductive. Thus, a conductive and a structural glue must be employed to ensure electrical connection and adhesion.

Finally, by placing multiple ladders, it is possible to create large active surfaces.

2. Silicon Detector Mechanical Characterization

The scope of the present section includes the mechanical characterization of the silicon detector material. The information of interest relates to stiffness and strength. Therefore, the retrieved parameters are the Young's modulus and the maximum stress and strain bearable by the silicon. In the present discussion, the material is thought to be isotropic and homogeneous (for a more detailed discussion on silicon directional properties please refer to [20–22]).

2.1. Test Samples

The test batch consisted of fourteen DSSDs with dimensions of $72.00 \times 41.40 \times 0.30 \text{ mm}^3$ (measured with 0.01 mm of accuracy). The crystal orientation was <111> for all specimens. The tested samples were spares from the AMS-02 experiment and therefore fully representative of real space hardware.

2.2. Test Description and Execution

Each sample underwent a *three-point bending* test, consisting of the application of a force perpendicular to the silicon surface while the sample rested on two supports. The force and the supports were round bars capable of exerting forces but not moments, thus constraining the perpendicular displacement but not the rotation. The test is detailed in Figure 5. The machine used for the test was a LLOYD LR30K.

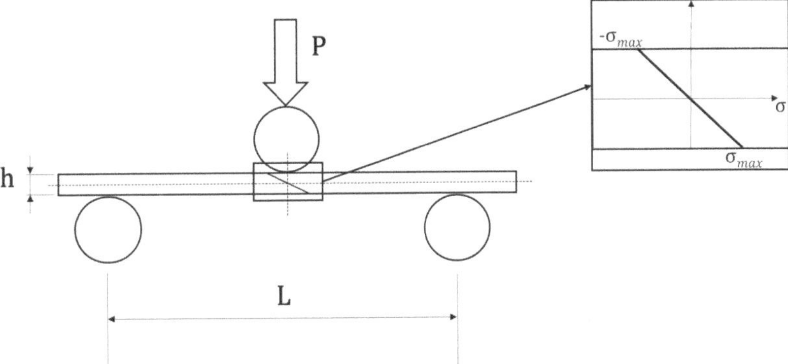

Figure 5. Three–point bending test scheme.

The illustration also presents the through-the-thickness stress profile, linked to the strain profile through the silicon elastic properties. The stress profile was induced in the SUT by the applied load P. Equation (1) relates the input force to the surface stress.

$$\sigma_{surf} = \frac{3PL}{2bh^2} \tag{1}$$

Concerning Figure 5, P and σ_{surf} are the applied force and the resulting surface stress, respectively, h is the specimen thickness, b is the specimen's width (through-the-paper dimension), and L is the free span (distance between supports), equal to 50 mm in the present case. Figure 6 portrays the test setup.

Starting from the rest position ($P = 0\,\text{N}$), the equipment moved the central rod downwards and measured the force opposed by the system under test (SUT). The motion had an initial engagement phase, where the rod filled the gap between the resting position and the SUT surface. This was followed by a strain test, where the force applied by the moving gauge increased, compensating for the opposing force of the deforming specimen. Eventually, the sample broke, and the force applied by the gauge dropped to zero. Using Equation (1), the stress in the silicon sample was extracted from the measurement of the gauge and was plotted in the time–stress graph shown in Figure 7. The circumstance that the stress increased only linearly indicated that the silicon only deformed elastically until the breaking point. The stress at the breaking point for this sample was 208.5 MPa.

Figure 6. Three-point bending test setup.

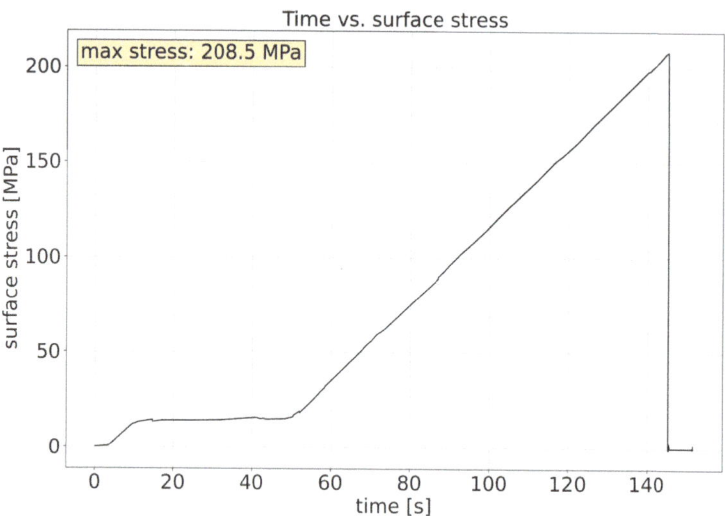

Figure 7. Silicon detector surface stress as a function of time.

The reported maximum stress values are presented in Table 1, while the strains are in Table 2.

Table 1. Sample data for max stress.

Sample	Max Stress [MPa]
1	187.4
2	305.1
3	272.8
4	220.2
5	384.3
6	236.3
7	208.8
8	293.6
9	151.5
Max	**384.37**
Min	**151.6**

Table 2. Sample data for Max Strain $\left[\frac{m}{m}\right]$.

Sample	Max Strain $\left[\frac{m}{m}\right]$
1	-0.0015
2	-0.0019
3	-0.0010
Max	**-0.00102**
Min	**-0.0019**

Having calculated the maximum stress bearable by the single tile, the focus shifted to the stiffness relating stress and strain through the relation of Equation (2).

$$E = \frac{\Delta\sigma}{\Delta\epsilon} \tag{2}$$

Since σ can be computed from the applied force, it was necessary to measure ϵ to determine E. For this reason, mono-directional strain gauges were installed on eight out of the fourteen specimens. Of the former, three samples were tested to failure, with five in the elastic region, thus allowing multiple repetitions on the same specimen (to verify the repeatability of the measurement). Finally, the Young's modulus was extracted from the strain–stress curves, like the one in Figure 8. The measured data are reported in Table 3.

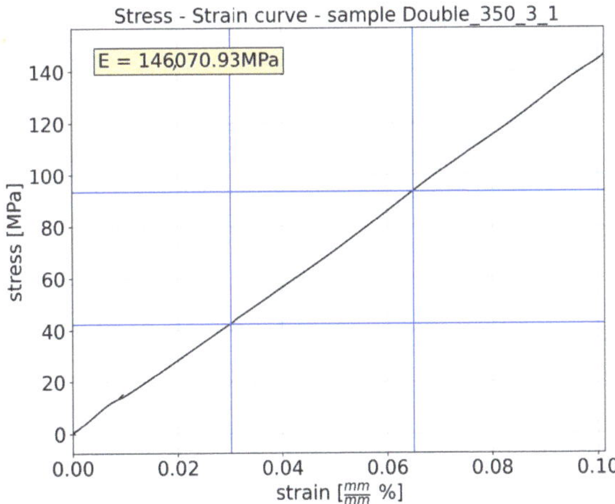

Figure 8. Stress–strain curve and Young's modulus estimation.

Table 3. Sample Data for E [MPa].

Sample	E [MPa]
1	142,996.0
2	151,949.7
3	131,619.0
4 rep1	159,680.2
4 rep2	151,574.8
4 rep3	154,870.3
4 rep4	159,155.2
4 rep5	151,675.6
5 rep1	135,986.4
5 rep2	139,668.3
5 rep3	135,712.0
5 rep4	134,508.5
5 rep5	139,795.8
6 rep1	121,552.2
6 rep2	124,688.9
6 rep3	121,231.5
6 rep4	116,789.3
6 rep5	118,367.0
7 rep1	146,106.7
7 rep2	146,901.6
7 rep3	145,125.7
7 rep4	151,382.1
7 rep5	150,104.5
Max	**159,680.2**
Min	**116,789.3**

Finally, another batch of 11 DSSD specimens, identical to the ones used for these tests, were weigh on a scale with accuracy 0.1 g. Through which it was possible to estimate the DSSD density (through the notorious Equation (3)).

$$\rho = \frac{m}{V} \tag{3}$$

2.3. Test Output Summary

In conclusion, the test campaign used fourteen silicon tile specimens, eight of which were equipped with strain sensors. The testing machine provided force data that were easily related to the through-the-thickness stress thanks to Equation (1). On the other hand, the strain measurements and Equation (2) led to the estimation of the Young's modulus and the maximum strain (for the three cases in which failure occurred). Additionally, the average density was computed through the weight measurements on eleven sensors. The test results are summarized in Table 4.

Table 4. Summary of silicon experimental campaign results.

Property	Mean (μ)	Standard Deviation (σ)	Sample Size
Density $[\frac{kg}{m^3}]$	2392	69.5	11
Max Stress (MPa)	251.15	70.60	9
Max Strain (%)	0.1457	0.0430	3
Young Modulus (GPa)	142.19	10.19	8 (23 reps.)

3. Dynamic Effect of the Adhesive Bond to the Structural Substrate

3.1. Overview of the Design Challenge

The present section discusses the effect adhesives have on the stresses experienced by silicon sensors when subjected to vibration. From the author's experience, two different classes of adhesives can be used for structural purposes: epoxy-based or silicon-based. The first being very rigid and the second more compliant. As explained later in this section, the second are more suitable for this application, since they introduces damping at the interface and mitigate shock loads. Nevertheless, the adhesive effect is different for each detector configuration, and some may be more susceptible to certain specific effects than others. A further complication is the requirement of a second, conductive glue, providing the bias to the backside of the sensors. This has to be epoxy-based and must be applied with a minimum area, to meet the electrical conductivity requirement. A stiff adhesive runs against the desire for the glue to dampen the transfer of vibrations. The challenge is to find a design combining a conductive glue with a softer silicon-based glue.The variable design parameters include the relative areas, the applied patterns, and the thickness of the glues, each of which is limited by further constraints

The criticality of the adhesive choice emerged during the shock test of the DAMPE [23] quarter plane prototype shown in Figure 9, where the tiles broke.

Figure 9. DAMPE quarter plane prototype.

In the present situation, the ladders, shown in Figure 9, were connected to the plane with the glue pattern shown in Figure 10.

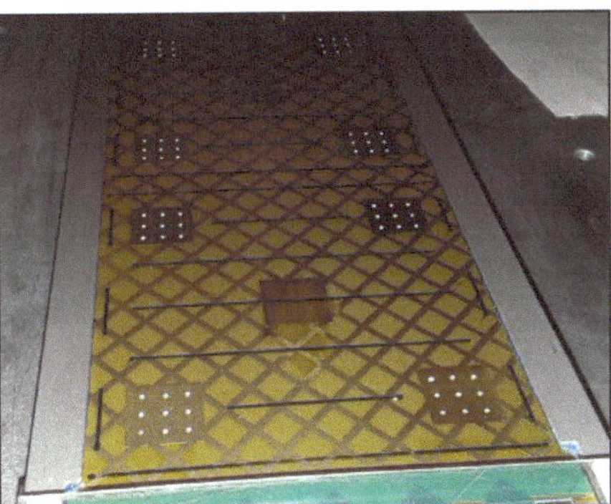

Figure 10. DAMPE gluing pattern.

The reader can notice two different adhesive patterns: thin dark lines of epoxy structural glue (not to be mistaken with the bottom brown copper cross pattern) and gray dots of electrically conducting glue (necessary to ensure the system's functionality).

As previously stated, epoxy glues are stiff and ensure a rigid connection between the ladder substrate and the supporting plane but at the same time provide little damping. The stiffness of the connection was considered the prime factor responsible for the failure registered during the shock test of the prototype (depicted in Figure 11).

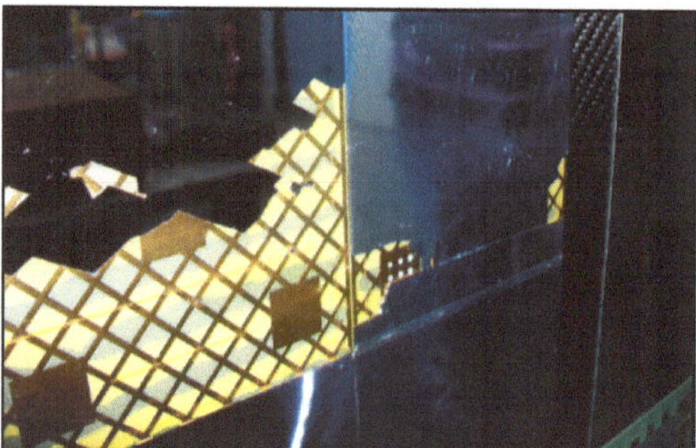

Figure 11. Silicon detector failure during the space qualification shock test.

For this reason, in the DAMPE flight model, structural duties were fulfilled by a compliant glue (silicon-based), replacing the much stiffer epoxy-based adhesive. For the sake of a better understanding, the difference in Young's modulus (E) of the two adhesives is very relevant, going from the 1 GPa of the epoxy-based to the 1 MPa for the silicon-based.

The successful launch and correct on-orbit operation of the DAMPE proved that this choice was effective. After this result, no further studies were performed and the silicon-based glue was made the standard for the adhesion of tiles to substrates.

Recently, it has been observed that the best electric glue used for non-structural purposes is an epoxy-based adhesive. This observation motivated a new study on the effect of the latter on tile mechanics, and this is the subject of the present section. To clarify, the change of structural glue was sufficient for the mentioned design. Although the conductive glue was still present, the performance increase resulting from the change was sufficient to pass the qualification test. Indeed, the amount of conductive glue was less than that of the structural glue. As a whole, the specific design was successful and no further investigations were performed. Conversely, the present study reopens this topic, to highlight this issue and provide general information to help future designs.

3.2. Glue Data

Before moving on with the test campaign, it is interesting to provide information on the gluing.

Starting with the involved adhesives, the epoxy glue used for this work was 3M scotch-weld epoxy adhesive 2216 gray, the Si-based was Dow Corning 3145 RTV MIL-A-46146 Gray, and the conductive glue was EPO-TEK® EJ2189. The glue was using a syringe. The deposition patterns were straight lines for structural glue (either Si or epoxy-based) and single dots for the conductive glue (as per Figure 10). The glue thickness was about 30 μm.

3.3. Experimental Campaign

The continuous effort towards the development of more efficient and/or compact detectors has led to new geometries, where the issue presented in the above can become relevant. Specifically, a geometry such as that of a mini-PAN with a single active surface, adhering to the substrate only on its boundaries, could increase epoxy-associated effects. Concurrently, this scenario is perfectly suited for tackling the topic.

Leveraging the former discussion, a test campaign was set up. The test samples were two mini-PAN trackers installed on the same mechanical interface (tracker module version 1) and then vibrated. The first was a tracker PCB with a dummy silicon detector (mechanically equivalent) glued only using structural glue (from now on, we will refer to the silicon-based glue as *structural glue*). The second was like the first (even the same PCB is used) with a different gluing: now both the structural and conductive adhesives were used. Both tests employed mechanical dummies. The detectors were equivalent to real ones mechanically but not electrically. Hence, there was no need to apply the bias(the conductive glue added for the bond to be mechanically relevant, not for electric purposes). Except for the adhesive bond, the two tests were the same. Hence, the following discussion applies to both.

The goal of the test was the estimation of the dynamic effect of the glue. In practice, by monitoring the overall damping of the assembly, it was possible to appreciate the effect of the adhesive.The damping was a good control parameter for our goal. Indeed, this study aimed to show the criticality of the glued connection and to qualitatively show the extent of the changes. A comprehensive discussion would require an extended discussion, foreseen for future studies. Instead, here, the authors would like to provide one lesson learned to guide designers facing similar issues.

It is important to attest that high-damping has a very beneficial effect on non-static loads and especially on shock loads [24]. There are several ways to compute the damping, most of which are experimental (that is the reason why this study relied on experiments) [25,26]. In this case, the choice was the method the commonly called the *3 dB method*. This method predicted the damping estimation from the frequency response function (FRF) peak amplitude drops [27,28]. Here, the damping of the first peak was considered (being associated with the flexural mode of the silicon tile). Thus, the damping value was a direct measure of how the glued interface filtered the inputs applied to the PCB (the adhesive was responsible for the load transfer between the PCB and the silicon). More information on the FRF is provided in the literature [29–31]. In the ideal case of undamped structures, the FRF amplitude at the peaks would be infinite (asymptotes in the function). In reality, all mechanical systems are damped and the FRF amplitude is finite. It can be proved that the damping is proportional to the peak width; narrower peaks have higher damping than wider ones. The *3 dB method* was used to estimate the damping from the peak aperture. For this aim, Equation (4) was used, where Ω_0 is the peak frequency, and ω_1 and ω_2 are the frequency points to the left and right of the peak with an amplitude 3 dB less than that of the peak (Figure 12).

$$\zeta = \frac{(\omega_2 - \omega_1)}{2\Omega_0} \tag{4}$$

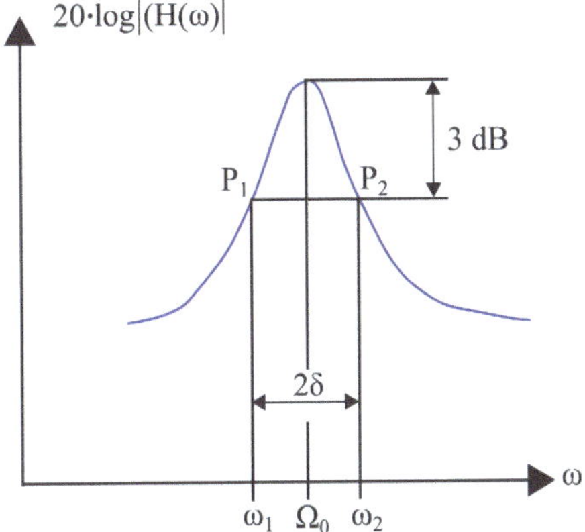

Figure 12. Illustrated explanation of the *3 dB* method.

Leveraging this method, it was possible to estimate the damping for both configurations. The analyzed cases were explanatory conditions chosen to present the issue and provide some indicative numbers. This part of the research did not provide a comprehensive report on silicon gluing. Indeed, it would be necessary to perform specific research considering different patterns, glues, and geometries. This is, of course, a prospect of the study. Conversely, the present research aimed to depict a serious mechanical issue to be considered during the design phases and to build a basis for more in-depth studies.

Detailing the test execution, both SUTs were installed on a shaking table and underwent a frequency sweep. To avoid any ambiguity, a shaker table is an experimental apparatus capable of generating dynamic loads on a platform called the *shaker head*. The provided load can be random or harmonic [32]. The first consists of a time-varying signal constituted by the superimposition of multiple sinusoidal functions with different frequencies and random phases. The second consists of the application of harmonic signals with fixed amplitude and time-varying frequencies, according to a predefined time-frequency law named the sweep-rate [33]. The main function of the shaker is to apply reference loads to verify the capability of the SUT to withstand a given load profile (either random or harmonic). Nevertheless, the shaker can also be used to determine the FRF at a certain point of the SUT. Once more, the FRF is the relation between the input (provided by the shaker in this case) and the output signal in a control point. If the structure is perfectly rigid, the measured signal is identical to the input signal and the FRF is 1 at all frequencies. Conversely, elastic structures resonate, thus the measurement at the resonance frequency will be higher than one (and dependent on the damping). As a whole, in this part of the research, the shaker was used to determine the FRF between the shaker profile and the center of the silicon. To do so, a low amplitude (to avoid damage to the structure) harmonic load was applied. The shaker used for the test was a Sentek L0315 reference.

Figure 13 gives a snapshot of the test setup.

Figure 13. Damping estimation setup.

Here, the shaker head (metallic disk below the mini-PAN tracker fixture) and two measurement points are visible: one in correspondence with the accelerometer, and the other a red dot. Indeed, while the first measurement was taken using standard techniques (piezoelectric accelerometer), for the second measure, a laser interferometer was employed. This choice was motivated by mass considerations: sensors should not affect the dynamics of the SUT. As a rule of thumb, the accelerometer should be 100 or 1000 times lighter than the tested component. Given the mass of the silicon surface, even the lightest accelerometer (0.2 g) would affect the results. For this reason, a laser interferometer was used. It is worth highlighting once more that there was a hole in the PCB in correspondence with the silicon surface, and only the sides of it were in contact with the PCB.

The final result of the study is presented in Figures 14 and 15, respectively, illustrating case 1 (only structural glue used) and case 2 (nominal bond: structural and electrical glue used).

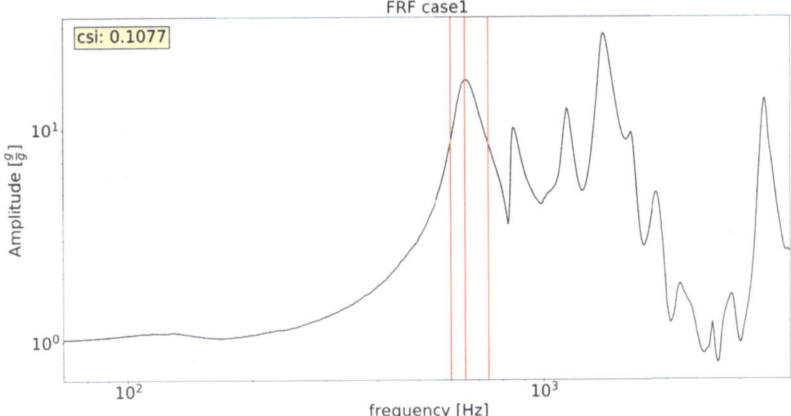

Figure 14. FRF and damping estimation case 1.

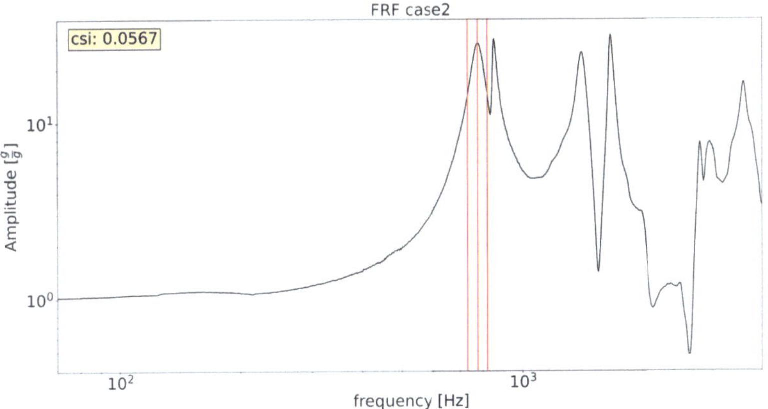

Figure 15. FRF and damping estimation case 2.

In both cases, the input profile was harmonic, with an amplitude of 4.905 $\frac{m}{s^2}$ ($\frac{1}{2}$ g), spanning between 20 Hz and 3.5 kHz. As specified before, the analysis focused on the first peak, which is considered the most relevant for the present discussion.

Discussing the results, the difference in damping was far from negligible. Indeed, the damping dropped from the $\xi = 0.11$ of the silicon case to the $\xi = 0.056$ of the stiffer condition. Thus, the test reported a damping drop of about 47%, proving that the glue effect on this kind of mechanical bond is far from negligible.

To conclude, the present study intended to stress the criticality of xSSD gluing, first noticed during DAMPE prototype shock tests. To attain this goal, two identical mini-PAN trackers were tested. The SUTs were identical in every aspect, except the adhesive configuration: one employed only silicon-based glue, and the other used both silicon-based and epoxy-based (necessary for electrical purposes) adhesives. The dynamic effect of the different gluing configurations was assessed through the measurement of the first-peak damping from the experimental FRFs. Given the contained mass of the silicon, it was necessary to perform the FRF measurements with a laser interferometer.

The result leads to the conclusion of the importance of gluing for the detectors' dynamics and the necessity of reducing the amount of conductive glue as much as possible if the objective is to have a soft and damped bond.

4. Wirebond Mechanical Studies

This part of the research concerns the mechanics of wirebonds. The discussion is split into two: an initial part collecting data from a wire pull test campaign, and a vibration campaign. The goals of the first were the assessment of wirebond connection quality and the sharing of experimental data with the scientific community. The goal of the second was to extend the heritage (coming from various successful space missions) of wirebond vibration endurance to more general cases. Indeed, the literature [34] advises the encapsulation of wirebonds, while various space missions have successfully employed unencapsulated connections. To conclusively prove the space suitability of unencapsulated bonds, the present study performed a random vibration campaign. Three different samples were vibrated at levels far above the vibrational space qualification levels requested by space standards. Additionally, a preliminary shock campaign was performed. The shock test aimed to enrich the picture and provide a more solid result. In any case, the latter can not be considered a nominal shock qualification campaign, because the required levels were not attained. Revisiting the earlier discussion, wirebonds are the electric connections between silicon tiles or between silicon tiles and front-end electronics. For the experiment to function properly, connections could not brake nor short-circuit (touching the surrounding wires). Thus, the dimensions of the connection were constrained by the spacing and the width of the strips. This necessitated the employment of 25 µm diameter wires (99% purity). The bonding procedure required a specific apparatus. Figure 16 presents a picture taken during the bonding process (bonding tool on the right, pale yellow).

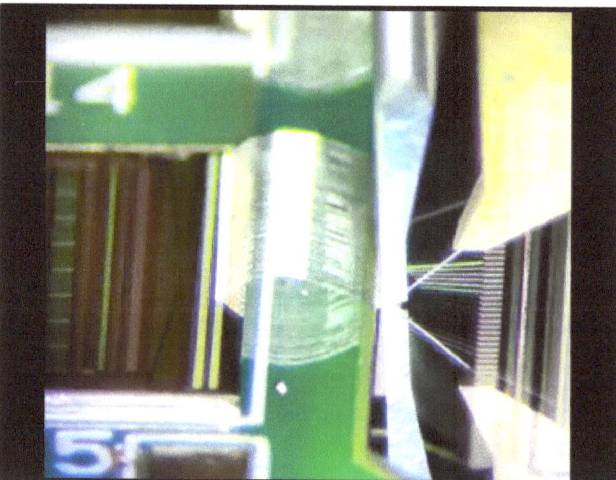

Figure 16. Bonding process.

4.1. Manufacturing Process Verification

To verify the mechanical strength of the electric connections, a set of 515 samples was tested. Each wire was pulled using a custom-made hooked dynamometer. The acceptance criteria mandated that only 15% of failures were padlifting. In the remaining cases, the wire broke either in the middle or at the base.

Detailing the setup, Figure 17 shows a picture taken during the test, whose structural configuration can be summarized with the scheme of Figure 18.

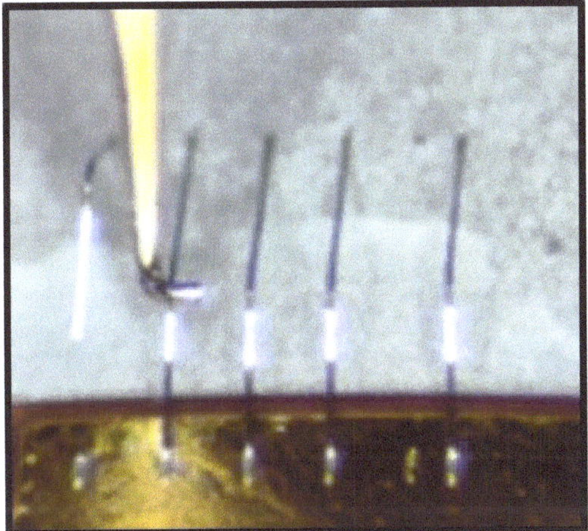

Figure 17. Wirebond pull test—test picture.

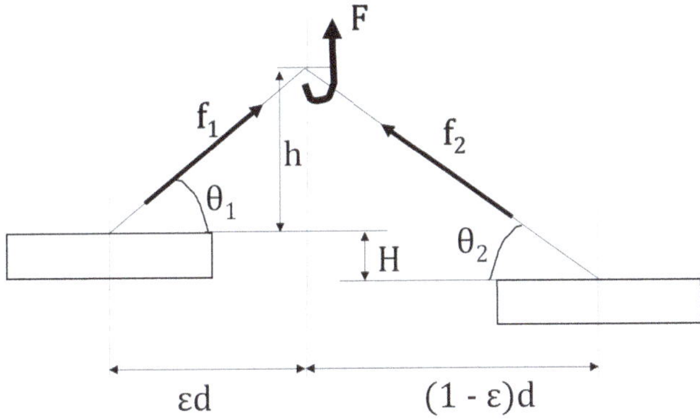

Figure 18. Wirebond pull test—structural scheme.

The latter presents the general situation of this kind of test. For the specific case, the parameters were

- $h = 750$ μm
- $H = 0$ μm
- $d = 1500$ μm
- $\epsilon = 0.5$

Resulting in $f_1 = f_2 = F\frac{\sqrt{2}}{2}$. The assumption for ϵ is quite strong: the application point depends on the operator's skills. Nevertheless, given the numerosity of the sample, we can assume that the uncertainty on ϵ was statistically mitigated.

The retrieved data are presented in Table 5.

Table 5. Summary of wirebond pull test data.

Property	Mean (μ)	Standard Deviation (σ)	Sample Size
Pull Force at break point [gf]	12.47	1.80	515

4.2. Vibrational Tolerance of Wirebonds

The majority of mechanical load experienced by a space mission is launcher-associated. Indeed, the on-orbit placement phase is by far the most critical for space systems.

Once again, the loads experienced in this phase are mainly dynamic and can be split into three categories: harmonic loads, random loads, and shock loads. The first comes from the launcher resonances; the second from the non-deterministic inputs from acoustic, aerodynamic, and thrust generation apparatus; and the third from instantaneous phenomena, such as booster and launcher stage separation.

For an object to be space qualified, it has to be subjected to a mechanical qualification campaign. Such campaigns involve the application of test loads with levels dependent on the specific case; based on the experiment mass, the configuration and position in the loads experienced during the flight can be very different.For random tests, common practice involves the use of a standard (American [35] or European [36]) or a launcher user manual (e.g., Falcon's [37]) for the profile definition. Instead, here, the test profile was intentionally more severe than that of the standards, to account for the possible dynamic amplification of the experiment structure. The study aimed to demonstrate the extreme tolerance of wirebonds to vibration and to prove the suitability of unencapsulated wirebonds for space applications. The second objective was quite relevant for the authors. Although the literature [34] suggests encapsulating the bonds for mechanical protection, this specific application experience proved this to not be mandatory; the space operative missions discussed in the introduction did not employ this solution. In any case, bond geometry severely affects the mechanical behavior, and success may be related to the specific cases. Hence, a dedicated analysis was deemed necessary to avoid future problems and to qualify the naked-wire approach.

To attain this goal, the SUTs were intentionally overtested. Not knowing the launch configuration, the objects were tested in different directions (X, Y, and Z). High-level random tests were the primary objective of this activity, and z-directed shock tests enriched the overall picture. In conclusion, the present activity aimed to cover all possible flight conditions, provide data applicable to future experiments, and definitively prove the space suitability of unpotted wirebonds.

4.2.1. SUTs and Setup

To produce general results not associated with specific configurations, three different test samples were selected. The first sample (Figure 19) comprised two AMS-02 DSSDs spaced 11 mm apart. It was not possible to electrically verify the object; thus, visual inspection was the only criterion available to judge the test results. The bond was intentionally longer for this campaign, to validate a sort of worst-case scenario. The detectors were glued on an FR-4 substrate and then on an aluminum plate (interface to the test apparatus).

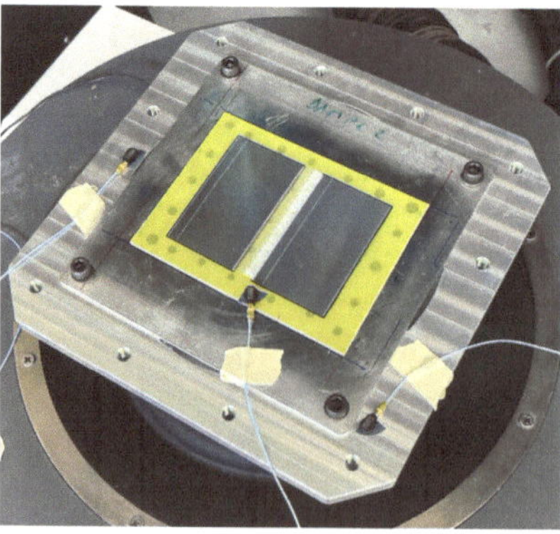

Figure 19. Test sample 1 *long bonds*.

Test sample no. 2 was a tile and electronics functional assembly from the AMS-02 L0 upgrade. Figure 20 presents the SUT 2. The test started with two pairs of front-end chips (the external ones from each side) not connected to the silicon sensor surfaces. Similarly to test sample 1, the assembly was glued on a metallic plate, working as a test interface.

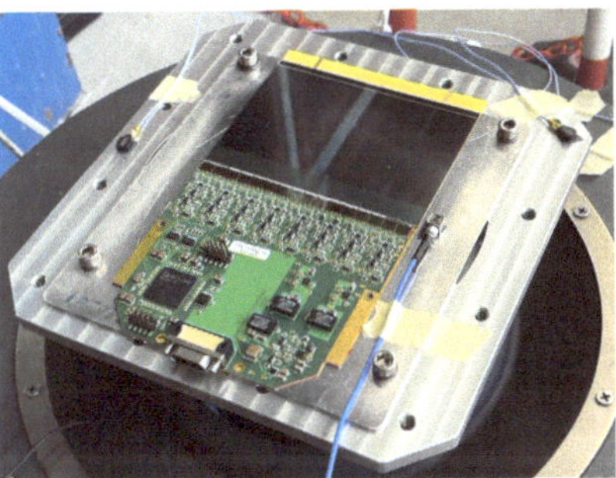

Figure 20. Test sample 2 AMS-02 L0 upgrade.

Test sample no 3 in Figure 21 employed a DAMPE silicon and front-end and it functioned like SUT 2. For all SUTs, silicon-based glue ensured the adhesion of the SUTs to the mechanical interface and the bonds are naked (no potting).

Figure 21. Test sample 3 DAMPE detector.

Each assembly underwent a random vibration along the X, Y, and Z axes (where Z is the direction normal to the silicon surface). No harmonic testing was performed, because the sine sweep space qualification excites frequencies below 100 Hz and in this range the tested objects behaved like rigid bodies; there was no dynamic amplification and the load experienced equaled the input. Instead, on higher portions of the spectrum, the component's dynamics played a relevant role, and the input loads were amplified. This is the critical part of the spectrum.

The random test profile ranged from 20 Hz to 2000 Hz with a plateau between 100 Hz and 600 Hz and a ramping profile elsewhere (the same shape as the one presented in Figure 22). The profile is provided in terms of PSD, as in the cited standards [35–37]. The interested reader can find additional information on the topic in the literature [15,29].

The severity of the random profile was measured through the root mean square (RMS) acceleration. All samples were tested at different levels (up to the maximum permitted by the apparatus i.e., 40 g RMS along the Z axis and 17 g RMS along X and Y) in the three directions. Concerning the equipment, the testing apparatus was the same Sentek L0315 shaker as in Section 3.3, with a different head expander for the X, Y, and Z tests. Figure 23 presents the X–Y test head expander, while Figure 24 presents that for Z.

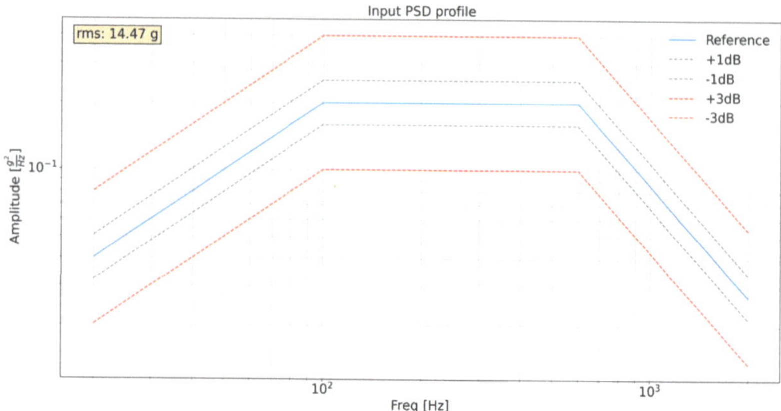

Figure 22. Random test profile of power spectra density from reference [35].

Figure 23. Head expander used for the X and Y tests.

Figure 24. Head expander used for the Z tests.

The Z fixture was necessary due to the dimensions of the assemblies, especially for test samples 2. Instead, the X–Y cubic fixture allowed the testing in the specified direction with a shaker force directed along Z (the SUTs were attached to the lateral face of the cube). Additionally, on-shaker Z-direction shock tests were performed on all samples.

4.2.2. Wirebond Vibration Tests

Extending the previous discussion, the campaign involved two types of test: random and shock (harmonic was not considered relevant since it excites lower frequencies). Random tests were performed in X, Y, and Z configurations. Shock tests were performed only along the Z direction. Concerning the test configuration, the Z setups for all test samples are visible in Figures 19–21. Conversely, Figure 25 introduces the X and Y configurations.

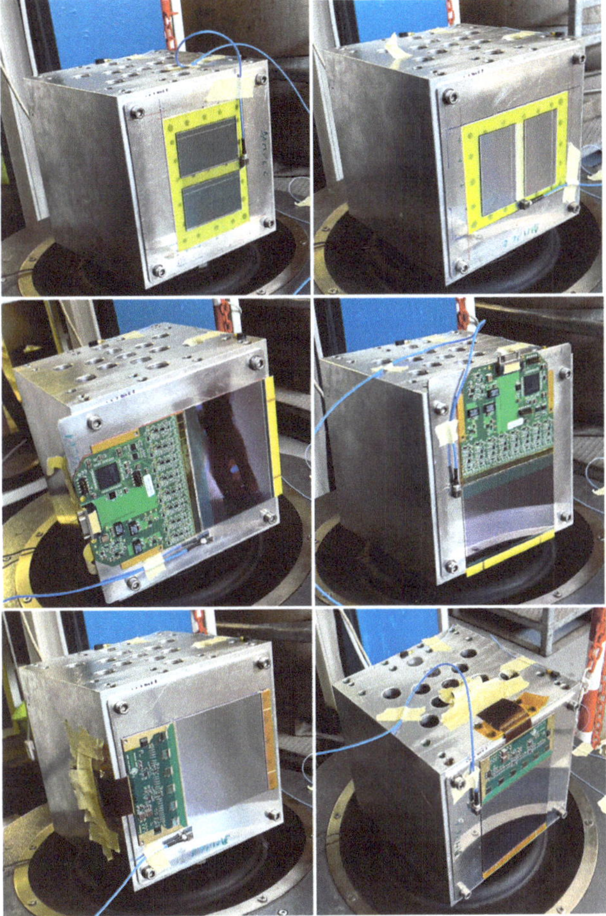

Figure 25. Head expander used for the X and Y test configurations.

On the topic of the specific tests, the sequence included an electrical continuity test (for samples 2 and 3) before and after each load cycle, along with a visual inspection.

Moving on to the load cycles, random and shock profiles were applied to the SUTs. The random span range was 20–2000 Hz, with a plateau between 100 Hz and 600 Hz and a ramped behavior (+3 dB, −5 dB) elsewhere. The random amplitude was dependent on the RMS. The test started from an RMS value of 14.5 g and proceeded to the maximum allowed by the shaker. Specifically, the PSD profile was not modified (same plateau). Instead, the RMS was increased. Thus, the input was translated upwards in such a way that the profile remained the same, while the area grew (hence the RMS).

Conversely, the shock profile was based on NASA GEVs (a curve in the bi-logarithmic plane) and linearly interpolated three points: 100 Hz—81.3 g, 625 Hz—500 g, and 5000 Hz—500 g. Again, the RMS gauged the random vibration severity and ranged from 14.4 g (GEVs nominal value) to a maximum of 40 g along the Z axis and 17 g along the X and Y axes (the fixtures mass constrained the maximum acceleration). On the other hand, the shock never reached the full level provided by the GEVs, and the severity was quantified by the percentage of the full level.

Another interesting aspect to consider was the input amplification due to non-ideal mechanical connections. Here, there were two connections: one between the shaker and the metallic plate, and one between the samples' substrate and the plate. This effect was quantified using the ratio of the measured RMS and the input RMS. The amplification values (measured before the failure of sample 1) were approximately 4.4, 1.8, and 1.95 for the three test samples. The higher value for test sample 1 was explained by the accelerometer position. Indeed, for cases 2 and 3, the measurement was taken on the metallic plate, thus not taking into account the adhesive bond. Instead, in case 1, the monitoring point was on the same substrate as the detectors. Hence, it is reasonable to assume that the glue amplification effect discussed in Section 3 occurred for all the samples and that the real value on the silicon substrate was four times higher than that of the input.

Post-test functional checks and visual inspection proved successful for samples 2 and 3, while for test sample 1 the visual inspection showed a loss of adhesive integrity in correspondence with the accelerometer position (check Figure 26).

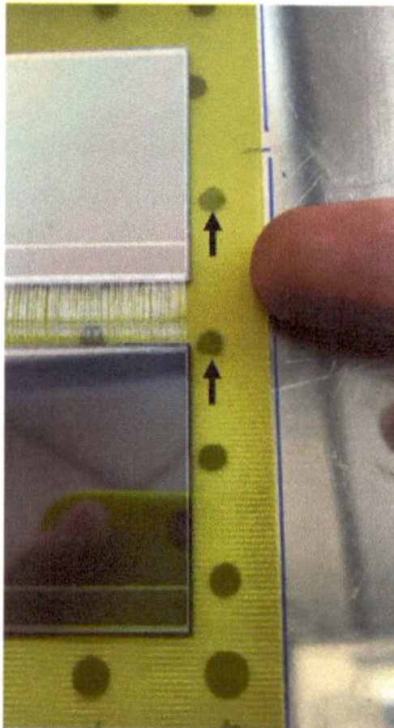

Figure 26. Long bond glue detachment points (marked with black arrows).

The latter event is reported here for the sake of comprehensiveness, but it was not concerning, since both the bonds and the silicon remained intact. Indeed, given the contained mass of the SUT, the accelerometer presence was not negligible and led to overtesting. Moreover, the fact that the assembly did not lose integrity when tested with an additional mass and to a very high level (RMS was more than double wrt to NASA requirements) proved the space-suitability of this technology. Additionally, the anomaly occurred after the Z random vibration. Therefore, the damaged component successfully underwent random X, Y, and shock Z.

Let us conclude this section with Table 6, summarizing the levels experienced by each test sample. Again the random intensity is quantified by the RMS (square of the integral of the PSD), while the shape of the PSD profile was not changed. Instead, the shocks were provided as percentages of the target value (500 g), which could not be attained in the present facility.

Table 6. Unpotted wirebond vibration and shock test summary.

Test Sample	Max. RMS X (g)	Max. RMS Y (g)	Max. RMS Z (g)	Max. Shock (% of NASA Reference)
1	17	17	30	90%—450 g
2	17	17	40	70%—350 g
3	17	17	40	90%—450 g

5. Result Summary and Conclusions

This work has covered various aspects related to spaceborne silicon detector mechanics. Namely, the mechanical characterization (extraction of elastic properties), the depiction of the gluing criticality, and wirebond strength (both in terms of wire strength and resistance to vibrations).

For the mechanical characterization, diverse AMS-02 spare detectors were tested on a three-point bending machine. Through these tests, it was possible to determine the elastic modulus, the maximum allowed stress, and the maximum allowed strain. Concurrently, another batch of AMS-02 spare detectors were weighed on a scale and measured with a caliper, leading to density estimation. A summary of the data is provided in Table 4. The characterization campaigns produced the inputs needed for the mechanical analysis of silicon detectors. Leveraging the presented data, a designer could set up a model (either static or dynamic) including silicon detectors as participating objects (not only as inert masses). This is permitted by the knowledge of elastic properties and density. Finally, failure data permit the estimation of safety factors and structural verification. As a whole, the provided data should help mechanical designers working with these detectors, by providing elastic, density, and failure data. As future developments, it would be interesting to consider the crystal orientation for a more precise assessment of mechanical properties. Although the present study provides good design inputs (in terms of mechanical properties), it could be interesting to extend the study to blank silicon wafers to give more general properties. Additionally, the three-point bending tests could be performed on operative detectors, to measure how bending affects the detector's properties. The latter investigation was not included in the present study given its focus on space applications, where the detector undergoes bending during the flight and not during its operative life. In any case, future studies could evaluate if the vibration induces a change in the silicon properties, changing its electrical properties. Section 3 details a critical mechanical aspect of silicon detector assembly: the gluing issue. Specifically, how the employment of different types of glue affects the load transfer between the substrate and the sensor. The lesson learned from a previous experiment had already resulted in changing the structural glue (from an epoxy base to a silicon base). In this document, the issue was described and further analyzed. In detail, the damping (used as a control parameter to provide qualitative information) of a nominal gluing assembly (conductive + structural glue) was compared to a non-nominal one (employing only structural glue). The negative impact of conductive glue (unavoidable and epoxy-based) was high. This study led to two conclusions: particular care should be taken when designing a sensors' gluing paths, and the conductive glue should be kept to a minimum required to ensure electrical functionality. Although each situation should be studied independently, the information contained here provides ideas and directions for preliminary mechanical analyses. The effect of different glues of the same kind is another prospect. The present research closed with a study on wirebonds (Section 4). The discussion was split in two: an initial part collecting data from a wire pull test campaign, and a vibration campaign. The first proved wirebond manufacturing quality and provided experimental data for the scientific community. The second (employing the bond manufacturing techniques validated before) proved the initial assumptions of high-vibration tolerance of unencapsulated bonds. The successful high-level vibration tests proved that the general wirebond tolerance to space vibrations is high. The relevant result was not the proven capabilities to withstand space vibration but the successful withstanding of vibrational levels far above the standard. Indeed, the space suitability of wirebonds was proven by on-orbit operation detectors (which successfully underwent a space qualification campaign and a real space flight). As a whole, the present study shows that, not only can unencapsulated bonds withstand space vibration, but the acceleration required to break them is at least one order of magnitude (the profile was increased to the maximum of the facility and the wirebonds did not break) above the mandated standard. The present study does not disprove the use of encapsulated bonds but states that it is not necessary in space applications. Different applications may consider using this solution based on other constraints, a higher handling damage risk related to a larger volume of components for example. The present study shows that wirebonds are very resistant to vibrations. Hence, there is no need to employ performance-increase solutions such as encapsulation. On the other hand, changing the manufacturing technology negatively affects project reliably (critical in space applications). Moreover, encapsulant addition negatively impacts the mass budget (especially given the length of the ladders) and introduces another manufacturing phase, affecting the time budget. As a whole, the advantages of encapsulation do not justify the replacing of a flight-proven approach (no encapsulation) with a possibly beneficial approach with no flight heritage (to our knowledge) and with reported [34] thermo-mechanical issues.

In conclusion, this research succeeded in procuring inputs and requirements for a more accurate design of spaceborne physics experiments involving silicon detectors. The collection of information and discussion provisioned by this paper should improve the accuracy of mechanical models and pave the way for new and fascinating solutions.

Author Contributions: On the single authors' contributions: Conceptualization E.M., Methodology E.M., L.M., G.M. and M.P.; validation E.M., M.I. and M.P.; investigation E.M., M.I., G.M., L.M., M.C., G.S. (Gianluigi Silvestre), R.P., A.S., L.F. and G.S. (Gianluca Scolieri); software G.S. (Gianluca Scolieri); writing E.M., M.P. and A.S.; review and editing E.M.; supervision G.A., F.C. (Filippo Cianetti), C.B. and L.T.; data curation E.M.; resources F.C. (Franck Cadoux), L.M., L.T. and G.A. All authors have read and agreed to the published version of the manuscript.

Funding: We would like to acknowledge the European Union—NextGenerationEU under the Italian Ministry of University and Research (MUR) National Innovation Ecosystem grant ECS00000041-VITALITY for funding one of the authors.

Conflicts of Interest: The authors declare no conflict of interest.

References

1. Pohl, M. Particle detection technology for space-borne astroparticle experiments. *arXiv* **2014**, arXiv:1409.1823.
2. Ambrosi, G. The AMS Silicon Tracker readout system: Design and performance. *Nucl. Instrum. Methods Phys. Res. Sect. A Accel. Spectrom. Detect. Assoc. Equip.* **1999**, *435*, 215–223. [CrossRef]
3. Alba, B. In-flight performance of the AMS-02 silicon tracker. *J. Phys. Conf. Ser.* **2013**, *409*, 012032. [CrossRef]
4. Ambrosi, G.; Burger, W.; Oliva, A. The AMS-02 silicon tracker: Recent results and current status. *Nucl. Instrum. Methods Phys. Res. Sect. A Accel. Spectrom. Detect. Assoc. Equip.* **2010**, *617*, 471–472. [CrossRef]
5. Chang, J.; Ambrosi, G. The DArk Matter Particle Explorer mission. *Astropart. Phys.* **2017**, *95*, 6–24. [CrossRef]
6. Azzarello, P.; Ambrosi, G.; Asfandiyarov, R.; Bernardini, P.; Bertucci, B.; Bolognini, A.; Cadoux, F.; Caprai, M.; De Mitri, I.; Domenjoz, M.; et al. The DAMPE silicon–tungsten tracker. *Nucl. Instrum. Methods Phys. Res. Sect. A Accel. Spectrom. Detect. Assoc. Equip.* **2016**, *831*, 378–384. [CrossRef]
7. Tykhonov, A.; Ambrosi, G.; Asfandiyarov, R.; Azzarello, P.; Bernardini, P.; Bertucci, B.; Bolognini, A.; Cadoux, F.; D'Amone, A.; De Benedittis, A.; et al. In-flight performance of the DAMPE silicon tracker. *Nucl. Instrum. Methods Phys. Res. Sect. A Accel. Spectrom. Detect. Assoc. Equip.* **2019**, *924*, 309–315. [CrossRef]
8. Azzarello, P.; Cadoux, F.; Favre, Y.; Iizawa, T.; Kole, M.; Lamarra, D.; Paniccia, M.; Stauffer, J.; Wu, X.; Xie, P.; et al. Development of a penetrating particle Analyzer for high-energy radiation measurements in deep space and interplanetary missions. In Proceedings of the International Astronautical Congress, IAC, Online, 12–14 October 2020.

9. Ambrosi, G.; Azzarello, P.; Bergmann, B.; Bertucci, B.; Cadoux, F.; Duranti, M.; Ionica, M.; Kole, M.; Paniccia, M.; Plainaki, C.; et al. The Penetrating particle ANalyzer (PAN) instrument for measurements of low energy cosmic rays. In Proceedings of the 2019 IEEE Nuclear Science Symposium and Medical Imaging Conference (NSS/MIC), Manchester, UK, 26 October–2 November 2019; pp. 1–8.
10. Perrina, C.; Ambrosi, G.; Azzarello, P.; Cadoux, F.; Catanzani, E.; Favre, Y.; La Marra, D.; Silvestre, G.; Wang, J.; Wu, X. The Tracking System of HERD. In Proceedings of the 36th International Cosmic Ray Conference—PoS(ICRC2019), Madison, WI, USA, 24 July–1 August 2019; Volume 358, p. 122. [CrossRef]
11. Dong, Y.; Zhang, S.; Ambrosi, G. Overall Status of the High Energy Cosmic Radiation Detection Facility Onboard the Future China's Space Station. In Proceedings of the 36th International Cosmic Ray Conference—PoS(ICRC2019), Madison, WI, USA, 24 July–1 August 2019; Volume 358, p. 062. [CrossRef]
12. Zhang, S.N.; Adriani, O.; Albergo, S.; Ambrosi, G. The high energy cosmic-radiation detection (HERD) facility onboard China's Space Station. In *Space Telescopes and Instrumentation 2014: Ultraviolet to Gamma Ray*; Takahashi, T., den Herder, J.W.A., Bautz, M., Eds.; SPIE: Bellingham, DC, USA, 2014. [CrossRef]
13. Adriani, O.; Altomare, C.; Ambrosi, G. Design of an Antimatter Large Acceptance Detector In Orbit (ALADInO). *Instruments* **2022**, *6*, 19. [CrossRef]
14. Battiston, R.; Bertucci, B. High precision particle astrophysics as a new window on the universe with an Antimatter Large Acceptance Detector In Orbit (ALADInO). *Exp. Astron.* **2021**, *51*, 1299–1330. [CrossRef]
15. Sarafin, T.; Larson, W. *Spacecraft Structures and Mechanisms: From Concept to Launch*; Space Technology Library, Springer: Cham, Switzerland, 1995.
16. Wijker, J. *Spacecraft Structures*; Springer: Berlin/Heidelberg, Germany, 2008.
17. Buckland, M. Development of the ITS3: A bent-silicon vertex detector for ALICE in the LHC Run 4. *Nucl. Instrum. Methods Phys. Res. Sect. A Accel. Spectrom. Detect. Assoc. Equip.* **2022**, *1039*, 166875. [CrossRef]
18. Dinu, N.; Fiandrini, E. Electrical characterization of the large amount of silicon sensors for alpha magnetic spectrometer (ams) tracker. In Proceedings of the 20th IEEE Instrumentation Technology Conference, Vail, CO, USA, 20–22 May 2003; Volume 1, pp. 786–791. [CrossRef]
19. Sauli, F. *Instrumentation in High Energy Physics*; World Scientific: Singapore, 1992. [CrossRef]
20. Thomsen, E.V.; Reck, K.; Skands, G.; Bertelsen, C.; Hansen, O. Silicon as an anisotropic mechanical material: Deflection of thin crystalline plates. *Sens. Actuators A Phys.* **2014**, *220*, 347–364. [CrossRef]
21. Newnham, R. *Properties of Materials: Anisotropy, Symmetry, Structure*; OUP Oxford: Oxford, UK, 2005.
22. Petersen, K. Silicon as a mechanical material. *Proc. IEEE* **1982**, *70*, 420–457. [CrossRef]
23. Bolognini, A. Development and Testing of the DAMPE Silicon Tracker Mechanical Structure. Ph.D. Thesis, Università degli Studi di Perugia, Perugia, Italy, 2015.
24. Rao, S. *Mechanical Vibration*, 3rd ed.; Addison-Wesley: Miami, FL, USA, 1995.
25. Tomac, I.; Slavič, J. Damping identification based on a high-speed camera. *Mech. Syst. Signal Process.* **2022**, *166*, 108485. [CrossRef]
26. Ruzzene, M.; Fasana, A.; Garibaldi, L.; Piombo, B. Natural Frequencies and Dampings Identification Using Wavelet Transform: Application to Real Data. *Mech. Syst. Signal Process.* **1997**, *11*, 207–218. [CrossRef]
27. Arora, V. Structural damping identification method using normal FRFs. *Int. J. Solids Struct.* **2014**, *51*, 133–143. [CrossRef]
28. Marchetti, F.; Ege, K.; Leclere, Q.; Roozen, N. Structural damping definitions of multilayered plates. In Proceedings of the ISMA2020-USD2020, Virtually, 7–9 September 2020; pp. 133–143.
29. Preumont, A. *Random Vibration and Spectral Analysis*, 1st ed.; Kluwer Academic Publisher: Dodrecht, The Netherlands, 2010.
30. Braccesi, C.; Cianetti, F.; Palmieri, M.; Zucca, G. The importance of dynamic behaviour of vibrating systems on the response in case of non-Gaussian random excitations. *Procedia Struct. Integr.* **2018**, *12*, 224–238. [CrossRef]
31. Palmieri, M.; Slavič, J.; Cianetti, F. Single-process 3D-printed structures with vibration durability self-awareness. *Addit. Manuf.* **2021**, *47*, 102303. [CrossRef]
32. Bendat, S.; Piersol, G. *Random Data: Analysis and Measurement Procedures*, 4th ed.; John Wiley & Sons: Hoboken, NJ, USA, 2010.
33. Palmieri, M.; Cianetti, F.; Zucca, G.; Morettini, G.; Braccesi, C. Spectral analysis of sine-sweep vibration: A fatigue damage estimation method. *Mech. Syst. Signal Process.* **2021**, *157*, 107698. [CrossRef]
34. Honma, A. Lessons Learned from the Module Production for the First CMS Silicon Tracker. *Instruments* **2022**, *6*, 73. [CrossRef]
35. Viens, M. *General Environmental Verification Standard (GEVS) for GSFC Flight Programs and Projects*; NASA Goddard Space Flight Center: Greenbelt, MD, USA, 2021.
36. The European Cooperation for Space Standardization (ECSS). *Space Engineering, Spacecraft Mechanical Loads Analysis Handbook*; ESA Requirements and Standards Division: Noordwijk, The Netherlands, 2004.
37. SPACE-X. Falcon 9 Launch Vehicle Payload User's Guide 2011. Available online: https://www.spacex.com/media/falcon-users-guide-2021-09.pdf (accessed on 1 November 2023).

Article

Double Photodiode Readout System for the Calorimeter of the HERD Experiment: Challenges and New Horizons in Technology for the Direct Detection of High-Energy Cosmic Rays

Pietro Betti [1,*], Oscar Adriani [1], Matias Antonelli [2], Yonglin Bai [3], Xiaohong Bai [3], Tianwei Bao [4], Eugenio Berti [5], Lorenzo Bonechi [5], Massimo Bongi [1], Valter Bonvicini [2], Sergio Bottai [5], Weiwei Cao [3], Jorge Casaus [6], Zhen Chen [3], Xingzhu Cui [4], Raffaello D'Alessandro [1], Sebastiano Detti [5], Carlos Diaz [6], Yongwei Dong [4], Noemi Finetti [5,7], Valerio Formato [8], Miguel Angel Velasco Frutos [6], Jiarui Gao [3], Francesca Giovacchini [6], Xiaozhen Liang [3], Ran Li [3], Xin Liu [4,9], Linwei Lyu [3], Gustavo Martinez [6], Nicola Mori [5], Jesus Marin Munoz [6], Lorenzo Pacini [5], Paolo Papini [5], Cecilia Pizzolotto [2], Zheng Quan [4], Junjun Qin [3], Dalian Shi [3], Oleksandr Starodubtsev [5], Zhicheng Tang [4], Alessio Tiberio [1], Valerio Vagelli [10,11], Elena Vannuccini [5], Bo Wang [3], Junjing Wang [4], Le Wang [12], Ruijie Wang [4], Gianluigi Zampa [2], Nicola Zampa [2], Zhigang Wang [4], Ming Xu [4], Li Zhang [4] and Jinkun Zheng [3]

1 Dipartimento di Fisica e Astronomia, Università degli Studi di Firenze and INFN Firenze, Via Sansone 1, 50019 Sesto Fiorentino, Italy; adriani@fi.infn.it (O.A.); bongi@fi.infn.it (M.B.); candi@fi.infn.it (R.D.); tiberio@fi.infn.it (A.T.)

2 INFN Trieste, via Valerio 2, 34127 Trieste, Italy; walter.bonvicini@ts.infn.it (V.B.); cecilia.pizzolotto@ts.infn.it (C.P.); gianluigi.zampa@ts.infn.it (G.Z.); nicola.zampa@ts.infn.it (N.Z.)

3 Xi'an Institute of Optics and Precision Mechanics of CAS, Xi'an 710019, China; baiyonglin@opt.ac.cn (Y.B.); bxh@opt.ac.cn (X.B.); caoweiwei@opt.ac.cn (W.C.); chenzhen@opt.ac.cn (Z.C.); gaojiarui@opt.ac.cn (J.G.); liangxiaozhen@opt.ac.cn (X.L.); liran@opt.ac.cn (R.L.); lvlinwei@opt.ac.cn (L.L.); qjj@opt.ac.cn (J.Q.); lotus@opt.ac.cn (D.S.); wbo@opt.ac.cn (B.W.); zhjink@opt.ac.cn (J.Z.)

4 Institute of High Energy Physics, Chinese Academy of Sciences, Beijing 100049, China; baotw@ihep.ac.cn (T.B.); cuixingzhu@ihep.ac.cn (X.C.); dongyw@ihep.ac.cn (Y.D.); xliu@ihep.ac.cn (X.L.); quanzheng@ihep.ac.cn (Z.Q.); tangzhch@ihep.ac.cn (Z.T.); wangjunjing@ihep.ac.cn (J.W.); wangrj@ihep.ac.cn (R.W.); wangzhg@ihep.ac.cn (Z.W.); mingxu@ihep.ac.cn (M.X.); zhangli@ihep.ac.cn (L.Z.)

5 INFN Firenze, Via Sansone 1, 50019 Sesto Fiorentino, Italy; berti@fi.infn.it (E.B.); lorenzo.bonechi@fi.infn.it (L.B.); bottai@fi.infn.it (S.B.); noemi.finetti@univaq.it (N.F.); mori@fi.infn.it (N.M.); lorenzo.pacini@fi.infn.it (L.P.); papini@fi.infn.it (P.P.); starodubtsev@fi.infn.it (O.S.); vannuccini@fi.infn.it (E.V.)

6 Ciemat, E-28040 Madrid, Spain; jorge.casaus@ciemat.es (J.C.); carlos.diaz@ciemat.es (C.D.); miguelangel.velasco@ciemat.es (M.A.V.F.); francesca.giovacchini@ciemat.es (F.G.); gustavo.martinez@ciemat.es (G.M.); jesus.marin@ciemat.es (J.M.M.)

7 Dipartimento di Scienze Fisiche e Chimiche, Università degli Studi dell'Aquila, Via Vetoio, Coppito, I-67100 L'Aquila, Italy

8 INFN—Sezione di Roma Tor Vergata, via della Ricerca Scientifica 1, I-00133 Roma, Italy; valerio.formato@roma2.infn.it

9 University of Chinese Academy of Sciences, Beijing 101408, China

10 Agenzia Spaziale Italiana, via del Politecnico s.n.c., I-00133 Roma, Italy; valerio.vagelli@asi.it

11 INFN Perugia, Via Alessandro Pascoli 23c, I-06123 Perugia, Italy

12 National Astronomical Observatories, Chinese Academy of Sciences, Beijing 100012, China; wangle@nao.cas.cn

* Correspondence: betti@fi.infn.it

Citation: Betti, P.; Adriani, O.; Antonelli, M.; Bai, Y.; Bai, X.; Bao, T.; Berti, E.; Bonechi, L.; Bongi, M.; Bonvicini, V.; et al. Double Photodiode Readout System for the Calorimeter of the HERD Experiment: Challenges and New Horizons in Technology for the Direct Detection of High-Energy Cosmic Rays. *Instruments* 2024, *8*, 5. https://doi.org/10.3390/instruments8010005

Academic Editor: Antonio Ereditato

Received: 12 October 2023
Revised: 13 December 2023
Accepted: 24 December 2023
Published: 22 January 2024

Abstract: The HERD experiment is a future experiment for the direct detection of high-energy cosmic rays and is to be installed on the Chinese space station in 2027. The main objectives of HERD are the first direct measurement of the *knee* of the cosmic ray spectrum, the extension of electron+positron flux measurement up to tens of TeV, gamma ray astronomy, and the search for indirect signals of dark matter. The main component of the HERD detector is an innovative calorimeter composed of about 7500 LYSO scintillating crystals assembled in a spherical shape. Two independent readout systems of the LYSO scintillation light will be installed on each crystal: the wavelength-shifting fibers system developed by IHEP and the double photodiode readout system developed by INFN and CIEMAT. In order to measure protons in the cosmic ray *knee* region, we must be able to measure energy release of about 250 TeV in a single crystal. In addition, in order to calibrate the system, we need to measure typical releases of minimum ionizing particles that are about 30 MeV. Thus, the readout systems should have a dynamic range of about 10^7. In this article, we analyze the development and the performance of the double photodiode readout system. In particular, we show the performance of a prototype readout by the double photodiode system for electromagnetic showers as measured during a beam test carried out at the CERN SPS in October 2021 with high-energy electron beams.

Keywords: cosmic rays; calorimeters; space instrumentation; large detector systems for particle and astroparticle physics

1. Introduction

Direct detection of cosmic rays is limited at high energy by the geometrical acceptance of space experiments. Indeed, the cosmic ray flux decreases with energy as $E^{-\gamma}$ with $\gamma \simeq 2.7$, limiting the number of particles at high energies. Thus, we need experiments with larger acceptances: a feature that contrasts with the high cost per weight of payloads and power consumption availability in space. The HERD (*High Energy cosmic-Radiation Detection facility*) [1,2] experiment is a new experiment for direct detection of high-energy cosmic rays that will be installed on the Chinese space station in 2027. HERD has an innovative design: with mass and power consumption comparable with that of the current experiments in orbit, it will have a very much larger geometric acceptance. Thanks to this, it will expand direct measurement of proton and nuclei fluxes up to the cosmic ray *knee* region (PeV/nucleon) and electron+positron flux up to tens of TeV. Thus, it will expand direct cosmic ray measurements more than one order of magnitude in energy with respect to the current experiments in orbit. In addition, HERD will perform gamma ray astronomy measurements, and with measurement of both electron+positron flux and gamma rays, it will search for indirect signals of dark matter.

The HERD detector is based on an innovative calorimeter geometry: it is surrounded on five faces by sub-detectors for tracking, charge measurement, and an anti-coincidence system. The calorimeter has a spherical shape and is composed of about 7500 three-centimeter cubic LYSO scintillating crystals, as shown in Figure 1. It is homogeneous, finely segmented, 3D, isotropic, and deep (about 55 X_0, and 3 λ_I). The first idea for this type of calorimeter was developed and studied by the CaloCube collaboration, which demonstrated the very large geometric acceptance that can be achieved with this type of space-borne calorimeter [3–8]. Indeed, thanks to its spherical shape, the HERD calorimeter has a very large acceptance. Considering that it is surrounded on five faces by sub-detectors, the experiment can detect particles arriving from five different directions (the only blind face is the one connected to the space station). The calorimeter has good energy resolution: about 2.5% for electromagnetic showers and less than 30% for hadronic showers. In addition, the cubic segmentation permits the 3D-reconstruction of events and good electron–hadron discrimination for particles coming from all directions. Thanks to these features, HERD's effective geometric factor is about 2.5 $m^2 sr$ for electrons and about 1 $m^2 sr$ for protons.

The scintillation light of the LYSO crystals is readout by two independent systems: one based on *Wavelength Shifting Fibers* (WLSFs) coupled to *Intensified scientific CMOS* (IsCMOS) developed by the Chinese Institute of High Energy Physics (IHEP), and the other one based on the use of two photodiodes with different active areas developed by INFN Florence, INFN Trieste, and CIEMAT Madrid. In order to calibrate the readout systems, we need to detect typical energy releases of minimum ionizing particles, which are about 30 MeV, in a crystal. In addition, we want to measure proton and nuclei fluxes up to the PeV/n energies. Since in a single LYSO cube the energy released by PeV/n particles can be as large as 250 TeV, our readout systems must have an extremely high dynamic range: larger than 10^7. Indeed, the saturation level of a single channel is more than 20 times higher than that in current experiments in orbit. In addition, the total number of channels of the HERD calorimeter will be about 20 times larger than in current calorimeters in orbit [9,10]. These characteristics raise challenges to maintain acceptable power consumption and to manage a higher number of channels.

In this article, we briefly describe the design of the double photodiode readout system. Then, we present the performance studies for electromagnetic showers that were measured on this system during a beam test at CERN SPS in October 2021. Finally, we introduce the new and latest update to the system with some hints about future tests.

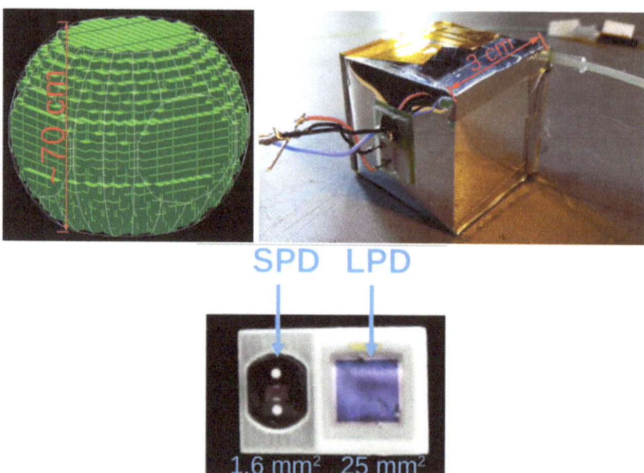

Figure 1. Top left: scheme of the structure of the calorimeter; about 7500 three-centimeter cubic LYSO crystals are assembled in a spherical shape. **Top right**: picture of a LYSO crystal with WLSF and PD readout systems installed; the crystal is covered with a reflective coating. A monolithic package with photodiodes is glued to front of the crystal. The WLSFs are glued on the top face of the crystal and are placed below a reflective coating; we can see the fibers coming out of the reflective coating in the upper right corner of the image. **Bottom**: an illustration of an in-house-built prototype of a monolithic package for the PD readout system, composed of LPD (*Large PhotoDiode*, 25 mm^2) and SPD (*Small PhotoDiode*, 1.6 mm^2).

2. The Double Photodiode Read-Out System

The design of the double photodiode readout system is described in detail in [11]. In this section, we recall only the basic elements. The system is based on the use of two photodiodes with different active areas: the *Large PhotoDiode* (LPD), model VTH2110, with an active area of about 25 mm^2; and the *Small PhotoDiode* (SPD), model VTP9412, with an active area of about 1.6 mm^2. Both PDs are produced by Excelitas Technologies. The use of PDs with different active areas permits an increase in the dynamic range of the system. Indeed, the LPD is sensitive to small signals that the SPD cannot detect, while the SPD is sensitive to large signals for which the LPD saturates the electronics. The LPD and SPD are glued in a plastic mask to assemble an in-house-built monolithic package (Figure 1). The monolithic package is then fixed with optical glue on a LYSO crystal surface (Figure 1). Finally, the crystal surface is covered by a reflective coating.

The main component of the *front-end electronics* is the HiDRA2 chip, based on the CASIS ASIC [12], that was developed by INFN Trieste specifically for the double photodiode readout system of HERD. The HiDRA2 chip has a high dynamic range (from a few fC to 52.6 pC), low noise, and low power consumption (about 3.73 mW per channel). To reach such a large dynamic range, an automatic gain selector for the *charge-sensitive amplifier* is implemented in the chip: the ratio between *high gain* and *low gain* is about 20. The chips are mounted on the HiDRA board, which is controlled by two other boards: the TROC2 that drives the HiDRA chips and the TROC1 that is the interface between the acquisition PC and the TROC2; the boards are developed by CIEMAT Madrid.

3. The SPS2021 Beam Test

3.1. Introduction

In October 2021, we carried out a beam test with a prototype of about 500 LYSO crystals at CERN SPS. Only 63 crystals were equipped with both the double photodiode and WLSF readout systems. The 63 crystals were arranged in 3 columns of 21 crystals each along the beam line (Figure 2), while all the other cubes were equipped with only WLSFs. For a detailed description of the prototype, see [11]. Prototypes of other HERD subsystems (tracker, anti-coincidence, etc.) were installed as well along the beam line upstream of the calorimeter prototype; however, in this article, only the calorimeter data acquired with the PD system are discussed.

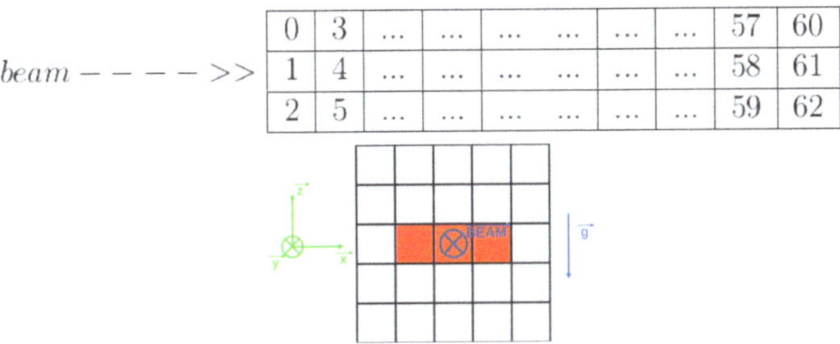

Figure 2. Top: scheme of the disposition of the crystals equipped with Double Photodiode read-out system, as seen from the sky point of view. **Bottom**: scheme of the crystal distribution in the prototype as seen from the beam's point of view; the crystals equipped with photodiodes are highlighted in red (the gravity-field direction is shown as reference).

During the beam test, different particle beams were used: muons at 250 GeV; electrons at 50, 100, 150, 200, and 250 GeV; and protons at 350 GeV. The LPDs were calibrated using the energy releases of 250 GeV muons, while the SPDs were calibrated through their correlation with the LPDs using high signals from showers induced by both electrons and protons. The results of the calibration and the characterization of the system during the SPS2021 beam test are discussed in [13]. Subsequently in this article, the energy is expressed in number of MIPs, as explained in [13].

In the following sections, we show the main results of the ongoing analysis of data acquired with the Double Photodiode read-out system for electromagnetic showers. In particular, we discuss the energy resolution and linearity of the calorimeter response for electromagnetic showers. Finally, we show the first measurement of the correlation between the photodiodes and the WLSF signals. In what follows, we consider only the data acquired with the beam hitting the central column of the calorimeter, as shown in Figure 2.

The following results have been reached analyzing only the calorimeter data acquired with the Double Photodiode read-out system. In future, this analysis could be improved using the data from all the detectors on the beam line.

3.2. Energy Resolution for Electromagnetic Showers

We estimate the energy deposited in the calorimeter with two different methods. In the first one, we sum the energy deposited in every crystal, while in the second one, we fit the longitudinal shower profile with a Gamma function. Indeed, the longitudinal profile of the energy deposit for an electromagnetic shower can be parametrized as [14]:

$$\frac{dE}{dt} = E_0 \cdot b \cdot \frac{(b \cdot t)^{a-1} \cdot e^{-b \cdot t}}{\Gamma(a)} \tag{1}$$

where E_0 is the energy of the particle that produced the shower, t is the length expressed in *radiation length* (X_0), a and b are parameters, and $\Gamma(a)$ is the Euler Γ function. Thus, by fitting this function to the shower's longitudinal profile, we can estimate the energy of the particle that has induced the shower. An example of this kind of fit is illustrated in Figure 3 (left) for a shower induced by a 250 GeV electron.

The two different estimates of the energy of the particle inducing the shower are compatible within less than 2%; thus, for the remainder of this paper, we consider the reconstructed energy to be the one given by the sum of the energy releases in the crystals.

We study the energy resolution of the calorimeter for electromagnetic showers with electron beams with the energies mentioned in Section 3.1. Considering all the events at the same beam energy, we build a histogram of the total energy release. We perform a fit with a logarithmic Gaussian [14] to estimate the peak position, and we use a confidence level method at 68% to estimate the distribution width. Finally, the energy resolution is given by the ratio between the distribution width and the peak position. In Figure 3 (right), the histogram and the fit result for 100 GeV electrons are shown.

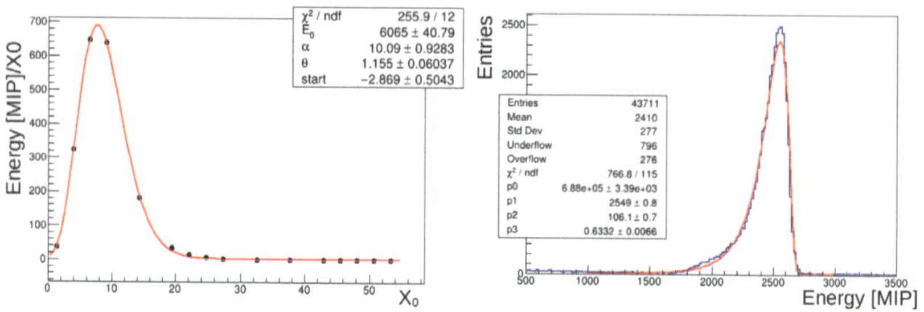

Figure 3. Left: fit with a Gamma function of the longitudinal shower profile for a 250 GeV electron shower (note that the energy is expressed in number of MIPs). **Right**: histogram of the total energy deposits for 100 GeV electron beam fitted with a logarithmic Gaussian.

The energy resolution estimated as a function of the beam energy is reported in Figure 4.

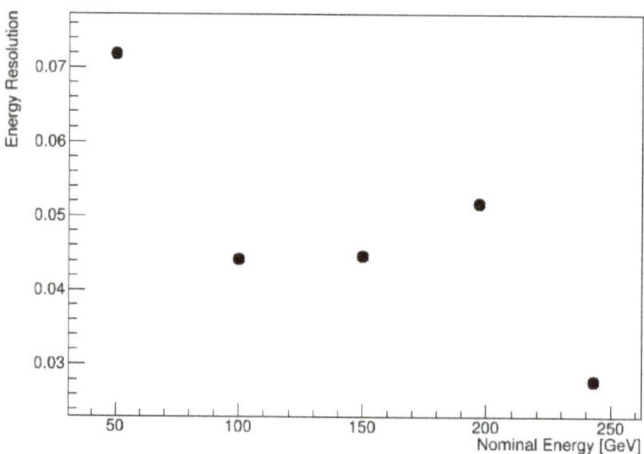

Figure 4. Energy resolution of the prototype for electromagnetic showers tested during SPS2021 beam test.

The energy resolution ranges from about 2.5% for 250 GeV electrons up to about 7% for 50 GeV electrons, and it does not monotonically decrease with energy. Instead, by a first Monte Carlo simulation study of the beam test, we expect the energy resolution to monotonically decrease with increasing of the energy from about 3.5% at 50 GeV to about 2.5% at 250 GeV. Thus, the decreases in the performance measured at certain energies are likely due to some experimental effects not implemented in the simulations. By a comparison with the Monte Carlo simulations, we found that if the beam is not parallel to the y-axis as shown in Figure 2 but is inclined in the yz-plane by less than 0.2 degrees with respect to that axis, it can cause a decrease in the energy resolution up to about 7%. Indeed, in this case, we have a lateral leakage of the shower in the vertical direction (z-axis) since we are considering only data acquired with one tray of crystals, because only one tray was equipped with the double photodiode readout system. Furthermore, inclination of the beam of this entity seems realistic considering the differences in beam shape that we monitored with the beam-line monitor when changing the energy of the electron beam and considering that the alignment procedure of the calorimeter was checked by eye with the help of a laser level, which is a procedure with a precision of a few mm. Finally, with the Monte Carlo simulations, we also checked that inclination with respect to the y-axis but in the xy-plane as compatible with the laser level alignment procedure cannot significantly influence the energy resolution (in this direction, indeed we have three columns of crystals and thus much better shower containment with respect to the z-axis).

3.3. Energy Linearity for Electromagnetic Showers

To measure the linearity of the prototype's response to electromagnetic showers, we build a graph, which has on the x-axis the nominal energy of the beam electrons and on the y-axis the energy measured as the sum of the energy releases in the crystals. Then, we perform a linear fit on the graph and estimate the deviation of every point from the fit, as illustrated in Figure 5. We can see in the figure that the non-linearity is less than 3%. This is quite a good result considering that this analysis does not make use of data from other subsystems like the particle tracker, and that due to the geometry of the set of cubes instrumented with PDs and because the beam structures vary with energy, we expect different lateral leakages along the vertical direction that varies with the energy.

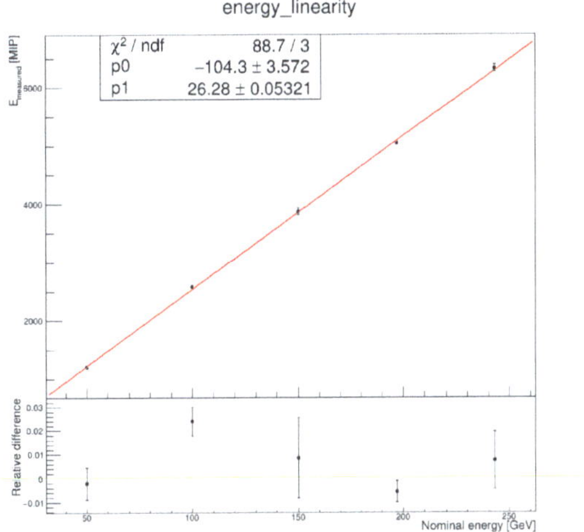

Figure 5. Linearity response for electromagnetic showers of the prototype tested during SPS2021 beam test. **Top**: energy measured in the calorimeter as a function of beam energy. The points are fitted with a straight line (red line). **Bottom**: the relative differences between the points and the fit result are plotted (for better visualization a green line corresponding to no differences is plotted).

3.4. Double Photodiode and WLSF Read-Out System Correlation

The beam test at SPS2021 is the first beam test in which the two readout systems for the scintillation light of LYSO crystals were completely integrated on the same crystals. Indeed, as described before, the 63 crystals we are considering for this analysis are equipped with both the WLSF and double photodiode readout systems. Thus, this is the first beam test during which we acquired signals induced by high-energy muons, electrons, and protons in the same crystals with the two independent systems. An example of the correlation of the signals of a crystal acquired with the LPD and with the WLSFs in high and low gain is shown in Figure 6. We clearly see in the figure that the two signals are correlated: this is the first measurement with high-energy particles of this correlation and demonstrates the possibility to use these two independent readout systems to collect light signals on a single cube and to crosscheck each other. The correlation for the single cubes has already proved a valuable tool during the beam test to monitor both the systems and to promptly spot possible problems in one of the two systems.

The correlation analysis was finalized only on single crystals. However in Autumn 2023, we performed a beam test at CERN PS and SPS with a 1000-crystal prototype for which all crystals were equipped with double photodiode and WLSF systems; so we are going to check the correlation between the two readout systems not only for the single cubes alone but also for aggregate variables like the total energy release.

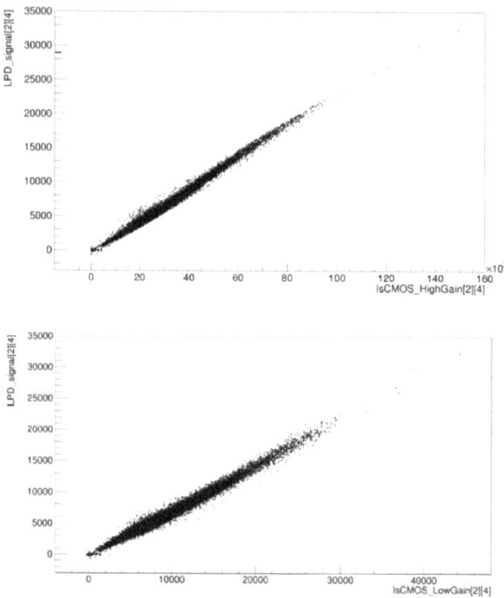

Figure 6. Correlation plots for signals acquired in the same crystals by double photodiode and WLSF readout systems. The y-axis shows the signal acquired by the LPD, while x-axis shows the signal acquired by the WLSFs in high gain (**top**) and in low gain (**bottom**). Both signals are expressed in ADC units. We clearly see that the two independent readout signals are correlated.

4. Development of a New Monolithic Package for Double Photodiode Read-Out System

As explained in [13], the first homemade prototype of the photodiode system (described in Section 2) does not have the final characteristics needed for the flight detector: indeed, the saturation level of the SPD is about 3.5 TeV instead of about 250 TeV. Therefore, after the characterization, we worked with Excelitas Technologies to produce a new version of the package with a modified SPD in order to reach the desired saturation level. The first project using this new photodiode package has already been presented in [13]. In this new version of the package, the SPD surface is covered with an inconel filter to attenuate the light entering its surface and to meet the requirement for the flight model in terms of maximum detectable energy release. In particular, the optical transmittance of the filter is about 1.5% for [410; 450] nm light, which is the LYSO's peak emission. In addition, LPDs and SPDs are directly assembled in the same FR4 package to form a monolithic package. Indeed, a homemade version of the package is very useful for the prototypal stage, but for the final sensors, we need an industrial version in order to keep the production under strong control and to minimize the variability between the packages. Over the course of the proceeding year, we finalized this first project and developed the first prototypes: a sketch of the homemade package, the project of the new package, and a prototype of the new package are illustrated in Figure 7.

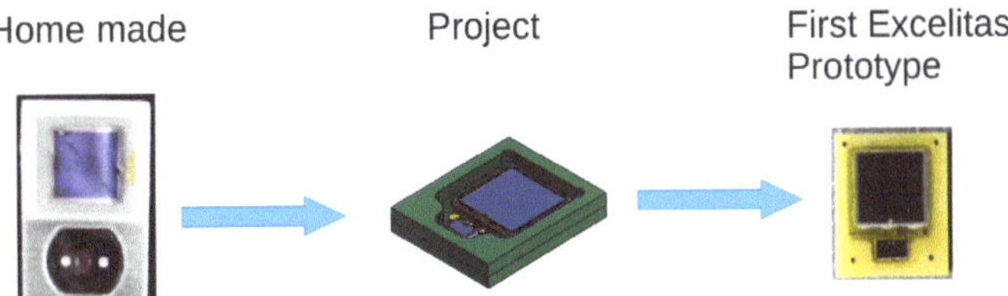

Figure 7. Sketch of the passage from the homemade package to the first Excelitas prototype.

This new version of the package has already been mounted on more than 1000 crystals to form a prototype of the calorimeter that was tested at PS and SPS in September and October 2023, respectively (as already mentioned in Section 3.4). All the crystals are equipped with both PD and WLSF systems.

The crystals were installed on 7 trays, every tray containing 7 columns with 21 crystals each. Thus, the prototype had a thickness of about 55 X_0 for particles parallel to the columns, as in the flight model. A picture of a tray of the prototype is reported in Figure 8.

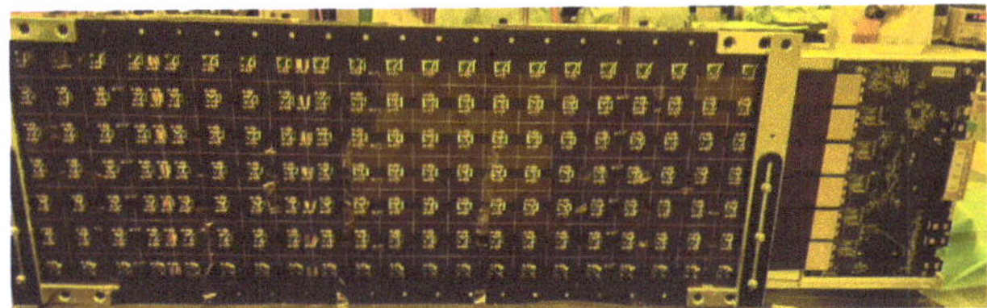

Figure 8. Picture of the bottom of one calorimeter tray for the prototype that was tested at CERN PS and SPS in September and October 2023. On the right is situated the *front-end electronics* board. The monolithic packages are glued to the crystals and come out from the bottom of the tray through some holes; they are connected to the front-end board via the brown cables that we can see in the figure. Both the cables and the front-end board are specifically designed for the double photodiode readout system.

The prototype was tested with muon, electron, proton, and nuclei beams. With respect to the SPS 2021 beam test, the other sub-detectors were also updated, and we acquired data on a common event-by-event basis. Thus, we have a preliminary kind of flight data type with all the info from every detector, allowing for a deep data analysis of the physics performance of the completed HERD detector.

5. Discussion

In this article, we have discussed the analysis of data acquired with a prototype of the HERD calorimeter with the double photodiode readout system at a beam test carried out at SPS in 2021. Specifically, we have analyzed the prototype's performance for electromagnetic showers. The energy resolution ranges from about 2.5% to about 7%. The 2.5% value is a good performance for the calorimeter prototype, while the higher value of the energy resolution is compatible with a small inclination of the beam that causes a lateral leakage of the shower in the vertical direction. With regard to the response linearity for electromagnetic showers, we measured a deviation from linearity of less than 3%, which is quite a good result considering the vertical leakage problem. Anyway, regarding this problem, strong improvement is expected when making use of the particle tracker information and instrumenting more cubes with PDs for better shower containment.

In addition, at the SPS 2021 beam test, we demonstrated the correlation between the WLSF and double photodiode readout systems on a crystal-by-crystal basis, and we already used this calorimeter feature to crosscheck the two systems. We expect to study the correlation on the full calorimeter using the global shower variables from the Autumn 2023 PS and SPS beam tests data.

The new prototype comprises about 1000 LYSO crystals, which are equipped with a new monolithic package that has been developed in collaboration with Excelitas Technologies. In this package, the LPD and SPD are directly assembled in the same FR4 package, and the SPD surface is covered with an optical filter in order to attenuate the signal and reach the desired dynamic range of the readout system.

In conclusion, with the calorimeter geometry and the double photodiode readout system, we are going to reach the desired calorimeter performance for the detection of electromagnetic showers. Moreover, a new monolithic package has been developed in order to extend the photodiode system's dynamic range. We are in the finalization phase of the system, which, step-by-step, is reaching the desired performance that will let the HERD experiment with its innovative calorimeter directly explore the unexplored high-energy range of cosmic rays up to the *knee* region for protons and nuclei and up to tens of TeV for electrons+positrons.

Author Contributions: Conceptualization, O.A., R.D. and J.C.; methodology, P.B., E.B., L.P. and N.M.; software, P.B., E.B., L.P., A.T. and N.M.; validation, P.B., E.B. and L.P.; formal analysis, P.B., E.B. and L.P.; investigation, O.S., P.B., E.B., L.P. and N.M.; resources, Y.B., T.B., W.C., X.C., J.G., R.L., X.L. (Xin Liu), L.L., Z.Q., J.Q., D.S., Z.T., B.W., J.W., R.W., Z.W., M.X., L.Z., J.Z., S.D., G.M., J.M.M., V.F., M.A.V.F., C.P., V.V., S.B., L.B., M.B., N.F., P.P., E.V., M.A., V.B., G.Z., N.Z., X.L. (Xiaozhen Liang), X.B., Z.C., L.W., C.D. and F.G.; data curation, P.B., E.B. and L.P.; writing—original draft preparation, P.B.; writing—review and editing, P.B. and N.M.; visualization, P.B. and L.P.; supervision, R.D.; project administration, O.A. and Y.D. All authors have read and agreed to the published version of the manuscript.

Funding: Grant No. 12027803 of the National Natural Science Foundation of China.

Data Availability Statement: The datasets presented in this article are not readily available because the data access is restricted to the members of the HERD collaboration. Requests to access the datasets should be directed to the authors.

Conflicts of Interest: The authors declare no conflicts of interest.

References

1. HERD The High Energy Cosmic Radiation Detection Facility. Available online: http://herd.ihep.ac.cn/ (accessed on 27 July 2022).
2. Pacini, L.; Adriani, O.; Bai, Y.L.; Bao, T.W.; Berti, E.; Bottai, S.; Cao, W.W.; Casaus, J.; Cui, X.Z.; D'Alessandro, R.; et al. Design and expected performances of the large acceptance calorimeter for the HERD space mission. In Proceedings of the 37th International Cosmic Ray Conference—PoS(ICRC2021), Berlin, Germany, 12–23 July 2021.
3. Vannuccini, E.; Adriani, O.; Agnesi, A.; Albergo, S.; Auditore, L.; Basti, A.; Berti, E.; Bigongiari, G.; Bonechi, L.; Bonechi, S.; et al. CaloCube: A new-concept calorimeter for the detection of high-energy cosmic rays in space. *Nucl. Instrum.* **2017**, *845*, 421–424. [CrossRef]
4. Adriani, O.; Agnesi, A.; Albergo, S.; Auditore, L.; Basti, A.; Berti, E.; Bigongiari, G.; Bonechi, L.; Bonechi, S.; Bongi, M.; et al. Calocube—A highly segmented calorimeter for a space based experiment. *Nucl. Instrum.* **2016**, *824*, 609–613.
5. Bongi, M.; Adriani, O.; Albergo, S.; Auditore, L.; Bagliesi, M.G.; Berti, E.; Bigongiari, G.; Boezio, M.; Bonechi, L.; Bonechi, S.; et al. CALOCUBE: An approach to high-granularity and homogeneous calorimetry for space based detectors. *J. Phys. Conf. Ser.* **2015**, *587*, 012029. [CrossRef]
6. Pacini, L.; Adriani, O.; Agnesi, A.; Albergo, S.; Auditore, L.; Basti, A.; Berti, E.; Bigongiari, G.; Bonechi, L.; Bonechi, S.; et al. CaloCube: An innovative homogeneous calorimeter for the next-generation space experiments. *J. Phys. Conf. Ser.* **2017**, *928*, 012013. [CrossRef]
7. Adriani, O.; Albergo, S.; Auditore, L.; Basti, A.; Berti, E.; Bigongiari, G.; Bonechi, L.; Bonechi, S.; Bongi, M.; Bonvicini, V.; et al. CaloCube: An isotropic spaceborne calorimeter for high-energy cosmic rays. Optimization of the detector performance for protons and nuclei. *Astropart. Phys.* **2017**, *96*, 11–17. [CrossRef]
8. Berti, E.; Adriani, O.; Albergo, S.; Ambrosi, G.; Auditore, L.; Basti, A.; Bigongiari, G.; Bonechi, L.; Bonechi, S.; Bongi, M.; et al. CaloCube: A new concept calorimeter for the detection of high energy cosmic rays in space. *J. Phys. Conf. Ser.* **2019**, *1162*, 012042. [CrossRef]

9.   Chang, J.; Ambrosi, G.; An, Q.; Asfandiyarov, R.; Azzarello, P.; Bernardini, P.; Bertucci, B.; Cai, M.S.; Caragiulo, M.; Chen, H.F.; et al. The DArk Matter Particle Explorer mission. *Astropart. Phys.* **2017**, *95*, 6–24. [CrossRef]

10.   Torii, S. The CALorimetric Electron Telescope (CALET): High Energy Astroparticle Physics Observatory on the International Space Station. In Proceedings of the 34th International Cosmic Ray Conference—PoS(ICRC2015), The Hague, The Netherlands, 30 July–6 August 2015.

11.   Adriani, O.; Antonelli, M.; Basti, A.; Berti, E.; Betti, P.; Bigongiari, G.; Bonechi, L.; Bongi, M.; Bonvicini, V.; Bottai, S.; et al. Development of the photo-diode subsystem for the HERD calorimeter double-readout. *JINST* **2022**, *17*, P09002. [CrossRef]

12.   Bonvicini, V.; Orzan, G.; Zampa, G.; Zampa, N. A Double-Gain, Large Dynamic Range Front-end ASIC with A/D Conversion for Silicon Detectors Read-Out. *IEEE Trans. Nucl. Sci.* **2010**, *57*, 2963–2970. [CrossRef]

13.   Betti, P.; Adriani, O.; Antonelli, M.; Bai, Y.; Bai, X.; Bao, T.; Berti, E.; Bonechi, L.; Bongi, M.; Bonvicini, V.; et al. Photodiode Read-Out System for the Calorimeter of the Herd Experiment. *Instruments* **2022**, *6*, 33. [CrossRef]

14.   Grupen C.; Shwartz B. *Particle Detectors*, 2nd ed.; Cambridge Monographs on Particle Physics, Nuclear Physics and Cosmology; Cambridge University Press: Cambridge, UK , 2008; pp. 230–272.

instruments

MDPI

Article

Design and Performance of a Low-Energy Gamma-Ray Trigger System for HERD

Luis Fariña *, Keerthana Lathika, Giulio Lucchetta, Monong Yu, Joan Boix, Laia Cardiel-Sas, Oscar Blanch, Manel Martinez and Javier Rico on behalf of the HERD Collaboration †

Institut de Física d'Altes Energies (IFAE), The Barcelona Institute of Science and Technology (BIST), 08193 Bellaterra, Barcelona, Spain; klathika@ifae.es (K.L.); glucchetta@ifae.es (G.L.); myu@ifae.es (M.Y.); jboix@ifae.es (J.B.); laia@ifae.es (L.C.-S.); blanch@ifae.es (O.B.); martinez@ifae.es (M.M.); jrico@ifae.es (J.R.)
* Correspondence: lfarina@ifae.es
† The full author list is included in the Appendix A.

Abstract: The High Energy cosmic-Radiation Detection (HERD) facility has been proposed as one of the main experiments on board the Chinese space station. HERD is scheduled to be installed around 2027 and to operate for at least 10 years. Its main scientific goals are the study of the cosmic ray spectrum and composition up to the PeV energy range, indirect dark matter detection, and all-sky gamma-ray observation above 100 MeV. HERD features a novel design in order to optimize its acceptance per weight, with a central 3D imaging calorimeter surrounded on top and on its four lateral sides by complementary subdetectors. A dedicated trigger, dubbed the ultra-low-energy gamma-ray (ULEG) trigger, is required to enable the detection of gamma rays down to ∼100 MeV. The ULEG trigger design is based upon the search for energy deposition patterns on the tracker and the anticoincidence shield, compatible with the conversion of a gamma ray within the tracker volume and resulting in enough tracker hits to allow for a good-quality gamma-ray direction reconstruction. We describe the current status of the design of the ULEG trigger system. We also characterize its performance in detecting gamma rays as inferred from Monte Carlo studies.

Keywords: HERD; trigger; gamma rays

Citation: Fariña, L.; Lathika, K.; Lucchetta, G.; Yu, M.; Boix, J.; Cardiel-Sas, L.; Blanch, O.; Martinez, M.; Rico, J., on behalf of the HERD Collaboration. Design and Performance of a Low-Energy Gamma-Ray Trigger System for HERD. *Instruments* **2024**, *8*, 31. https://doi.org/10.3390/instruments8020031

Academic Editors: Matteo Duranti and Valerio Vagelli

Received: 6 February 2024
Revised: 23 April 2024
Accepted: 26 April 2024
Published: 4 May 2024

1. Introduction

The High Energy cosmic-Radiation Detection (HERD) facility is a future detector of charged cosmic rays and gamma rays scheduled to be installed aboard the Chinese space station. The experiment will begin operations around 2027, and will run for at least 10 years. As a cosmic ray detector, it aims to produce detailed spectra of the different cosmic ray species up to the knee energies and to search for dark matter signatures in these spectra. As a gamma ray detector, it will monitor the whole gamma-ray sky thanks to its unprecedented field of view [1].

HERD is designed as a multi-directional detector in order to efficiently utilize its mass budget (see Figure 1b). At its center lies a 3D-segmented calorimeter (CALO) surrounded by the rest of the subdetectors which are arranged in five active faces. From the inside out, HERD features a fiber tracker (FIT) for track reconstruction, a plastic scintillation detector (PSD) used for gamma identification and charge reconstruction, and a silicon charge detector (SCD) that provides precise charge reconstruction. On one of the lateral faces, a transition radiation detector will be used to calibrate the CALO.

The FIT (see also Figure 1a) is divided into five sectors, each covering one face of the cube. Each of these sectors consists of several layers of scintillating fibers, spanning the whole length of the side they are on. These layers are arranged in tracking pairs or double layers, with the fibers in adjacent layers running in perpendicular directions, forming a total of seven double layers. Each individual layer is segmented into modules, each containing a fiber mat and its readout. In our reference geometry, modules are ∼10 cm wide, each of them producing a signal that can be used for triggering. The top face contains 12 modules in every layer, while the lateral faces contain, in alternating layers, 10 vertical and 8 horizontal modules.

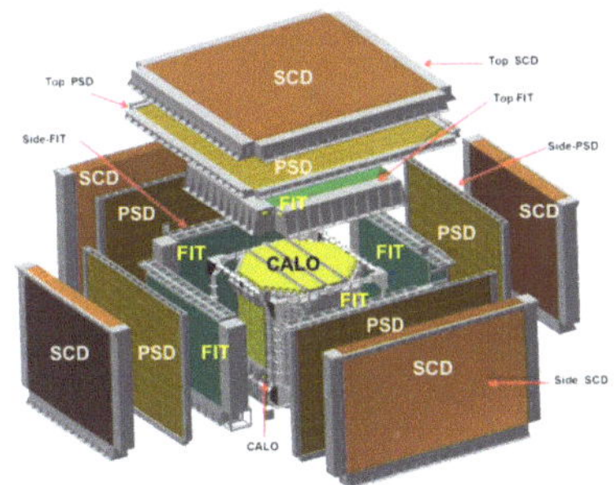

(**a**) HERD design and subdetectors.

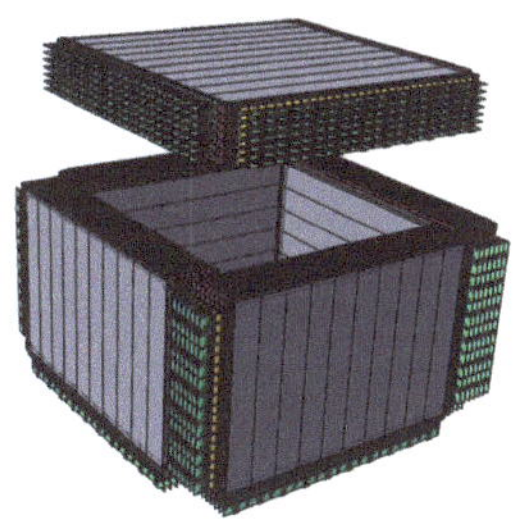

(**b**) FIT segmentation.

Figure 1. HERD and FIT models.

As for the PSD, we consider a reference geometry where it consists of two staggered layers of scintillating square tiles of size 10×10 cm^2. The top face contains 14×14 tiles, and each of the lateral faces contain 13×9 tiles. Each of these tiles is read out individually, and they produce a signal that can be used by the different triggers.

Since HERD aims to measure the fluxes of various species of cosmic ray particles, it features a set of triggers specifically designed for different particle species and energy ranges [2]. Among them, HERD includes a baseline trigger for gamma rays above $\sim$500 MeV, requiring a certain energy deposition in the CALO plus a combination of signals from other subdetectors. We know that HERD is sensitive to gamma rays down to the few-tens-of-MeV range (ultra-low-energy gamma, ULEG, rays) [3], but in the Low Earth Orbit, where the HERD will operate, gamma rays are outnumbered by charged particles by up to five orders of magnitude. Lowering the energy deposition threshold at the CALO would rapidly saturate its readout, so a different approach is needed.

The trigger systems in previous generations of pair-production gamma-ray space detectors operate under the designs similar to each other: first, a trigger signal is produced when a group of hits appears in spatial/temporal coincidence. Dependence on patterns containing several hits prevents triggering by noise and allows for preselecting the direction of the primary particle. For this purpose, a specific subdetector can be used (in EGRET, two layers of scintillating material are placed above and below the tracker for this purpose), or the tracker itself can be used (this is the case for AGILE, which requires hits in three out of four possible in adjacent double layers, and also in Fermi-LAT, where the main trigger condition requires hits in three adjacent double layers). In order to reject charged primaries, an additional subetector is used (the anticoincidence shield), which must show no hits, consistently with the primary being neutral, and converting only afterwards. In both AGILE and Fermi-LAT, this subdetector is segmented into smaller elements, and information from the spatial location of the triggering hits is used to constrain a region of interest to be checked instead of the whole subdetector. Finally, the calorimeter can be checked to verify that the shower intersects it, if needed, and then instrument can be read out and the information on the event's interactions saved.

The ULEG trigger is designed in these three stages:

- Level 0 (L0): initial information of patterns of three aligned FIT modules (three-in-a-row (3IR) patterns), and of PSD elements hit by the shower.
- Level 1 (L1): the PSD information is checked against the region(s) of interest defined by the triggered elements in the FIT.
- Level 2 (L2): the energy deposited in the CALO is checked. In this study, we consider a threshold at 100 MeV.

It is important to understand the rates at which the individual elements of HERD's subdetectors are activated and the rates at which these L0, L1, and L2 signals are produced in order to ensure that the readouts do not saturate. The limiting factor is the CALO readout, which can operate at about 800 Hz [4]. However, the ULEG trigger needs to coexist with the others without hindering the achievement of HERD's scientific goals, so its rate should be well below this number.

In this study, our goal is to develop and optimize the design of the ULEG trigger, considering its efficiency on the target sample of gamma rays and its trigger rate under realistic environmental conditions.

2. Materials and Methods

2.1. Monte Carlo Simulations

The response of the detector to incident particles is simulated with the help of Herd-Software, a framework for simulation and data analysis, which contains HERD-specific detector models and analysis algorithms. Particle generation and interactions with the detector volumes are simulated using GGS ("Generic Géant4 Simulation", [5]), a package used to carry out fast simulations with Géant4 [6]. The body of the space station is not physically simulated due to its computational complexity; instead, a volume approximating its shape is used as an exclusion region. The simulation of the detector response to these interactions is carried out using a custom EventAnalysis code available in HerdSoftware.

The goal is to study the performance of the ULEG trigger under different metrics, specifically the gamma-ray detection efficiency, and the trigger rates at different trigger levels and of different designs. The trigger rate for a given species of particle can be calculated as a convolution of its flux $\Phi(E, \theta, \phi)$ and trigger acceptance:,

$$\frac{\mathrm{d}N}{\mathrm{d}t} = \int \frac{\mathrm{d}^2\Phi(E, \theta, \phi)}{\mathrm{d}\Omega \mathrm{d}E} A_{\mathrm{eff}}(E, \theta, \phi)\mathrm{d}\Omega \mathrm{d}E, \tag{1}$$

where $A_{\mathrm{eff}}(E, \theta, \phi)$ is the effective area which can be estimated with a Monte Carlo study from

$$A_{\mathrm{eff}}(E, \theta, \phi) = S_{\mathrm{gen}}(\theta, \phi)\frac{N_{\mathrm{sel}}(E, \theta, \phi)}{N_{\mathrm{gen}}(E, \theta, \phi)}, \tag{2}$$

where $S_{\mathrm{gen}}(\theta, \phi)$ is the generating surface perpendicular to the (θ, ϕ) direction and $N_{(\mathrm{gen,sel})}$ refer. respectively, to the number of particles generated and to the number of particles that pass the selection cuts.

For this study, the samples of primary particles for the Monte Carlo study consist of 10^7 events per species detailed (namely protons, electrons, positrons, and alpha particles—fluxes of other particle species are negligible), generated with isotropic spatial distribution and log-uniform energy distribution between 1 MeV and 100 GeV. We consider the expected fluxes at their highest, i.e., at the highest geomagnetic latitudes seen by HERD, as reconstructed from various models and empirical data from previous missions (NINA-2, AMS-01, PAMELA, MARYA, see [7]). We assume that fluxes are independent of pointing direction except from the change at the Earth's limb, located at a polar angle of ~108.3° at the orbital altitude of HERD (above the Earth's limb the fluxes contain primary and secondary cosmic rays; below, there are only secondaries). In particular, we do not consider any smoothing of this boundary, nor do we consider any azimuthal dependence in the arrival direction of charged particles.

For the case of testing the trigger efficiency on gamma rays, we define a fiducial sample of gamma rays as the subset of gamma rays that

- travel downwards, i.e., up to 90° from the zenith,
- do not interact with the detector (namely they do not comptonize) before undergoing pair conversion,
- convert within the FIT,

- after conversion, both the electron and the positron produce hits or interactions in at least three consecutive FIT double layers in both tracking directions within each double layer (for a total of at least six hits), and
- the total energy deposited in the CALO is at least half of that of the primary.

These conditions are chosen as an approximate parametrization of the set of gamma rays that can be distinguishable from cosmic rays and the albedo background gamma rays, and for which adequate track reconstruction and energy resolution can be obtained.

2.2. Three-in-a-Row Trigger Design

The ULEG trigger is based on the same three-in-a-row concept implemented in Fermi-LAT [8]. To produce a valid trigger signal, an event must produce at least three hits in consecutive FIT double layers, and in both tracking directions within each double layer. This responds to a minimal prerequisite for acceptable track reconstruction of the gamma-ray's direction.

In Fermi-LAT, the tracker is segmented into an array of 4×4 towers, each producing independent triggers. HERD's tracker is more finely segmented (10 cm wide modules to Fermi's 40×40 cm^2 tower-like modules). Additionally, because of its 5-face design, the normal geometrical cross-section of the FIT is larger than that of the CALO, and so the edges of the tracker are involved only for particles with off-axis incidence angles. Thus, a significant fraction of the particles of interest interacts with three fiber modules that are not vertically aligned. For this reason, we also allow trigger patterns that end in modules adjacent to the first one, as long as they pass through either of the two modules between them (see Figure 2).

This extra degree of freedom in the trigger pattern causes the inclusion of some patterns that actually decrease trigger performance, as they increase the trigger rate and/or the complexity of the trigger, without meaningful marginal increases to its scientific performance. Therefore, we remove these patterns from the trigger design. They are:

- non-CALO-intersecting: the geometric cross-section of the tracker is larger than that of the calorimeter. Some groups of 3 modules, located near the edges of the tracker, trigger primarily in events in which the shower is directed away from the calorimeter. The only information available in these cases comes from the shower development in the tracker, leading to poor energy reconstruction.
- upwards-pointing: in the lateral faces, some patterns respond primarily to particles travelling upwards, from the direction of the Earth.

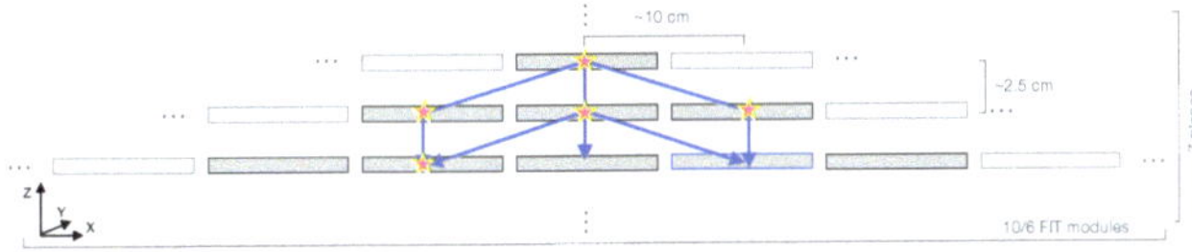

Figure 2. Schematic view of the 3IR trigger patterns.

2.3. Veto Design

At Level 1, the ULEG trigger cross-references the signals from the FIT 3IR and the PSD, producing the PSD-vetoed signals. Here, we present the performance of this veto strategy when checking the whole PSD, both as a baseline and as the simplest possible implementation. This incurs a significant loss of efficiency at high energy due to vetoing on the backsplashed part of the shower, which could be mitigated by adopting a strategy where the 3IR pattern is used to restrict the PSD elements that can produce a veto to those within a given region of interest.

3. Results

3.1. Individual Element Activation Rates

Every element capable of producing a signal relevant to the trigger is simulated independently, so Equations (1) and (2) can be applied not just to the different trigger conditions, but also to the activation of each of the individual sensitive elements that generate the signals used in the ULEG (the PSD tiles and the FIT mats). Since the sizes, shapes, location and numbers of these elements are subject to some modification as the design of the detector is finalized, we report here the activation rates per element per unit area.

We find that the activation rate of the individual elements is highly dependent on their location within HERD. An example is provided in Figure 3 for the case of the PSD. The outer layer is more exposed than the inner one due to the flux of $\sim$1 MeV particles that do not penetrate to the deeper parts of the detector. In the inner layer, some elements are partially uncovered due to the staggering of the two layers and have activation rates similar to those of the outer layer. On the other hand, higher energy particles (above $\sim$1 GeV) produce more interactions in the deeper layers due to increased backsplash and are responsible for the hotspot at the center of the top face, where the CALO is located. Shadowing due to the body of the space station causes the top face to be less exposed than the lateral faces, and in the latter it causes the rates to decrease as the elements move closer to the space station mount.

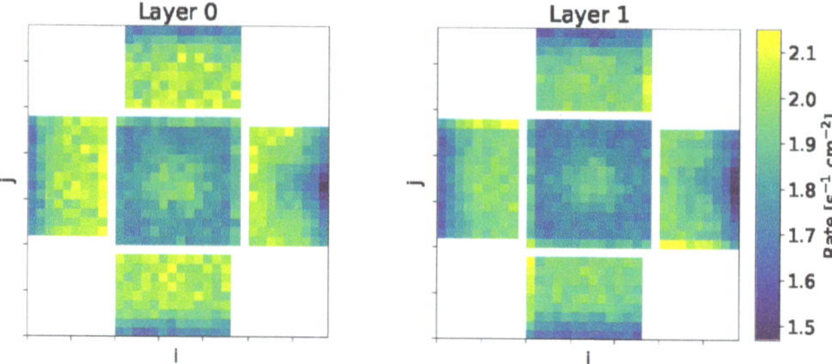

Figure 3. Estimated total activation rates of the individual tiles for the outer (0, left) and inner (1, right) layers of the PSD. Each of the pixels represents a single PSD tile. The five sections in each plot correspond to the five faces of HERD: at the center is the top face, and from the right and in counterclockwise order the other faces are designated X+, Y+, X−, Y−, and they are aligned with the space station directions forward, port, aft, and starboard, respectively. In this representation, the lateral faces are rotated so that the pixels closer to the top face correspond to the upper part of the face they are on.

According to our simulations, the activation rates are $\sim$2.2 cm^{-2}s^{-1} and $\sim$1.7 cm^{-2}s^{-1} for the most exposed PSD tile and FIT module. The flux models considered have large uncertainties for energies below $\sim$10 MeV due to the lack of precise measurements; for the estimated worst-case-scenario (higher flux) model, the activation rates could be as high as $\sim$5.2 cm^{-2}s^{-1} and $\sim$3.5 cm^{-2}s^{-1}, respectively, about a factor two higher. The increase is higher for the PSD, as it is the outermost detector involved in the ULEG and these particles have low penetrative power.

3.2. Trigger Performance and Rates

The total estimated trigger rates for the studied detector geometry are $\sim$2.6 · 10^4 s^{-1}, $\sim$2 · 10^2 s^{-1}, and $\sim$60 s^{-1} for 3IR, L1 and L2, respectively (see Figure 4). The bulk of the trigger rate at L1 and L2 corresponds to events that pass along directions corresponding to the edges of the detector (see Figure 5), which indicates that this number is affected by inaccuracies in the description of HERD used for the simulations. Note that here, we consider only the limited information available at the speed at which the triggers have to operate. L2 events are still susceptible to further cleaning after the data are stored. In the offline analysis, more information is available, including, e.g., the full resolution of the FIT (the trigger uses only mat-level resolution) and supplementary vetoing with subdetectors outside the PSD.

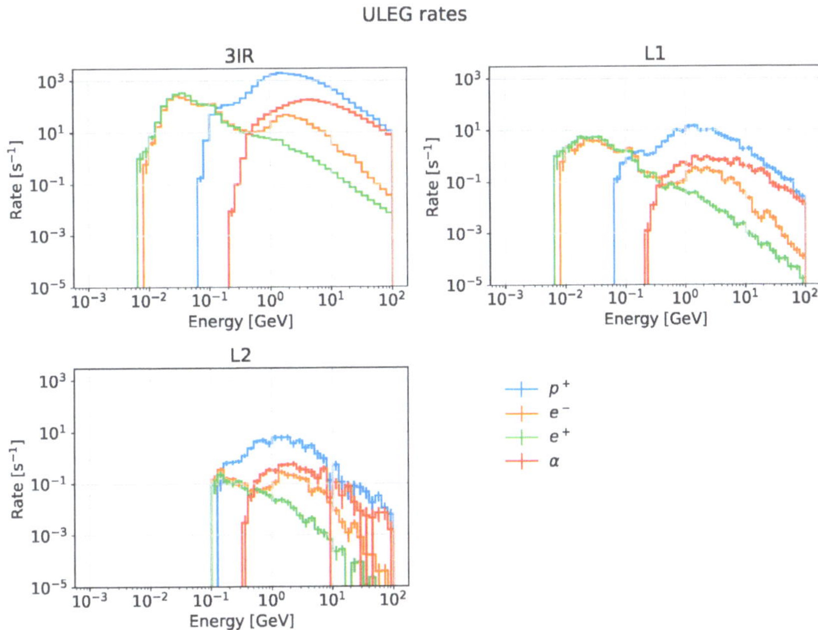

Figure 4. Expected particle rates for the different backgrounds under study and for ULEG trigger levels 3IR, L1 and L2.

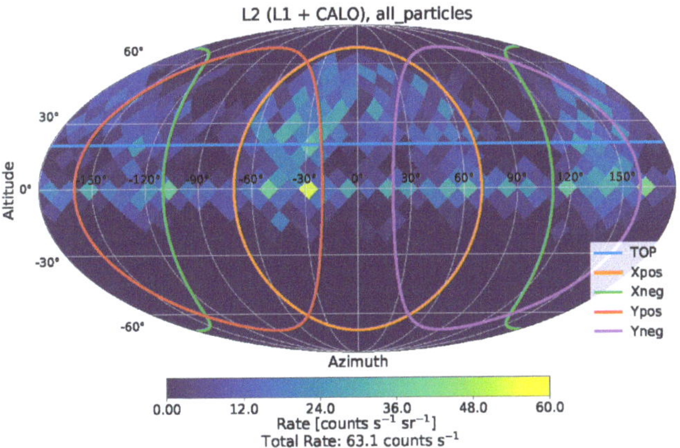

Figure 5. L2 trigger rates according to the direction of the primary as seen from HERD. The boundary of a putative 70° field of view is drawn for each face as a visual aid. Note that the boundary of the top face's field of view is a circle corresponding to the 20° N parallel.

In order to estimate the contribution from background events that are physically indistinguishable from gamma rays, we also consider a model of HERD with a completely hermetic PSD, finding an irreducible rate of $\sim 0.2\,\mathrm{s}^{-1}$ background events comparable to the expected rates from the averaged galactic diffuse gamma-ray background (see Figure 6), which is the dominant gamma-ray diffuse background.

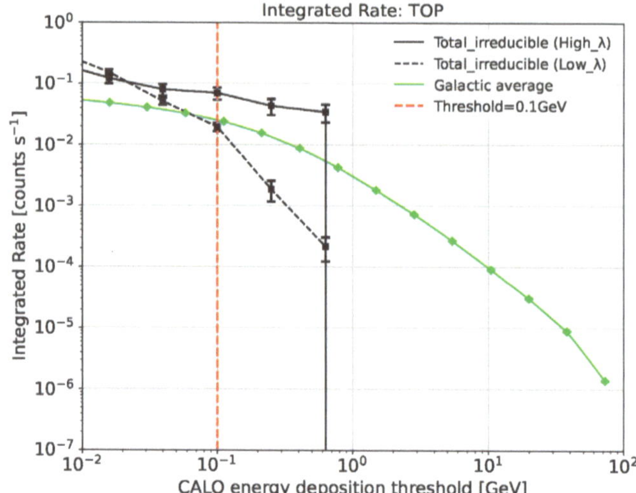

Figure 6. Trigger rates from the irreducible background compared with the rates expected with the diffuse background, as inferred from simulations with a fully hermetic HERD model.

As for the trigger efficiency, we find that the 3IR design is extremely efficient in the energy range we consider, but the PSD veto introduces a loss of efficiency due to backsplash-induced veto, with ∼30% (∼70%) events being vetoed at ∼1 GeV (∼10 GeV) (Figure 7). This could be mitigated by adopting a veto strategy based on a region of interest determined by the combination of modules that produce the trigger. Other HERD triggers provide accessory detection capabilities above ∼500 MeV and above ∼15 GeV [9], and the ULEG trigger is designed to attain high efficiency in the lower end of the spectrum.

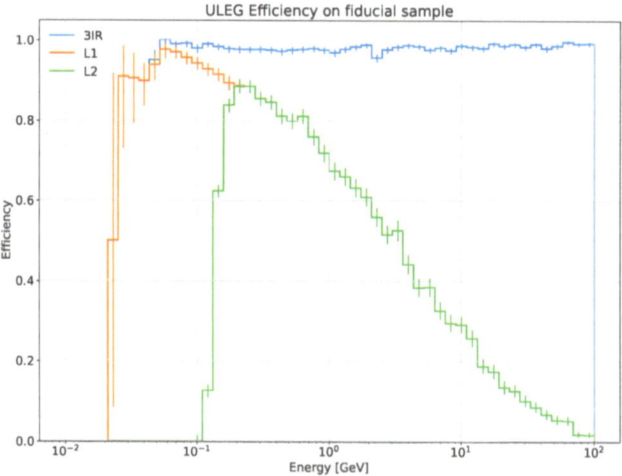

Figure 7. Efficiency of the 3IR, L1 and L2 levels on the fiducial gamma-ray sample.

4. Discussion

We present a preliminary design for the HERD ULEG trigger and its expected performance in the detection of gamma rays in the 100 MeV–100 GeV energy range. The performance of this design is within the limits imposed by the mission's objectives, but improvements are expected. The final design will feature a region-of-interest- and time-of-flight-based veto strategy. A proof-of-concept version of the hardware implementation of this trigger is undergoing validation at beam tests at PS and SPS at CERN in 2023 and 2024.

Author Contributions: Conceptualization: L.F., K.L., G.L., M.Y., J.B., L.C.-S., O.B., M.M. and J.R.; Data curation: L.F., K.L., G.L., M.Y., J.B., L.C.-S. and J.R.; Formal analysis: L.F., K.L., G.L., M.Y. and J.R.; Investigation: L.F., K.L., G.L., M.Y., J.B. and L.C.-S.; Software: L.F., K.L., G.L., M.Y., J.B. and L.C.-S.; Validation: L.F., K.L., G.L. and M.Y.; Visualization: L.F., K.L., G.L. and M.Y.; Writing—original draft: L.F.; Writing—review & editing: K.L., G.L., M.Y., J.B., L.C.-S., O.B., M.M. and J.R.; Funding acquisition: J.R.; Project administration: J.R.; Supervision: J.R. All authors have read and agreed to the published version of the manuscript.

Funding: This work is supported by Agencia Estatal de Investigación, Ministerio de Ciencia e Investigación: PID2020-116075GB-C22/AEI/10.13039/501100011033, PRE2019-091232, PRE2021-097518; European Union NextGenerationEU: PRTR-C17.I1; Generalitat de Catalunya.

Data Availability Statement: Dataset available on request from the authors

Conflicts of Interest: The authors declare no conflict of interest.

Abbreviations

The following abbreviations are used in this manuscript:

HERD	High Energy cosmic-Radiation Detection facility
CALO	CALOrimeter
FIT	FIber Tracker
PSD	Plastic Scintillation Detector
SCD	Silicon Charge Detector
TRD	Transition Radiation Detector
3IR	Three in a row
RoI	Region of Interest
ULEG	Ultra-Low-Energy Gamma
CSS	Chinese Space Station

Appendix A. The HERD Collaboration

O. Adriani[1,2], F. Alemanno[3,4,a], C. Altomare[5], G. Ambrosi[6], M. Antonelli[7], P. Azzarello[8], X.H. Bai[9], Y.L. Bai[9], T.W. Bao[10], M. Barbanera[6], F.C.T. Barbato[3,4], F. Bernard[11], P. Bernardini[12,13], E. Berti[2], B. Bertucci[6,14], P. Betti[1,2], X.J. Bi[10,15], G. Bigongiari[16,17], O. Blanch[18], J. Boix[18], M. Bongi[1,2], V. Bonvicini[7], S. Bottai[2], P. Brogi[16,17], C. Brugnoni[6,14], F. Cadoux[8], I. Cagnoli[3,4], H.Y. Cai[10,15], D. Campana[19], W.W. Cao[9], L. Cardiel-Sas[18], J. Casaus[20], E. Casilli[12,13], R. Catala[21], E. Catanzani[6,14], P.W. Cattaneo[22], D. Cerasole[5,23], L. Chang[24], H. Chen[10,15], K. Chen[25], L. Chen[26], M.L. Chen[10], P.D. Chen[27], R. Chen[25], Y.D. Cheng[10,15], F. Cianetti[6,14], A. Comerma[28], X.Q. Cong[29], P. Coppin[8], X.Z. Cui[10], R. D'Alessandro[1,2], D. D'Urso[6,30], C. Díaz[20], C. Dai[31], I. De Mitri[3,4], F. de Palma[12,13], C. De Vecchi[22], V. Di Felice[32], A. Di Giovanni[3,4], M. Di Santo[3,4], L. Di Venere[5], Y.W. Dong[10], G. Donvito[5], Y.J. Du[33], M. Duranti[6], A. Espinya[21], K. Fang[10], L. Fariña[18], Y. Favre[8], H.B. Feng[31], M. Fernandez Alonso[3,4], N. Finetti[2,34], G. Fontanella[3,4], V. Formato[32], J.M. Frieden[11], Y. Fu[33], P. Fusco[5,23], J.R. Gao[9], F. Gargano[5], D. Gascón[21,35], D. Gasparrini[32], E. Ghose[12,13,b], F. Giovacchini[20], S. Gómez[21,28], K. Gong[10], M.H. Gu[10], D. Guberman[21], C. Guerrisi[5,23], R. Guida[36], D.Y. Guo[10], J.H. Guo[37], H.L. He[10,15], H. Hu[10], H.J. Hu[31], Y.M. Hu[37], Z.X. Hu[29], G.S. Huang[27], W.H. Huang[38], X.T. Huang[38], Y.G. Huang[33], M. Ionica[6], F. Jia[31], J.S. Jia[33], F. Jiang[31], X.W. Jiang[10], Y. Jiang[6,14], P. Jiao[33], A. Kotenko[8], D. Kyratzis[3,4,c], D. La Marra[8], K.R. Lathika[18], L. Li[10], M.J. Li[38], M.X. Li[25], Q.Y. Li[39], Q.Y. Li[40], R. Li[9], S.L. Li[10,15], T. Li[29], T. Li[38], X.Q. Li[10], X.Q. Li[41], Y.Y. Li[39], Z.H. Li[10,15], M.J. Liang[10,15], X.Z. Liang[9], C.L. Liao[10,15], F. Licciulli[5], Y.J. Lin[29], B.H. Liu[24], D. Liu[38], H. Liu[26], H.B. Liu[31], H.W. Liu[10], X. Liu[10,15], X.J. Liu[10], X.W. Liu[31], Y.Q. Liu[10], F. Loparco[5,23], S. Loporchio[5,23], L. Lorusso[5,23], B. Lu[10], R.S. Lu[10,15], Y.P. Lu[10], G. Lucchetta[18], J.G. Lv[10], L.W. Lv[9], P. Maestro[16,17], E. Mancini[6], R. Manera[21], J. Marin[20], P.S. Marrocchesi[16,17], G. Marsella[42,43], G. Martinez[20], M. Martinez[18], J. Mauricio[21], M.N. Mazziotta[5], G. Morettini[6,14], N. Mori[2], L. Mussolin[6,14], S. Nicotri[5], Y. Niu[38], A. Oliva[44], D. Orlandi[4], M. Orta[21,35], G. Osteria[19], L. Pacini[2], B. Panico[19,45], F.R. Pantaleo[5,23], G. Panzarini[5,23], S. Papa[36], P. Papini[2], A. Parenti[3,4], M. Pauluzzi[6,14], W.X. Peng[10], F. Perfetto[19], C. Perrina[11], G. Perrotta[36], R. Pillera[5,23], A. Pinard[11], C. Pizzolotto[7] S. Qian[10], R. Qiao[10], J.J. Qin[9], X.B. Qiu[29], Z.Y. Qu[40], Z. Quan[10], A. Rappoldi[22], G. Raselli[22], F. Renno[36], J. Rico[18], M. Rossella[22], A. Sanmukh[21,35], A. Sanuy[21,35], E. Savin[44], M. Scaringella[2], V. Scotti[19,45], D. Serini[5], D.L. Shi[9], Q.Q. Shi[38], I. Siddique[3,4,b], L. Silveri[3,4], G. Silvestre[6], A. Smirnov[3,4], P.G. Song[33], O. Starodubtsev[2], Y.Z. Su[41], D. Sukhonos[8], J.Y. Sun[10,15], X.L. Sun[10], Z.T. Sun[10], A. Surdo[13], Z.C. Tang[10], A. Tiberio[1,2], A. Tykhonov[8], V. Vagelli[6,46], E. Vannuccini[2], M. Velasco[20], R. Walter[47], A.Q. Wang[38], C. Wang[33], D. Wang[25], G.F. Wang[27], H. Wang[25], J.J. Wang[10], J.W. Wang[33], M. Wang[38], R.J. Wang[10], W.S. Wang[10], X.P. Wang[41], Y. Wang[33],

Z.G. Wang[10], Z.H. Wang[26], J.J. Wei[37], P. Wei[31], Y.F. Wei[27], B.B. Wu[10], C. Wu[29], L.B. Wu[3,4,d], Q. Wu[10,15], X. Wu[8], H.Q. Xie[41], X. Xing[9], Y.W. Xing[33], M. Xu[10], S.H. Xu[9], Z.X. Yan[10,15], H.B. Yang[41], H.T. Yang[10,15], X.G. Yang[10,15], Y. Yang[31], P.F. Yin[10], Y.H. You[10,15], M. Yu[18], G. Zampa[7], N. Zampa[7], D.L. Zang[10], C. Zhang[10], F. Zhang[10], F.Z. Zhang[10], H.M. Zhang[10], J. Zhang[33], J. Zhang[39], S.D. Zhang[9], S.N. Zhang[10,15], X. Zhang[33], X.T. Zhang[10], C.X. Zhao[41], R. Zhao[33], J.K. Zheng[9], Y.J. Zheng[26], F.R. Zhu[26], J.Y. Zhu[41], K.J. Zhu[10], L.K. Zou[40].

[1] Department of Physics, University of Florence, Via Sansone 1, I-50019 Sesto Fiorentino, Firenze, Italy

[2] Istituto Nazionale di Fisica Nucleare, Sezione di Firenze, Via Sansone 1, I-50019 Sesto Fiorentino, Firenze, Italy

[3] Gran Sasso Science Institute (GSSI), Viale Crispi 7, I-67100 L'Aquila, Italy

[4] Istituto Nazionale di Fisica Nucleare, Laboratori Nazionali del Gran Sasso, Via Acitelli 22, I-67100 Assergi, L'Aquila, Italy

[5] Istituto Nazionale di Fisica Nucleare, Sezione di Bari, via Orabona 4, I-70126 Bari, Italy

[6] Istituto Nazionale di Fisica Nucleare, Sezione di Perugia, Via Alessandro Pascoli 23c, I-06123 Perugia, Italy

[7] Istituto Nazionale di Fisica Nucleare, Sezione di Trieste, via A. Valerio 2, I-34127 Trieste, Italy

[8] Département de Physique Nucléaire et Corpusculaire (DPNC), Université de Genève, 24 quai Ernest-Ansermet, CH-1211 Genève 4, Switzerland

[9] Xi'an Institute of Optics and Precision Mechanics of CAS, No.17 Xinxi Road, New Industrial Park, Xi'an Hi-Tech Industrial Development Zone, Xi'an 710019, China

[10] Institute of High Energy Physics, Chinese Academy of Sciences, 19B Yuquan Road, Shijingshan District, Beijing 100049, China

[11] Institute of Physics, Ecole Polytechnique Fédérale de Lausanne (EPFL), Bâtiment PH, Station 3, CH-1015 Lausanne, Switzerland

[12] Dipartimento di Matematica e Fisica "E. De Giorgi", Università del Salento, I-73100 Lecce, Italy

[13] Istituto Nazionale di Fisica Nucleare, Sezione di Lecce, Via per Arnesano, I-73100 Lecce, Italy

[14] Università degli Studi di Perugia, Piazza Università 1, I-06123 Perugia, Italy

[15] University of Chinese Academy of Sciences, No.1 Yanqihu East Rd, Huairou District, Beijing 101408, China

[16] Department of Physical Sciences, Earth and Environment, University of Siena, via Roma 56, I-53100 Siena, Italy

[17] Istituto Nazionale di Fisica Nucleare, Sezione di Pisa, Largo B. Pontecorvo 3, I-56127 Pisa, Italy

[18] Institut de Física d'Altes Energies (IFAE), The Barcelona Institute of Science and Technology (BIST), E-08193 Bellaterra, Barcelona, Spain

[19] Istituto Nazionale di Fisica Nucleare, Sezione di Napoli, Via Cintia, I-80126 Napoli, Italy

[20] Centro de Investigaciones Energéticas, Medioambientales y Tecnológicas (CIEMAT), E-28040 Madrid, Spain

[21] Departament de Física Quàntica i Astrofísica (FQA) and Institut de Ciències del Cosmos (ICCUB), Universitat de Barcelona (UB), E-08028, Barcelona, Spain

[22] Istituto Nazionale di Fisica Nucleare, Sezione di Pavia, Via Bassi 6, I-27100 Pavia, Italy

[23] Dipartimento di Fisica "M. Merlin" dell'Università e del Politecnico di Bari, via Amendola 173, I-70126 Bari, Italy

[24] North Night Vision Technology Co., Ltd., Hongwai Road 5, Kunming 650217, China

[25] PLAC, Key Laboratory of Quark & Lepton Physics (MOE), Central China Normal University, Wuhan 430079, China

[26] School of Physical Science and Technology, Southwest Jiaotong University, No.999, Xi'an Road, Chengdu 611756, China

[27] Department of Modern Physics, University of Science and Technology of China, Hefei 230026, China

[28] Serra Hunter fellow at Polytechnic University of Catalonia (UPC), Electronics Department, E-08019 Barcelona, Spain

[29] North Night Vision Science & Technology (Nanjing) Research Institute Co., Ltd, Kangping Street 2, Nanjing 211100, China

[30] Università degli Studi di Sassari, Piazza Università 21, I-07100 Sassari, Italy

[31] Guangxi Key Laboratory for Relativistic Astrophysics, Guangxi University, Daxue East Road 100, Nanning 530004, China

[32] Istituto Nazionale di Fisica Nucleare, Sezione di Roma Tor Vergata, via della Ricerca Scientifica 1, I-00133 Roma, Italy

[33] Institute of Special Glass Fiber & Optoelectronic Functional Materials, China Building Materials Academy, Guanzhuang Dongli 1, Chaoyang district, Beijing 100024, China

[34] Department of Physical and Chemical Sciences, University of L'Aquila, Via Vetoio, Coppito, I-67100 L'Aquila, Italy

[35] Institut d'Estudis Espacials de Catalunya (IEEC), E-08034 Barcelona, Spain

[36] Dipartimento di Ingegneria Industriale, Università degli Studi di Napoli Federico II, P.le Tecchio 80, I-80125 Napoli, Italy

[37] Purple Mountain Observatory, CAS, No.10 Yuanhua Road, Qixia District, Nanjing 210023, China

[38] Shandong University (SDU), 72 Binhai Road, Jimo, Qingdao 266237, China

[39] Shandong University (SDU), 27 Shanda Nanlu, Jinan, Shandong 250100, China

[40] Shandong Institute of Advanced Technology (SDIAT), 1501, Panlong Road, Jinan, Shandong 250100, China

[41] Institute of Modern Physics, CAS, 509 Nanchang Rd., Lanzhou 730000, China

[42] Dipartimento di Fisica e Chimica "E. Segrè", Università degli Studi di Palermo, via delle Scienze, I-90128 Palermo, Italy

[43] Istituto Nazionale di Fisica Nucleare, Sezione di Catania, Via Santa Sofia 64, I-95123 Catania, Italy

[44] Istituto Nazionale di Fisica Nucleare, Sezione di Bologna, Viale C. Berti Pichat 6/2, I-40127 Bologna, Italy

[45] Università di Napoli Federico II, Dipartimento di Fisica "Ettore Pancini", Via Cintia, I-80126 Napoli, Italy

[46] Agenzia Spaziale Italiana, via del Politecnico s.n.c., I-00133 Roma, Italy
[47] Département d'Astronomie, Université de Genève, Chemin d'Ecogia 16, CH-1290 Versoix, Switzerland
[a] Now at Dipartimento di Matematica e Fisica "E. De Giorgi", Università del Salento, I-73100, Lecce, Italy and Istituto Nazionale di Fisica Nucleare, Sezione di Lecce, Via per Arnesano, I-73100 Lecce, Italy.
[b] Also at Dipartimento di Fisica, Università di Trento, via Sommarive 14, I-38123 Trento, Italy.
[c] Now at Département de Physique Nucléaire et Corpusculaire (DPNC), Université de Genève, 24 quai Ernest-Ansermet, CH-1211 Genève 4, Switzerland.
[d] Now at Institute of Deep Space Sciences, Deep Space Exploration Laboratory, Hefei 230026, China.

References

1. Gargano, F. The High Energy cosmic-Radiation Detection (HERD) facility on board the Chinese Space Station: Hunting for high-energy cosmic rays. In Proceedings of the 37th International Cosmic Ray Conference—PoS (ICRC2021), Berlin, Germany, 15–22 July 2021 ; p. 26. [CrossRef]
2. Wu, Q. Status of the HERD trigger design. In Proceedings of the 38th International Cosmic Ray Conference (ICRC2023)—Cosmic-Ray Physics (Direct, CRD), Nagoya, Japan, 26 July–3 August 2023; p. 126. [CrossRef]
3. Fariña, L.; Jouvin, L.; Rico, J.; Mori, N.; Gargano, F.; Formato, V.; de Palma, F.; Pizzolotto, C.; Casaus, J.; Mazziota, M.N.; et al. Gamma-ray performance study of the HERD payload. In Proceedings of the 37th International Cosmic Ray Conference (ICRC2021)—GAD–Gamma Ray Direct, Berlin, Germany, 15–22 July 2021; p. 651. [CrossRef]
4. Liu, X.; Adriani, O.; Bai, X.H.; Bai, Y.L.; Bao, T.W.; Berti, E.; Betti, P.; Bottai, S.; Cao, W.W.; Casaus, J.; et al. Double read-out system for the calorimeter of the HERD experiment. In Proceedings of the 38th International Cosmic Ray Conference (ICRC2023)—Cosmic-Ray Physics (Direct, CRD), Nagoya, Japan, 26 July–3 August 2023; p. 97. [CrossRef]
5. Mori, N. GGS: A Generic Geant4 Simulation package for small- and medium-sized particle detection experiments. *Nucl. Instrum. Methods Phys. Res. Sect. A Accel. Spectrometers Detect. Assoc. Equip.* **2021**, *1002*, 165298. [CrossRef]
6. Agostinelli, S.; Allison, J.; Amako, K.; Apostolakis, J.; Araujo, H.; Arce, P.; Asai, M.; Axen, D.; Banerjee, S.; Barrand, G.; et al. Geant4—A simulation toolkit. *Nucl. Instrum. Methods Phys. Res. Sect. A Accel. Spectrometers Detect. Assoc. Equip.* **2003**, *506*, 250–303. [CrossRef]
7. Lucchetta, G.; Fariña, L.; Yu, M.; Lathika, K.R.; Rico, J.; Cardiel, L.; Boix, J.; Martínez, M.; Casaus, J.; Mauricio, J.; et al. Gamma-ray performance of the High Energy cosmic-Radiation Detection (HERD) space mission. In Proceedings of the 38th International Cosmic Ray Conference (ICRC2023)–Gamma-ray Astronomy (GA), Nagoya, Japan, 26 July–3 August 2023; p. 691. [CrossRef]
8. Ackermann, M.; Ajello, M.; Albert, A.; Allafort, A.; Atwood, W.B.; Axelsson, M.; Baldini, L.; Ballet, J.; Barbiellini, G.; Bastieri, D. The Fermi Large Area Telescope On Orbit: Event Classification, Instrument Response Functions, and Calibration. *Astrophys. J. Suppl.* **2012**, *203*, 4. [CrossRef]
9. Velasco Frutos, M.A.; Bao, T.; Berti, E.; Bonvicini, V.; Casaus, J.; Giovacchini, F.; Liu, X.; Marco, R.; Marín, J.; Martínez, G.; et al. The High Energy cosmic Radiation Detector (HERD) Trigger System. In Proceedings of the 37th International Cosmic Ray Conference (ICRC2021)—CRD—Cosmic Ray Direct, Berlin, Germany, 15–22 July 2021; p. 62. [CrossRef]

MDPI

Article

The PMT Acquisition and Trigger Generation System of the HEPD-02 Calorimeter for the CSES-02 Satellite

Marco Mese [1,2,*], Antonio Anastasio [1], Alfonso Boiano [1], Vincenzo Masone [1], Giuseppe Osteria [1], Francesco Perfetto [1], Beatrice Panico [1], Valentina Scotti [1,2] and Antonio Vanzanella [1] on behalf of the CSES-Limadou Collaboration

[1] National Institute for Nuclear Physics, Sezione di Napoli, Via Cintia, I-80126 Naples, Italy; antonio.anastasio@na.infn.it (A.A.); boiano@na.infn.it (A.B.); vincenzo.masone@na.infn.it (V.M.); osteria@na.infn.it (G.O.); perfetto@na.infn.it (F.P.); bpanico@na.infn.it (B.P.); scottiv@na.infn.it (V.S.); antonio.vanzanella@na.infn.it (A.V.)

[2] Department of Physics, University of Naples Federico II, Via Cintia 21, I-80126 Naples, Italy

* Correspondence: mese@na.infn.it

Abstract: This contribution describes the acquisition and trigger system for the HEPD-02 calorimeter that will be used onboard the CSES-02 satellite for the CSES/Limadou mission. This mission arises from the collaboration between the Chinese Space Agency (CNSA) and the Italian Space Agency (ASI) and plans the realization of a constellation of satellites which will monitor ionospheric parameters supposed to be related to earthquakes. It will also monitor the solar activity and the interaction with the magnetosphere and will study the cosmic rays in low energy ranges, extending data from PAMELA and AMS. The CSES-02 satellite will be equipped with various instruments, including the High-Energy Particle Detector (HEPD-02), which was designed to measure the energy of particles coming from Van Allen belts. Signals from the HEPD-02 are acquired and digitized by an electronic board that also produces the trigger for the experiment. A new generation ASIC (CITIROC) for the amplification, shaping and memorization of signals from PMTs will be used on this board. The new ASIC allows the use of the peak detector feature, optimizing the acquisition of signals with different temporal characteristics. Along with this, new algorithms for trigger generation have been developed, providing trigger pre-scaling, concurrent trigger masks and Gamma Ray Burst detection. Using pre-scaled concurrent triggers will allow the study of very sensitive regions of a satellite's orbit such as the South Atlantic Anomaly and polar regions and to detect rare events such as GRBs while still monitoring particle bursts. In this contribution, the progress status of this work will be presented along with the measurements and tests made to finalize the flight model of the board.

Keywords: trigger; pmt; cses; limadou; calorimeter; earthquake; cosmic; rays; detector; GRB

Citation: Mese, M.; Anastasio, A.; Boiano, A.; Masone, V.; Osteria, G.; Perfetto, F.; Panico, B.; Scotti, V.; Vanzanella, A., on behalf of the CSES-Limadou Collaboration. The PMT Acquisition and Trigger Generation System of the HEPD-02 Calorimeter for the CSES-02 Satellite *Instruments* **2023**, *7*, 53. https://doi.org/10.3390/instruments7040053

Academic Editor: Antonio Ereditato

Received: 15 October 2023
Revised: 20 November 2023
Accepted: 6 December 2023
Published: 11 December 2023

1. Introduction

The CSES (China Seismo-Electromagnetic Satellite) mission is an Italian–Chinese space mission developed by the China National Space Administration (CNSA) and the Italian Space Agency (ASI). Various Italian universities and research centers contribute to the mission.

The main objective is the investigation of the upper ionosphere phenomena to obtain information about the correlation between seismic events and perturbations of physical quantities such as the electric and magnetic field of the Earth, the plasma frequency, the composition of the ionosphere and the flux of particles precipitating from Van Allen belts [1,2]. Italy contributes to the mission in the context of the LIMADOU program.

The CSES-Limadou program will be a multi-satellite mission that foresees the development of a constellation of satellites that will be equipped with various detectors specifically developed for the study of ionospheric parameters. The first satellite of the constellation, named CSES-01, was launched on 2 February 2018 and is still in operation [3,4].

From CSES-01 first scientific data, it was possible to obtain different results that include the monitoring of the G3 geomagnetic storm, which happened on 26 August 2018 [5], the study of the solar modulation of the cosmic rays [6] and the trapped proton fluxes inside the South Atlantic Anomaly (SAA) [7].

Furthermore, a retrospective analysis of high-magnitude seismic events, such as the Palu earthquake on 28 September 2018 ($M_W = 7.5$) and the Papua New Guinea earthquake on 14 May 2019 ($M_W = 7.6$) [8], has been made using data collected by the CSES-01 detectors. From these studies, several ionospheric anomalies that could be related to the preparation phase of these seisms can be deduced [6,7,9,10].

The second satellite of the constellation, named CSES-02, will be launched in early 2024 and will be placed in an orbit with a 180° phase shift with respect to CSES-01 to increase the temporal resolution. Thanks to the improvements made on the HEPD-02 detector, the CSES-02 will operate also in polar regions and the SAA, while CSES-01 was designed to be operative only for latitude between ±65° and outside the South Atlantic Anomaly (SAA).

The instruments that will be onboard the CSES-02 satellite, are listed in Table 1 with a brief description of their observation targets [11].

Table 1. CSES-02 instruments and observation targets.

Category	Payload Name	Observation Target
Particle energy	High-Energy Particle Detector (HEPD)	Electrons: 3 to 100 MeV Protons: 30 to 200 MeV
	Medium Energetic Electron Detector (MEED)	Electrons: 25 keV to 3.2 MeV
Electromagnetic field	Electric Field Detector (EFD)	Electric field: DC 3.5MHz
	High-Precision Magnetometer (HPM)	Magnetic field: 10 Hz to 20 kHz
	Search Coil Magnetometer (SCM)	Magnetic field: 10 Hz to 20 kHz
In situ plasma	Plasma Analyzer Package (PAP)	Composition: H^+, He^+, O^+ N_i: 5×10^2 to 1×10^7 cm^{-3} T_i: 500 to 10,000 K
	Langmuir Probe (LP)	N_i: 5×10^2 to 1×10^7 cm^{-3} T_i: 500 to 10,000 K
Plasma profile construction	GNSS Occultation Receiver	TEC by transmit VH/U/L signal
	Tri-Band Beacon	TEC by transmit VH/U/L signal
	Ionospheric	O_2 135.6 nm and N_2 LBH airglow

The responsibility of the realization of the High-Energy Particle Detector (HEPD-02) and the Electric Field Detector (EFD-02), for the second satellite, has been assigned to the Italian collaboration.

The HEPD-02 will measure the flux of particles and their energies and will also be able to detect Gamma Ray Bursts in the range from MeV to tens of MeV [12] and will be described in detail in the next sections.

The EFD-02 will measure the components of the electric field of the Earth, but its description is outside the scope of this article.

2. The HEPD-02 Detector

The High-Energy Particle Detector realized for the CSES-02 satellite (HEPD-02) is designed to detect electrons, protons and light nuclei in the energetic ranges from 30 to 200 MeV for protons and from 3 to 100 MeV for electrons.

The detector's dimensions are $403.6 \times 530 \times 382.5$ mm^3, and its mass is 50 kg. The power consumption, at the highest reachable trigger rate, is 43 W and the data budget is less than 100 Gbit per day.

It is structured in four main components:

1. The direction detection;
2. The trigger system;
3. The calorimeter;
4. The veto system.

These components are represented in Figure 1a [13], and a brief description of the detector's sections follows.

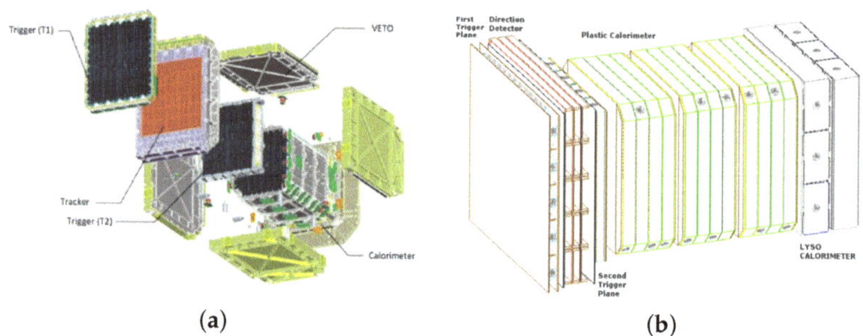

(**a**)   (**b**)

Figure 1. (**a**) Exploded view of the HEPD-02 detector, (**b**) structure of the trigger system and the calorimeter.

2.1. The Direction Detector

The direction detector is an innovative tracker made with Monolithic Active Pixel Sensors (MAPSs) and designed to measure the entrance angle of the particles.

It is composed of five turrets made by three sensitive planes that mount ten MAPS chips. These chips are composed of 512×1024 pixels with a size of 29.24×26.88 μm^2 and can reach a spatial resolution of approximately 4 μm.

The turrets are acquired by a dedicated electronic board called T-DAQ (Tracker Data Acquisition) [14,15].

2.2. The Trigger System and the Calorimeter

The energy of particles entering inside the HEPD-02 is converted into light signals using several scintillators that differ in material and dimension depending on their function.

The light produced is then acquired by two photomultiplier tubes by Hamamatsu (R9880-210 PMTs), placed on the opposite sides of each scintillator, for a total of 64 PMTs.

The trigger system is composed of two segmented planes of plastic scintillators (EJ-200), displaced orthogonally with respect to each other, and surrounding the direction detector.

Five segments of EJ-200 constitute the first layer of the trigger system and are aligned with the turrets of the direction detector ($154.6 \times 32.5 \times 2$ m^3). The second layer is realized with four segments ($150 \times 36 \times 8$ m^3) placed orthogonally with respect to TR1 [16].

A range calorimeter follows the trigger system, and it is made of twelve planes of plastic scintillators ($150 \times 150 \times 10$ m^3) and two segmented planes of LYSO scintillators, which extend the energetic range of the detector thanks to their high density.

The LYSO planes are segmented in three bars ($150 \times 49 \times 25$ m^3) arranged orthogonally to each other (see Figure 1b).

The plastic scintillators are named RAN_1 to RAN_12, while the LYSO planes are EN_1 and EN_2.

2.3. The Veto System

To identify particles that are not contained inside the detector or enter from the side or the bottom of it, the calorimeter is contained inside five planes of EJ-200 scintillators which constitute the veto system.

The lateral planes are called LAT, while the bottom scintillator is called BOT.

2.4. Electronics

The electronic boards used for HEPD-02 are listed below:

- HV-CTRL (High-Voltage Control), which is designed for the control of the high voltages of the PMTs;
- LV-CTRL (Low-Voltage Control), which controls the power of the other boards;
- T-DAQ (Tracker Data Acquisition), which is used for the readout of the MAP sensors of the tracker;
- PMT&T (PMT readout and Trigger), which is designed for the acquisition and digitization of the PMT signals and the generation of the trigger signal for the detector;
- DPCU (Data Processing and Control Unit), which controls all the subsystems and manages the communication toward the satellite.

All the subsystems provide HOT/COLD redundancy and communicate with DPCU via SpaceWire Lite protocol.

In this contribution, the PMT&T board will be described in detail, while a brief description of the DPCU and T-DAQ boards follows.

2.5. T-DAQ Board

The T-DAQ board is based on XC7A100T FPGA (Xilinx, San Jose, CA, USA) and acquires data from the direction detector, performing the following operations: configures the MAPS chips and manages the acquisition and their calibration, manages signals from and to the PMT & T board, and packs the data and transfers it to the DPCU board.

2.6. DPCU

The DPCU is based on a Zynq XC7Z7045 FPGA (Xilinx, San Jose, CA, USA) and is designed to control the other subsystems (T-DAQ, LV/HV-CTRL and PMT&T) and communicate with the satellite platform with CAN BUS and RS422 protocols. It also manages the HEPD-02 modes (stand-by, safe mode and nominal mode) and the calibration or acquisition operations. When the PMT&T and T-DAQ boards have data, the DPCU reads it and transmits it to the satellite.

In case of malfunctioning of any of the HEPD-02 subdetectors, the DPCU also has the task of providing recovery procedures.

3. The PMT Readout and Trigger System

The signals produced by the calorimeter's PMTs are digitized by a dedicated electronic board, called PMT&T, that mounts two CITIROC readout chips made by Weeroc that are managed by an A3PE3000L FPGA by Microsemi (Chandler, AZ, USA) [16,17].

The PMT&T board configures CITIROCs parameters, starts and stops the acquisition, produces the trigger for the detector if certain logic conditions on PMT signals, called "trigger masks", are met, and sends the data to the DPCU.

A schematic description of the board is shown in Figure 2b.

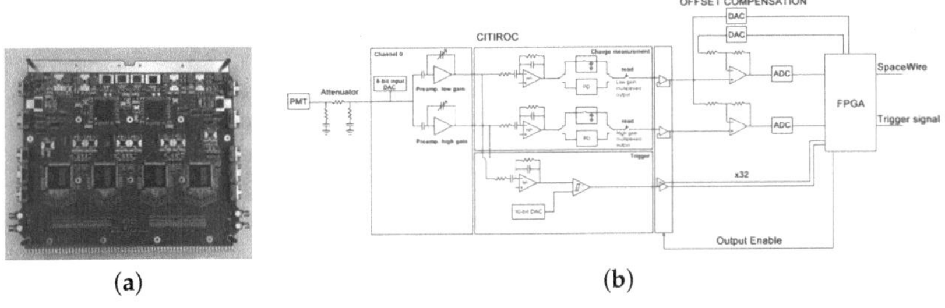

(a)   (b)

Figure 2. (**a**) The electronic board used for PMT readout and trigger system, (**b**) schematic description of the trigger board.

Since CITIROCs are designed for SiPMs, which have positive polarity, the PMT signals are collected from the last dynode and not from the anode, making it possible to avoid the use of inverters and lowering the board consumption. To match the input range of the CITIROCs, the dynode signals are also attenuated with "pi pad" attenuators.

3.1. CITIROCs

The internal structure of CITIROCs is shown in Figure 3.

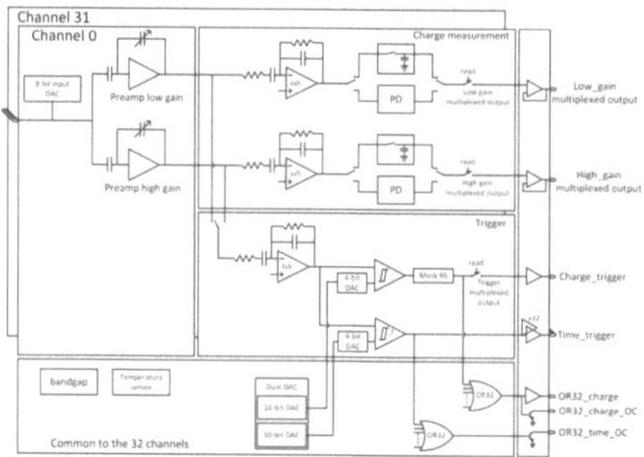

Figure 3. CITIROC internal structure.

An amplification stage is present on each channel, featuring two independent charge preamplifiers with different gain ranges: the high-gain preamplifier, with a gain from 10 to 600 and the low-gain preamplifier, with a gain from 1 to 60. The presence of two independent acquisition chains ensures a wide dynamical range, which is essential in the case of HEPD-02 due to the wide energy range required by design.

A configurable shaper follows the preamplifier with shaping times that can be selected between 12.5 and 87.5 ns with 12.5 ns steps.

The shaper's output can be stored in analogue memory circuits which can work in two operating modes: the Track and Hold mode, which samples the output of the shaper at a specific instant, and the Peak Detection mode, which follows the shaper's output in a defined time window.

Since the scintillators of HEPD-02 have very different timing characteristics, the PMT&T board uses the Peak Detection mode.

The FPGA uses the "Time_trigger" signals produced by the two CITIROCs to generate the trigger for the experiment. This enables the analogue output of the CITIROCs and starts the ADC conversions.

3.2. Scientific Data

The PMT&T board produces 240 bytes of data at each trigger and stores it in a 62×40 bytes FIFO. Since a single packet occupies four locations, the maximum amount of trigger events that the FIFO can store is 10.

The data content is described in Table 2.

A Finite State Machine (FSM) packs the data in the FIFO as soon as the ADC finishes the conversions and if there is enough room in memory to store it. Another independent FSM unpacks data into 62 SpaceWire registers when the DPCU is ready to read it (see Figure 4).

Table 2. Data packet description.

Name	Length (bytes)	Description
Trigger counter	4	Number of events acquired
Timestamp	4	Time from the power-on (16 µs resolution)
Trigger ID	1	Identification number for the trigger configuration
ADC data	192	ADC conversion of the two CITIROCs' output
Lost trigger	2	Triggers counted during dead time
Alive time	4	Alive time counter (5 µs resolution)
Dead time	4	Dead time counter (5 µs resolution)
Trigger flags	8	Flags indicating over-the-threshold channels
Turret flags	1	Flags indicating which tracker turret has been hit
Turret counters	20	Signals counted for each turret
Total	240	

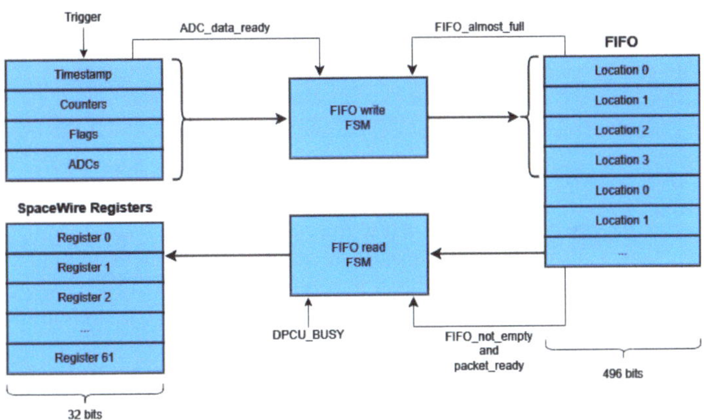

Figure 4. Data packet finite state machines.

3.3. Boards Interwork

The communication between the PMT&T and the other subsystems is managed by a register-based SpaceWire protocol, which is described in Figure 5.

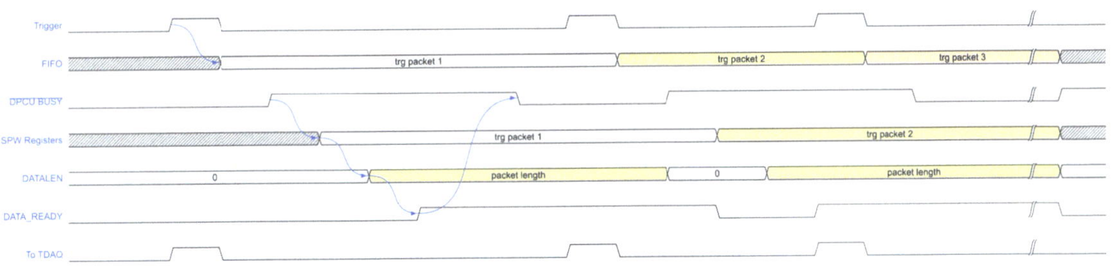

Figure 5. Timing diagram describing the dataflow inside the FPGA and the signals used to communicate with DPCU and TDAQ, the arrows indicate a cause-effect relationship between signals.

At each trigger event, the PMT&T board sends the trigger signal to the T-DAQ and the DPCU and starts the ADC conversions. Once the conversions finish, the FIFO is filled with data (first arrow in Figure 5).

If the DPCU is not busy (the $\overline{\text{DPCU\_BUSY}}$ signal, as shown in Figure 5, is high), the first packet in the FIFO is transferred to a series of SpaceWire registers and the total length of the data is stored in a separate register (second and third arrow of the timing diagram).

When the PMT&T board asserts the DATA_READY signal, the DPCU starts reading all the data registers.

3.4. Trigger Masks

The detector acquisition is conditioned by boolean expressions that are called "trigger masks".

The PMT&T board implements three classes of trigger masks: the first class is named "event acquisition masks" and validates events produced by particles that are contained inside the detector. The "event monitor masks" class, instead, is used for efficiency measurements and to report particles that escape from the bottom or the sides of the calorimeter. The last class, named "GRB detection masks", represents an innovation for the mission and is designed for detecting Gamma Ray Bursts in the 2 to 20 MeV energy range.

The event acquisition masks are defined as follows:

$$M_1 = TR1And \tag{1}$$

$$M_2 = TR1 \cdot TR2 \tag{2}$$

$$M_3 = TR1 \cdot TR2 \cdot RAN\_02 \tag{3}$$

where the names of the detector sections (i.e., $TR1$, $TR2$, RAN_xx and EN_xx) identify the OR of the signals produced by the PMTs. The "And" suffix is used when the AND of the two PMTs connected to the same scintillator is used.

These three masks are designed for particles with gradually increasing energies: M_1 will produce a trigger for particles that lose all their energy in the first trigger plane. The other two masks allow the detection of particles that reach the TR2 scintillator (M_2) and the RAN_02 slab (M_3).

The following masks belong to the event monitor class:

$$M_4 = RAN\_01 \cdot RAN\_07 \cdot RAN\_12 \tag{4}$$

$$M_5 = TR2 \cdot BOT \tag{5}$$

$$M_6 = BOT \cdot EN1 \cdot EN2 \cdot \overline{TR1 + TR2 + LAT} \tag{6}$$

$$M_7 = (RAN\_05And + RAN\_06And + RAN\_07And + RAN\_08And) \cdot \\ (\overline{RAN\_04 + RAN\_09}) \tag{7}$$

The M_4 masks allows for efficiency measurements if compared with M_1, M_2 or M_3. Masks M_5, M_6 and M_7 are designed for not contained particles.

Lastly, there are two GRB detection masks:

$$M_8 = (EN1And + EN2And) \cdot (\overline{RAN\_12 + LAT + BOT}) \tag{8}$$

$$M_9 = (RAN\_05And + RAN\_06And + RAN\_07And + RAN\_08And) \cdot \\ (\overline{RAN\_04 + RAN\_09 + LAT}) \tag{9}$$

The M_8 mask will trigger when signals are produced only in the LYSO crystals and not in the preceding plastic scintillators. The M_9 mask, instead, is triggered by signals produced only in the central stages of the calorimeter and not in the surrounding scintillators.

In addition to these, a "generic trigger mask" can be configured to obtain the AND of any scintillator.

Six of these trigger masks can be selected to operate in concurrency and can be pre-scaled. This allows the control of the bandwidth occupation between different physics channels and to acquire data in regions of the orbit where the particle rate, especially for low energies, significantly increases and can reach hundreds of megahertz.

To obtain these features, six multiplexers are connected to the output of the 10 trigger masks, and their outputs can be selected via SpaceWire registers.

The outputs of four of these multiplexers are connected to counters that ignore a configurable number of triggers, actually accomplishing the pre-scaling.

3.5. First and Second-Level Triggers

All the trigger masks that include anticoincidences, such as the GRB masks, are prone to produce spurious triggers if signals present delays between them.

A second-level trigger system has been developed to minimize the occurrence of these problems. This system allows the sampling of the trigger masks twice, assuring that the overlap between signals is long enough to exclude spurious triggers.

If this overlap lasts at least 30 ns, a "valid trigger" signal is produced and the FPGA starts the acquisition and the ADC conversion. If not, the CITIROCs analogue memories are cleared and the peak detection circuit is reset.

4. PMT&T Board Measurements

Several measurements have been conducted on the PMT&T board to define operative parameters that optimize the signal acquisition.

For these measurements, which will be described in detail in the following sections, a Python script has been developed to automatize the configuration, the acquisition, and the analysis of data.

4.1. Optimization of the Input Signal Conditioning Circuit

A series of measurements were performed to find the minimum value of the input signal attenuation that ensures a good input dynamic range for the CITIROCs.

In Figure 6a, it is possible to see that with a 3x attenuation, a compression for signals greater than 3 V is present.

This compression is caused by the protection diode inside the CITIROC's inputs and would reduce the available input dynamic range since PMTs' signals can extend up to 8 V.

The channel with the 6x attenuation, instead, shows a good linearity over the amplitude range of the input signal.

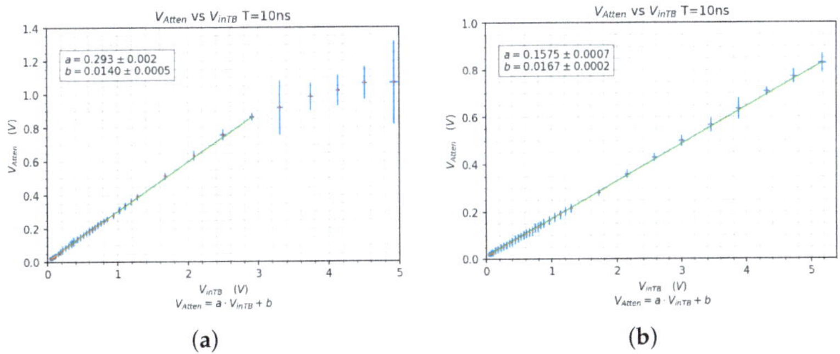

Figure 6. Output of the input signal attenuators: (**a**) 3x attenuator, (**b**) 6x attenuator.

From these considerations, a good value for the attenuators would be 6x, but from the calibration curves that will be presented in Section 4.2, it has been observed that doubling the attenuation and using higher gains for the preamplifiers allows the CITIROCs to work in a more linear region.

For these reasons, the final values of the attenuations have been chosen considering an attenuation of 12x for signals produced by almost all the scintillators except for those of T1. These scintillators, being thinner than all the others, produce smaller signals that would be wiped by the 12x attenuation, and therefore, the value 6x is used instead.

Table 3 shows the attenuation values for all the PMT&T channels.

Table 3. Gains and attenuation factors for PMT readout and trigger board.

Board Channels	Scintillators	Attenuation	HG	LG
0–4, 32–36	T1	6	20	2
5–20, 27–31, 37–52, 59–63	T2, RAN BOT, LAT	12	20	2
21–26, 53–58	EN	12	10	1.5

4.2. CITIROC's Preamplifiers and Shapers Calibration

As mentioned before, the CITIROC's preamplifiers and shapers can be configured with different values for the amplification and the shaping time.

To select the optimal combination of these parameters, a series of measurements have been conducted using square pulses with different amplitudes and durations.

Three classes of signals were used, considering the FWHM of the typical signal produced by the PMTs for different types of scintillators: 5 ns wide pulses for T1 scintillators, 10 ns for the calorimeter and 50 ns for the LYSO crystals.

Figure 7 shows, as an example, the mean of ADC counts versus the amplitude of calorimeter-like signals, while Figure 8 shows the same plots for LYSO-like signals.

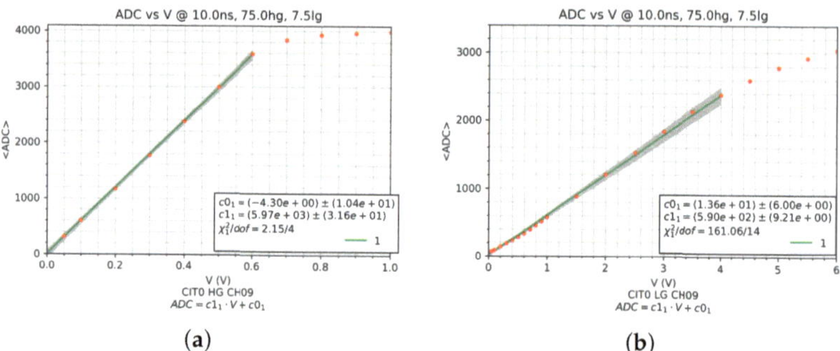

(a)   (b)

Figure 7. ADC means vs. V for calorimeter-like signals: (**a**) high gain = 75, (**b**) low gain = 7.5. The red dots represent experimental points, the green line the linear fit.

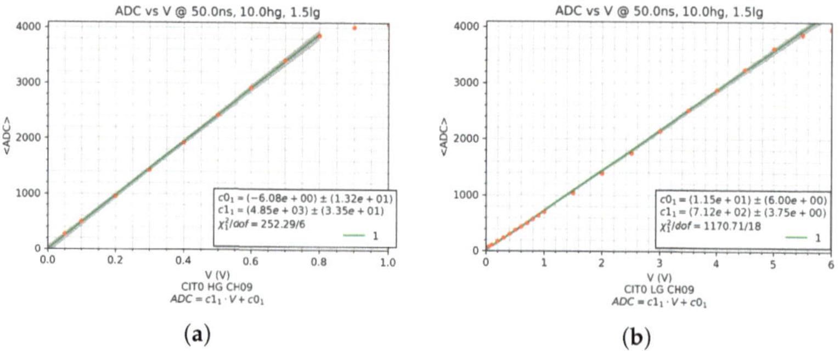

(a)   (b)

Figure 8. ADC vs. V for LYSO-like signals: (**a**) high gain = 10, (**b**) low gain = 1.5. The red dots represent experimental points, the green line the linear fit.

At first, an amplification of 75 for high-gain preamplifiers and 7.5 for low gain was used, but after the detector integration, crosstalk between channels was observed for particles that produced high-amplitude signals in a large number of scintillators.

This effect was less evident for lower amplifications and led to a change in the selected parameters for the T1 and calorimeter preamplifiers.

The final values for preamplifiers gain are shown in Table 3.

4.3. Threshold Calibration

The correspondence between the DAC value set for the threshold and the minimum signal that produces a trigger has been obtained with threshold scans at different input signal amplitudes.

An example of an S-curve made using 40 mV and 50 ns pulses, with a frequency of 1 kHz, is shown in Figure 9a.

The threshold calibration curves have been obtained repeating this measurement for different signal amplitudes and using the DAC values that reduce the trigger efficiency to 50%.

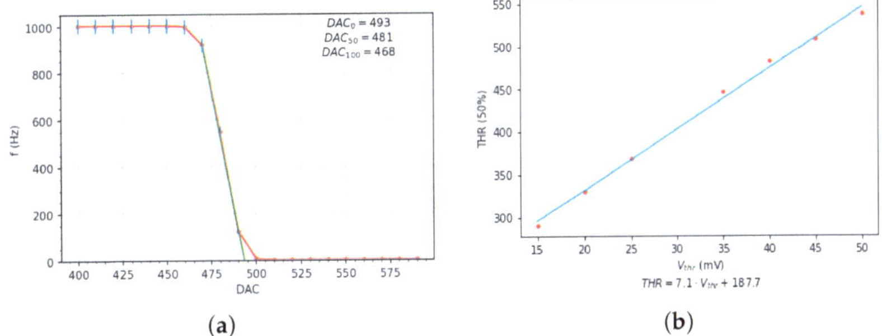

(a)   (b)

Figure 9. (**a**) S-curve acquired with a signal of 40mV, 50ns at 1kHz, (**b**) threshold DAC values (THR) as a function of the input signal (V_{thr}). The green line represents the linear fit of points in the descending region of the curve.

5. Results

The working parameters for the PMT&T board, which include the preamplifier's gain and the attenuation values for the input signal conditioning circuit, are reported in Table 3.

The flight model of the HEPD-02 detector successfully used these values during the environmental tests and the beam tests.

As an example of the acquisition made by the Flight Model of the HEPD-02, with the proposed parameters, Figure 10 shows the ADC distribution produced by all the scintillators of the calorimeter, during a beam test with 228 MeV protons. For each scintillator, only one of the two PMTs has been taken into account.

The path of the protons can be followed by observing the peak of the ADC distributions: in TR1, most of the particles passed in the second segment (TR1_2_1), while the peaks in nearby scintillators (TR1_1_1 and TR1_3_1) are less pronounced, and almost no signal is produced in TR1_4_2 and TR1_5_2.

In TR2, the central slabs are mainly affected by the beam particles, as can be seen by the peak distribution of TR2_2_2 and TR2_3_2.

The presence of the peaks in all the RAN scintillators indicates that the energy of the particle is sufficiently high to pass through the whole detector.

The distributions from the last two segmented planes of LYSO crystal show that the particles exit mainly through EN1_1_1 and EN1_2_1 and EN2_2_1.

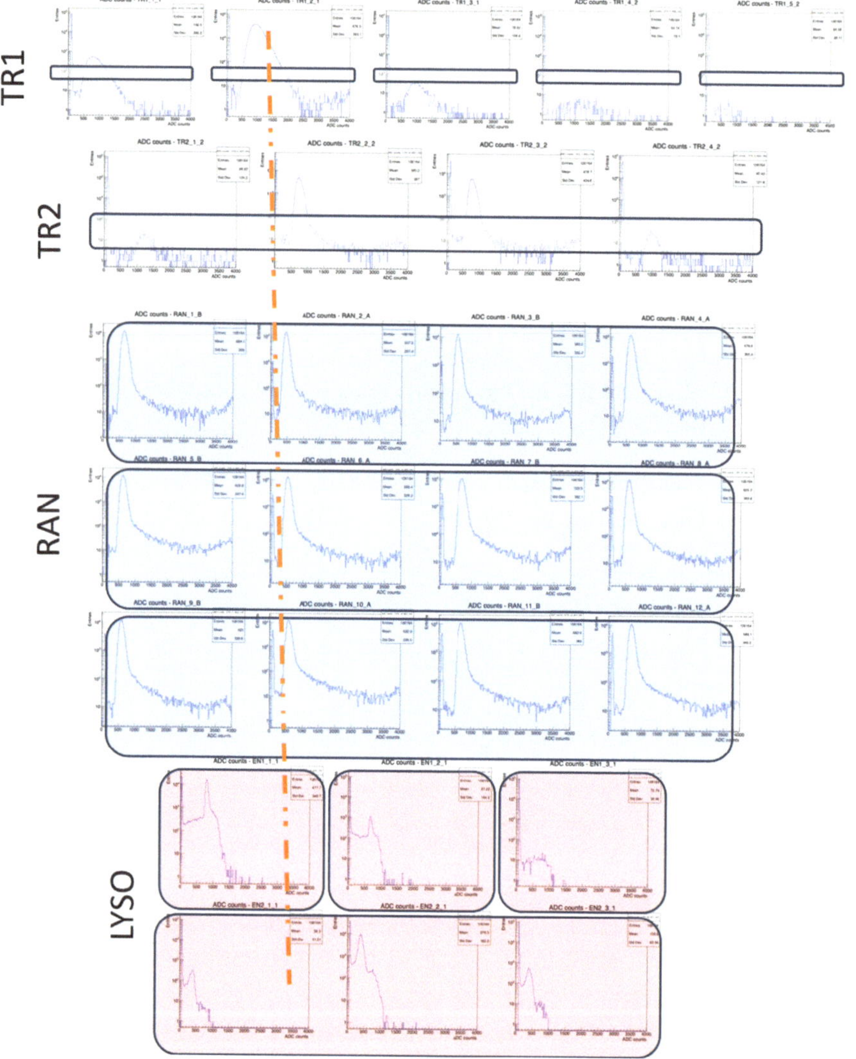

Figure 10. Protons with kinetic energy of 228 MeV acquired by HEPD-02.

6. Conclusions

The PMT&T board, developed for HEPD-02, brings improvements to the flexibility of the instrument and allows the detector to operate in regions such as SAA and poles, extending the scientific significance of the mission.

Concurrent and prescaled triggers allow the use of different masks per orbital zone and also make possible the development of the GRB detection algorithm.

The added functionalities have been intensively tested and work as expected.

Author Contributions: Conceptualization, M.M., G.O. and V.S.; methodology, M.M., G.O. and V.S.; software, M.M.; validation and resources, A.A., A.B., V.M. and A.V.; data curation, F.P. and B.P.; writing—original draft preparation, M.M.; writing—review and editing, G.O. and V.S.; supervision, G.O. and V.S. All authors have read and agreed to the published version of the manuscript.

Funding: This work was supported by the Italian Space Agency in the framework of the "Accordo Attuativo 2020-32.HH.0 Limadou Scienza+" (CUP F19C20000110005) and the ASI-INFN agreement n.2014-037-R.0, addendum 2014-037-R-1-2017.

Data Availability Statement: The data presented in this study are available on request from the corresponding author.

Acknowledgments: We would like to thank the staff of Electronics and Detectors Service (SER) of the INFN Division of Naples, in particular V. Masone, A Boiano, A Vanzanella and A. Anastasio, for their strong and undivided support throughout this project.

Conflicts of Interest: The authors declare no conflict of interest.

References

1. Pulinets, S.; Boyarchuk, K. *Ionospheric Precursors of Earthquakes*; Springer: Berlin/Heidelberg, Germany, 2004.
2. Picozza, P.; Battiston, R.; Ambrosi, G.; Bartocci, S.; Basara, L.; Burger, W.J.; Campana, D.; Carfora, L.; Casolino, M.; Castellini, G. Scientific goals and in-orbit performance of the High Energy Particle Detector on board the CSES satellite. *Astrophys. J. Suppl. Ser.* **2019**, *243*, 16. [CrossRef]
3. Ambrosi, G.; Bartocci, S.; Basara, L.; Battiston, R.; Burger, W.J.; Campana, D.; Caprai, M.; Carfora, L.; Castellini, G.; Cipollone, P.; et al. The electronics of the High-Energy Particle Detector on board the CSES-01 satellite. *Nucl. Instrum. Methods Phys. Res. Sect. A* **2021**, *1013*, 165639. [CrossRef]
4. Ambrosi, G.; Bartocci, S.; Basara, L.; Battiston, R.; Burger, W.J.; Campana, D.; Carfora, L.; Castellini, G.; Cipollone, P.; Conti, L.; et al. Beam test calibrations of the HEPD detector on board the China Seismo-Electromagnetic Satellite. *Nucl. Instrum. Methods Phys. Res. Sect. A* **2020**, *974*, 164170. [CrossRef]
5. Palma, F.; Sotgiu, A.; Parmentier, A.; Martucci, M.; Piersanti, M.; Bartocci, S.; Battiston, R.; Burger, W.J.; Campana, D.; Carfora, L.; et al. The August 2018 Geomagnetic Storm Observed by the High-Energy Particle Detector on Board the CSES-01 Satellite. *Appl. Sci.* **2021**, *11*, 5680. [CrossRef]
6. Bartocci, S.; Battiston, R.; Burger, W.J.; Campana, D.; Carfora, L.; Castellini, G.; Conti1, L.; Contin, A.; Donato, C.D.; Persio, F.D. Galactic Cosmic-Ray Hydrogen Spectra in the 40–250 MeV Range Measured by the High-energy Particle Detector (HEPD) on board the CSES-01 Satellite between 2018 and 2020. *ApJ* **2020**, *901*, 8. [CrossRef]
7. Martucci, M.; Bartocci, S.; Battiston, R.; Burger, W.J.; Campana, D.; Carfora, L.; Conti, L.; Contin, A.; De Donato, C.; De Santis, C.; et al. New results on protons inside the South Atlantic Anomaly, at energies between 40–250 MeV in the period 2018–2020. from the CSES-01 satellite mission.ck *Phys. Rev. D* **2022**, *105*, 062001. [CrossRef]
8. Piersanti, M.; Materassi, M.; Battiston, R.; Carbone, V.; Cicone, A.; D'Angelo, G.; Diego, P.; Ubertini, P. Magneto-spheric–Ionospheric–Lithospheric Coupling Model. 1: Observations during the 5 August 2018 Bayan Earthquake. *Remote Sens.* **2020**, *12*, 3299. [CrossRef]
9. Ambrosi, G.; Bartocci, S.; Basara, L.; Battiston, R.; Burger, W.J.; Carfora, L.; Castellini, G.; Cipollone, P.; Conti, L.; Contin, A.; et al. The HEPD particle detector of the CSES satellite mission for investigating seismo-associated perturbations of the Van Allen belts. *Sci. China Technol. Sci.* **2018**, *61*, 643–652. [CrossRef]
10. Available online: https://cses.web.roma2.infn.it/ (accessed on 13 October 2023).
11. Shen, X.; Zhang, X.; Yuan, S.; Wang, L.; Cao, J.; Huang, J.; Zhu, X.; Piergiorgio, P.; Dai, J. The state-of-art of the China Seismo-Electromagnetic Satellite mission. *Sci. China Technol. Sci.* **2018**, *61*, 634–642. [CrossRef]
12. Perciballi, S.; Iuppa, R.; Mese, M.; Nozzoli, F.; Osteria, G.; Scotti, V.; Ubertini, P. Performance of the HEPD-02 LYSO calorimeter and expected sensitivity to GRBs detection. *Proc. Sci.* **2021**, *395*, 583. [CrossRef]
13. De Santis, C.; Ricciarini, S. The High Energy Particle Detector (HEPD-02) for the second China Seismo-Electromagnetic Satellite (CSES-02). *PoS* **2021**, *ICRC2021*, 058. [CrossRef]
14. Iuppa, R.; Beolé, S.; Coli, S.; De Cilladi, L.; Gebbia, G.; Ricci, E.; Ricciarini, S.B.; Serra, E.; Zuccon, P. The innovative particle tracker for the HEPD space experiment onboard the CSES-02 satellite. *Proc. Sci.* **2021**, *395*, 70. [CrossRef]
15. Ricciarini, S.B.; Beolé, S.; de Cilladi, L.; Gebbia, G.; Iuppa, R.; Ricci, E.; Zuccon, P. Enabling low-power MAPS-based space trackers: a sparsified readout based on smart clock gating for the High Energy Particle Detector HEPD-02. *Proc. Sci.* **2022**, *395*, 71. [CrossRef]
16. Mese, M.; Osteria, G.; Scotti, V.; Anastasio, A.; Collaboration, C.L. The HEPD-02 trigger and PMT readout system for the CSES-02 mission. *Proc. Sci.* **2021**, *ICRC2021*, 063. [CrossRef]
17. Scotti, V.; Mese, M.; Osteria, G. A Versatile Readout and Trigger System for the High Energy Particle Detector Onboard the Satellite CSES-02. In Proceedings of the 2019 IEEE Nuclear Science Symposium and Medical Imaging Conference (NSS/MIC), Manchester, UK, 26 October–2 November 2019. [CrossRef]

 *instruments*

Article

A Compact Particle Detector for Space-Based Applications: Development of a Low-Energy Module (LEM) for the NUSES Space Mission

Riccardo Nicolaidis [1,2,*,†], Francesco Nozzoli [1,2,†], Giancarlo Pepponi [3,†] and on behalf of the NUSES Collaboration [‡]

[1] Physics Department, University of Trento, Via Sommarive 14, 38123 Trento, Italy; francesco.nozzoli@unitn.it
[2] INFN-Trento Institute of Fundamental Physics and Applications, Via Sommarive 14, 38123 Trento, Italy
[3] Fondazione Bruno Kessler, Via Sommarive 18, 38123 Trento, Italy; pepponi@fbk.eu
[*] Correspondence: riccardo.nicolaidis@unitn.it
[†] These authors contributed equally to this work.
[‡] Collaborators of the NUSES Collaboration are indicated in Supplementary Materials.

Abstract: NUSES is a planned space mission aiming to test new observational and technological approaches related to the study of relatively low-energy cosmic rays, gamma rays, and high-energy astrophysical neutrinos. Two scientific payloads will be hosted onboard the NUSES space mission: Terzina and Zirè. Terzina will be an optical telescope readout by SiPM arrays, for the detection and study of Cerenkov light emitted by Extensive Air Showers generated by high-energy cosmic rays and neutrinos in the atmosphere. Zirè will focus on the detection of protons and electrons up to a few hundred MeV and to 0.1–10 MeV photons and will include the Low Energy Module (LEM). The LEM will be a particle spectrometer devoted to the observation of fluxes of relatively low-energy electrons in the 0.1–7-MeV range and protons in the 3–50 MeV range along the Low Earth Orbit (LEO) followed by the hosting platform. The detection of Particle Bursts (PBs) in this Physics channel of interest could give new insight into the understanding of complex phenomena such as eventual correlations between seismic events or volcanic activity with the collective motion of particles in the plasma populating van Allen belts. With its compact sizes and limited acceptance, the LEM will allow the exploration of hostile environments such as the South Atlantic Anomaly (SAA) and the inner Van Allen Belt, in which the anticipated electron fluxes are on the order of 10^6 to 10^7 electrons per square centimeter per steradian per second. Concerning the vast literature of space-based particle spectrometers, the innovative aspect of the LEM resides in its compactness, within $10 \times 10 \times 10$ cm^3, and in its "active collimation" approach dealing with the problem of multiple scattering at these very relatively low energies. In this work, the geometry of the detector, its detection concept, its operation modes, and the hardware adopted will be presented. Some preliminary results from the Monte Carlo simulation (Geant4) will be shown.

Keywords: low energy module; NUSES; particle bursts; silicon detectors; PIPS; cosmic rays; particle identification; ΔE-E telescope

Citation: Nicolaidis, R.; Nozzoli, F.; Pepponi, G.; on behalf of the NUSES Collaboration. A Compact Particle Detector for Space-Based Applications: Development of a Low-Energy Module (LEM) for the NUSES Space Mission. *Instruments* **2023**, *7*, 40. https://doi.org/10.3390/instruments7040040

Academic Editor: Antonio Ereditato

Received: 26 October 2023
Revised: 3 November 2023
Accepted: 8 November 2023
Published: 13 November 2023

1. Introduction: The NUSES Space Mission

NUSES is a space mission aimed at testing innovative approaches for studying cosmic rays, gamma rays, and astrophysical neutrinos. The satellite will host two payloads, named Terzina [1,2] and Zirè [3–7]. Terzina will be a pathfinder for future missions devoted to observing Ultra High Energy Cosmic Rays (UHECRs) and neutrino astronomy using space-based instruments [8–11].

Zirè is a particle detector that tests novel instruments for the detection of γ-rays while monitoring fluxes of charged particles, such as electrons, protons, and light nuclei with kinetic energy from a few to hundreds MeV. The detector's primary goal is to count the trapped particles precipitating out of the Van Allen Belts (VABs) and look for any anomalies that might arise in the vicinity of tectonic events, including earthquakes or lithosphere-volcanic eruptions.

Monitoring solar activity and its cyclical cycle, which lasts roughly 11 years, is another crucial science goal of Zirè. Monitoring the incidence of phenomena like Solar Flares (SFs) or Coronal Mass Ejections (CMEs) during solar maximum is especially helpful [12]. The Zirè instrument will allow an online monitoring of the magnetospheric environment, useful for space weather characterizations Moreover, the study of energetic particles in the magnetosphere will advance our knowledge of the acceleration mechanisms at work during those occurrences. In conclusion, detecting photons with energy up to tens of MeV is the other science goal. This enables the investigation of some of the most intense and violent occurrences in astrophysics, known as Gamma Ray Bursts (GRBs) [13,14], which are fast and powerful gamma-ray pulses originating from extremely distant sources.

An additional detector, the Zirè-Low Energy Module (LEM), will cooperate with Zirè. The LEM is going to be inserted into the outer structure of the NUSES bus. Figure 1a shows the NUSES satellite. The on-board payloads are labeled at the edges of the figure. The LEM sub-detector has been designed to fit within a $10 \times 10 \times 10$ cm^3 volume. The LEM's goal will be to accomplish event-based particle identification (PID) for particles with relatively low kinetic energy, such as sub-MeV electrons and MeV protons.

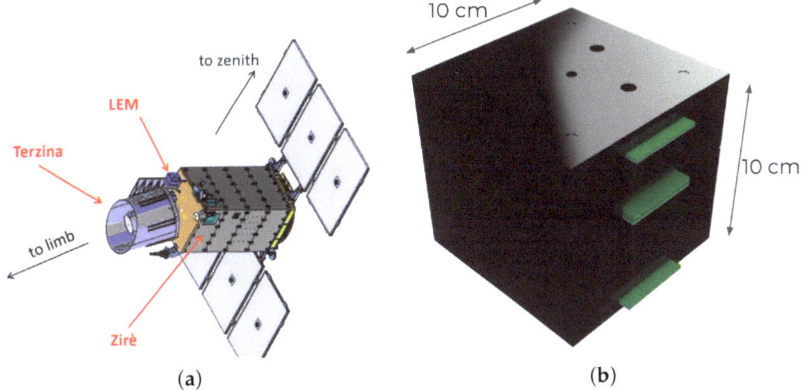

Figure 1. (a) The NUSES platform can be visualized in 3D. The Terzina detector faces the Earth's limb and measures fluorescence light from Extensive Air Showers (EAS) or Upgoing Air Showers (UAS) that are caused by the decay of the neutrino τ. The Zirè-LEM is shown as a small purple box. The detector is positioned outside the satellite's tray and is pointing towards the zenith. Lastly, Zirè is located inside the tray of the satellite and has three windows facing external space. (b) A visual representation of the Zirè-LEM detector. The 8 mm thick aluminum shield is depicted in the picture as a dark surface. Its goal is to reduce the occupancy of the veto scintillators, absorbing a large fraction of sub-MeV electrons. The particles that are the target are those whose incident directions allow them to enter the detector through the five holes shown in the picture.

2. The Need for a Low Energy Module

One of NUSES's key missions, as previously stated, is to monitor particle precipitation from the Van Allen Belts (VABs) and investigate eventual correlations to seismic activity, all while validating models for Lithosphere–Atmosphere–Ionosphere–Magnetosphere interactions. Several processes can result in the release of trapped charged particles from radiation belts. While our understanding of these processes is still limited, electromagnetic fluctuations within the radiation belt are widely assumed to be a significant factor. Geomagnetic/solar storms, thunderstorms, but also human-generated electromagnetic emissions, and seismic events can all cause these electromagnetic fluctuations. As highlighted in [15], measuring the fluxes of trapped charged particles can improve our understanding of the connection between the lithosphere, atmosphere, ionosphere, and magnetosphere. As pointed out in [16], there is statistical evidence indicating a temporal correlation between particle precipitation from the Van Allen Belts and major seismic events. These findings stimulate interest in more accurate measurements of electron fluxes with energies spanning in the range of 0.1–7 MeV, which could be a candidate channel for identifying hypothetical seismic precursors.

As a result, a Low-Energy Module (LEM) has been included in the design of the NUSES satellite in order to expand the observed energy window by Zirè. The Low Energy Module will be a small spectrometer with a volume of $10 \times 10 \times 10$ cm^3 and mass less than 2 kg (Figure 1b for the volume envelope of the LEM), capable of measuring the kinetic energy, arrival direction of low energy charged particles down to 0.1 MeV for electrons.

The primary goal of this detector is to observe the magnetosphere and ionosphere surroundings. Furthermore, the LEM instrument will investigate particle composition in the challenging conditions of the South Atlantic Anomaly (SAA), to quantify the isotopic ratios of H and He and potentially determine the proportion of heavier nuclei.

3. Geometry and Detection Concept of the LEM

The direction of a particle in a particle spectrometer, such as the Zirè, is usually established using tracking techniques. When dealing with relatively low-energy particles, however, the conventional tracking method is not feasible due to significant multiple scattering within the first sensitive element of the direction detector. A collimation technique, as discussed in [17], is required to determine the arrival direction of relatively low-energy particles. This approach requires utilizing a well-constructed passive shield with adequate thickness for blocking energetic particles coming from "unidentified" or random directions.

The passive collimator makes it simpler to detect particles arriving from "accepted" directions. To avoid the need for large and heavy passive protections, the LEM spectrometer employs a technique known as "active collimation". This method employs shaped plastic scintillators as ACD, successfully distinguishing particles that pass through a relatively lightweight passive shield.

The LEM features a 0.8-cm thick aluminum barrier to maintain bearable occupancy levels for the veto detectors in places such as the South Atlantic Anomaly (SAA) and the inner radiation belt adjacent to the poles, enabling consistent particle composition observations in those regions.

The LEM's particle identification capabilities rely on the long-established ΔE-E spectrometric method, which is discussed in references [18–20]. The approach employs five couples of silicon detectors (Passivated Implanted Planar Silicon or PIPS) positioned in a telescopic arrangement. The PIPS detector has a typical resolution of about 10 keV.

In Figure 2, a representation of the detector shows the components and its preliminary assembly. On the other hand, in Figure 3, a schematic view of the instruments explains its adopted detection scheme. A particle, when approaching the detector, can enter the instrument through the holes in the top aluminum structure, preventing the detection by the perforated top ACD (made of plastic scintillator). Depending on the particle's direction, charge and kinetic energy are determined by one of the five ΔE-E spectrometers. Each spectrometer consists of a thinner silicon detector (100 μm thick) placed on top of a thicker silicon detector (300 μm thick).

To improve the LEM's particle detection capabilities across a wider energy range, a calorimeter is placed beneath the PIPS detectors. This calorimeter is made of a plastic scintillator (2 cm thick) that can detect electron fluxes up to 10 MeV. As a result, there is expected to be a reasonable overlap with the Zirè flux data, as noted in [21]. A bottom ACD (made of plastic scintillator) is also used to identify high-energy particles that are not completely contained by the plastic scintillator calorimeter.

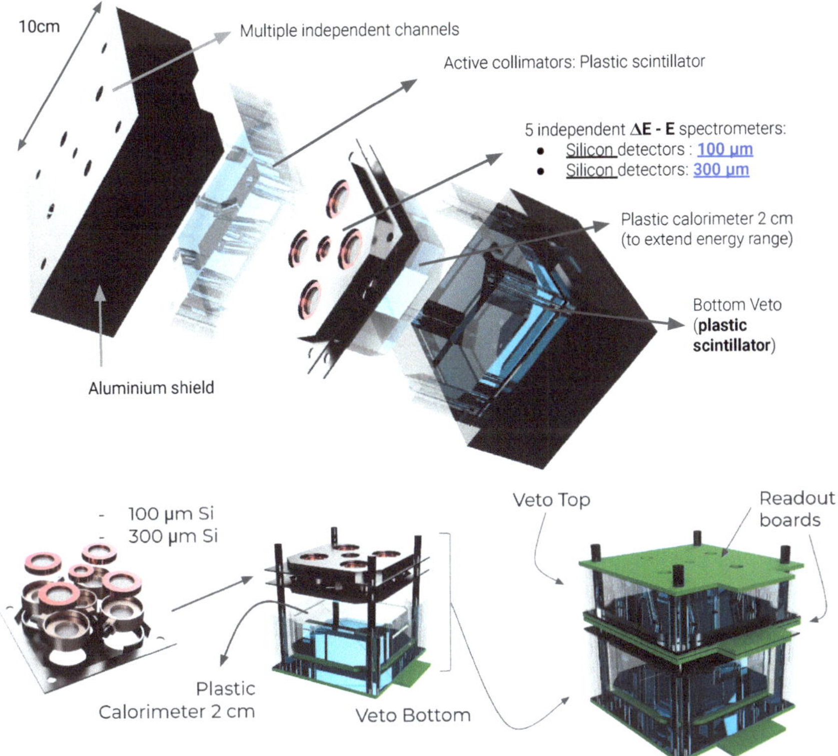

Figure 2. In the upper part of the figure, the expanded visualization of the LEM detector and its internal components. From the upper-left part of the figure: the aluminum shield has five holes/channels for the detector. The five channels are then evident in the active collimator, which is made of a plastic scintillator. In the core of the detector, the 5 ΔE-E detectors are positioned. After that, a calorimeter made of plastic scintillator is added to expand the energy range. In the lower part of the detector, there is an ACD made of plastic scintillator, followed by the bottom section of the aluminum shield. Below, the images describe the assembling and the geometry of the Low Energy Module (LEM) detector.

By considering a non-relativistic, low-energy charged particle traversing the ΔE-E telescope, the deposited energy in the thinner silicon detector, $\Delta E \propto \frac{Z^2}{\beta^2}$, and the particle's kinetic energy, $E_k \approx \frac{1}{2}m(\beta c)^2$, depends on the particle's velocity. If we combine these parameters, it is possible to define a particle classifier as

$$\text{PID} = \log_{10}\left(\frac{\Delta E}{1\,\text{MeV}} \cdot \frac{E_k}{1\,\text{MeV}}\right) \approx \log_{10}\left(Z^2 m\right) + \text{constant}. \tag{1}$$

This Particle IDentification (PID) classifier, dimensionless by definition, is primarily determined by the particle's mass, denoted as m, and its charge in modulus, represented by Z. As a result, this PID classifier partially (and approximately) does not depend on the velocity (and therefore on the kinetic energy) of the particle.

For the characterization of the detector's performances, a GEANT4 (version 11.0.3) Monte Carlo simulation [22] was appositely developed. The Physics List adopted in our application is the standard `FTFP_BERT`. For the simulation of the geometry reported in Figure 2, developed with a parametric computer-aided design (FreeCAD 0.20) software, we adopted the Geometry Description Markup Language (GDML) [23]. We generated the

GDML file (compatible with the GEANT4 toolkit) using the GDML Workbench [24] for FreeCAD 0.20. Figure 4 shows the event display generated by the GEANT4 application specifically developed for characterizing the detector. The cross-sectional view enables the reader to observe the tessellated solids used in the instruments.

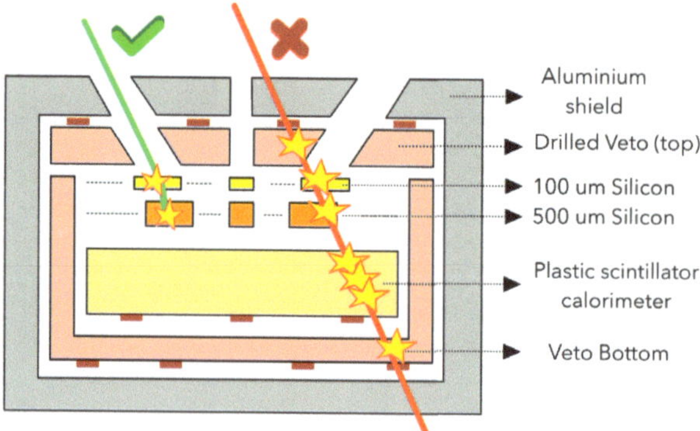

Figure 3. Schematics of the LEM detection approach; the green track is an example of good events (fully confined), and the red track represents an event to be rejected (not fully confined). The yellow star markers represent the energy deposited by the charged particle into the sensitive elements of the detector. Good events are characterized by a partial energy deposit in the thinner SD (100 μm) and a complete energy release in the thicker SD (300 μm) or, eventually, in the plastic calorimeter. In the second case, since the energy resolution of the plastic scintillator is worse, energy measurement will be affected by a larger uncertainty. Nevertheless, only when the energy release caused by the particle is confined within the detector an accurate PID is possible. Events to be rejected are characterized by an energy release in at least one of the two ACDs, or in more than two SDs not aligned on the same axis (e.g., two SDs that belong to different independent channels). Nonetheless, MIP particles (e.g., atmospheric muons on the ground), corresponding to crossing particles, will be used for calibration purposes.

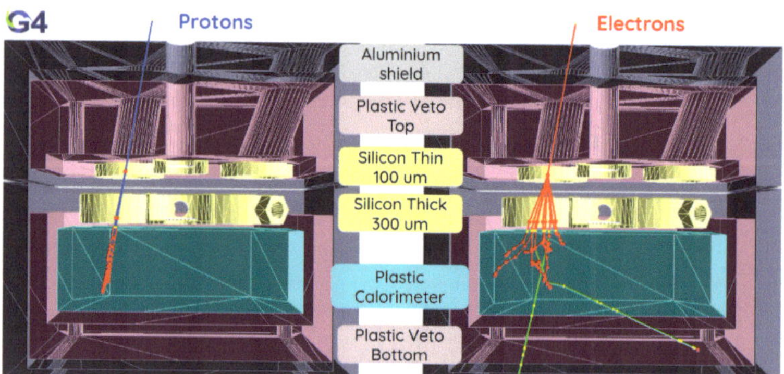

Figure 4. The picture shows the event display of the GEANT4 simulation of the Zire-LEM detector. (**Left Panel**) Visualization of 10-proton events (particle's trajectory is depicted by the blue line) with kinetic energy uniformly extracted between 3 and 50 MeV. (**Right Panel**) Visualization of 10-electron events (particle's trajectory is depicted by the red lines) with kinetic energy randomly extracted between 0.1 and 5 MeV. The green lines are photons produced during the electron's bremsstrahlung. Only in one case (displayed on the right-hand side of the left panel), the photon is re-absorbed via the photoelectric effect. It is possible to see that for electrons, the multiple scattering phenomenon is more impacting. This provides a graphical visualization of the need for an innovative active collimation technique for detecting the particle's direction at relatively low energy.

The PID classifier is shown in Figure 5a. It's interesting to note that the non-relativistic assumption is not valid for electrons. Nonetheless, they can still be identified since the electron's mass is $\approx 1/2000$ times the proton's mass. However, the previously mentioned limited energy resolution for the plastic scintillator calorimeter can cause a reduction in PID capabilities and performance at relatively high energies, in particular when particles traverse the thicker silicon detector and deposit their energy in the plastic calorimeter.

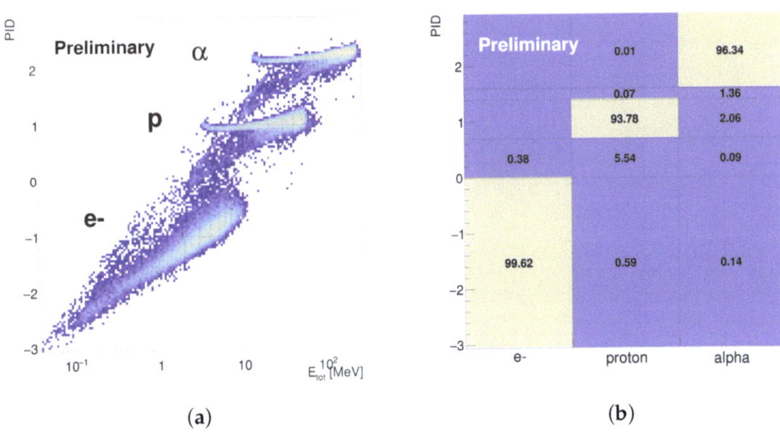

(a)   (b)

Figure 5. (**a**) PID capability for events impinging on the top 100 μm silicon detector and fully contained within the LEM. (**b**) Particle tagging efficiency for the three families of particles: electrons, protons, and alpha particles. For each particle family (Monte Carlo truth) reported on the horizontal axis, the tagging efficiency is reported on each histogram bin.

To estimate the particle identification efficiency, it is possible to define some specific intervals for the PID for each particle: $(-3, 0)$ for electrons, $(0.7, 1.4)$ for protons, and $(1.6, 2.5)$ for alpha particles. In the table shown in Figure 5b, the particle identification tagging efficiencies, for the three families of particles, are higher than 90 % in the three respective PID proxy intervals. In particular, we observed that more than 90% of particles, for each of the three generated classes (electrons, protons, and alpha), were correctly tagged.

The angular resolution and FOV of the detector for protons and electrons are characterized in Figure 6. The scatter plot depicts a projection on the plane of the particle's incident direction (at the Monte Carlo truth level). The origin of the plot is assumed to represent the zenith direction, which is the axis perpendicular to the front drilled aluminum surface of the LEM. The color indicates which ΔE-E channel has been triggered. The entire LEM FoV is around 45°. The RMS angular resolution for protons and alpha particles is around 6°. We acquired a lower resolution ($\approx 12°$) for electrons. Interactions with the inner edges of the LEM openings on the top Aluminum shield were expected to cause such an observable effect. The LEM geometric factor is in the range 0.1–0.3 cm^{2}sr for electrons in the 0.2–12-MeV energy window, for protons in the 3–70 MeV energy window, and for alpha particles in the 15–280 MeV energy window. The estimation was carried out assuming the definitions and methods described in [25]. Figure 7 displays the estimated geometric factor for electrons, protons, and alpha particles. Knowing the orbit parameters of the NUSES mission (Sun-synchronous, 97 degrees, LEO 550 km), a preliminary map of the expected rates of the LEM can be evaluated using the model International Radiation Environment Near Earth AE9/AP9 (IRENE-AE9/AP9) [26]. In the LEO environment, the most impacting populations of charged particles are trapped protons and electrons. With IRENE-AE9/AP9 we could estimate the differential omnidirectional/isotropic fluxes of those particles.

Figure 8 shows that the LEM will encounter a significant acquisition rate ($\approx$50 kHz) in the South Atlantic Anomaly (SAA). Therefore, a dual data transmission strategy is being developed. An "event-based" approach will be adopted for rates below 1 kHz. On the other hand, a "histogram-based" approach will be adopted for higher rates. This will ensure proper usage of the assigned data bandwidth to the LEM.

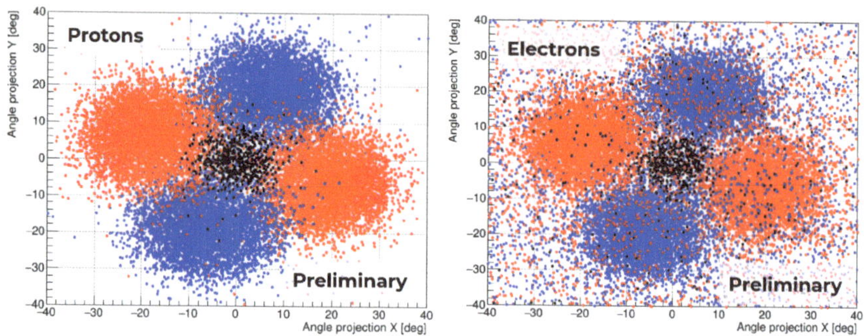

Figure 6. Angular resolution and Field Of View (FOV) of the LEM for protons on the left and electrons on the right. The different colors encode the pair in the ΔE-E spectrometer that is triggered. Since the detector exhibits axial symmetry, different colors are used to distinguish between adjacent lateral channels (in blue or red) and the central channel (in black). It is possible to see, on the right panel, the important effect of the electron's multiple scattering. Some electrons hitting the aluminum shield are then scattered in the direction encoded by one of the five channels.

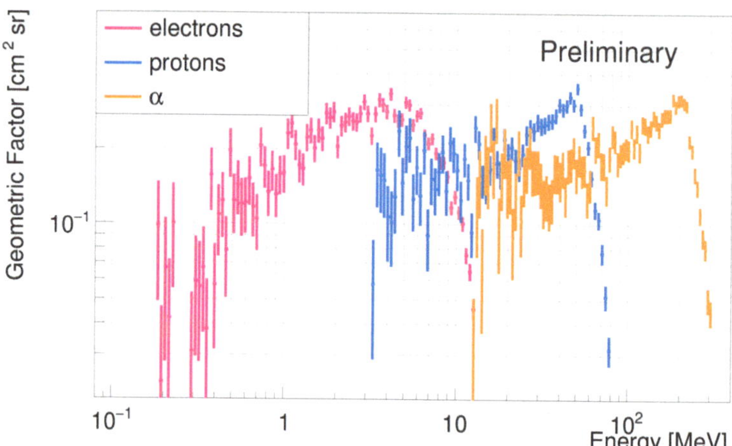

Figure 7. Geometric factor estimation for the LEM.

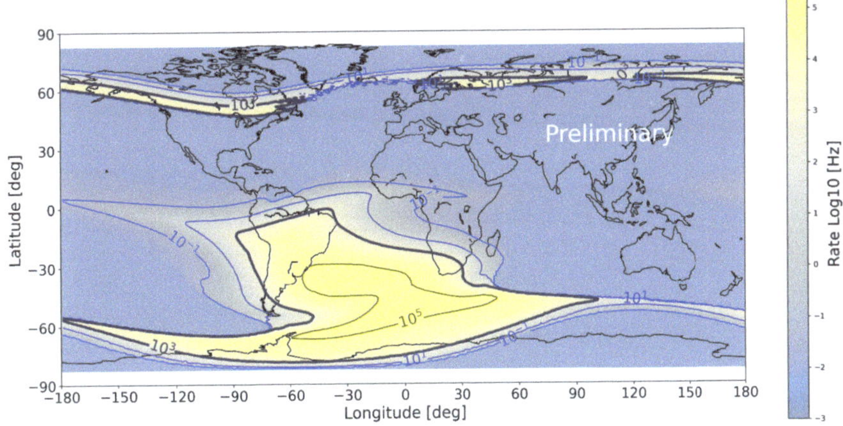

Figure 8. Estimated rate map having considered a satellite polar orbit, Sun-synchronous 97° and 550 km of altitude. For the conversion from isotropic fluxes to rates we used an $\simeq$0.2 cm^{2}sr geometric factor. The thicker contour in the map represents the region inside which the LEM will operate in the histogram-based mode.

4. Preliminary Test on PIPS Sensors

The core of the LEM detector is constituted by the ΔE-E spectrometers, which comprise five detectors. Four of these pairs of detectors have a circular shape with an area of 150 mm^2. The central pair of detectors has an area of 55 mm^2 (as depicted in Figure 9). The smaller diameter of the central PIPS detector was chosen to ensure consistent geometric acceptance across all five channels. The top sensors, each with a thickness of 100 μm, will be the R-series (ruggedized) PIPS detectors produced by ORTEC/AMETEK [27].

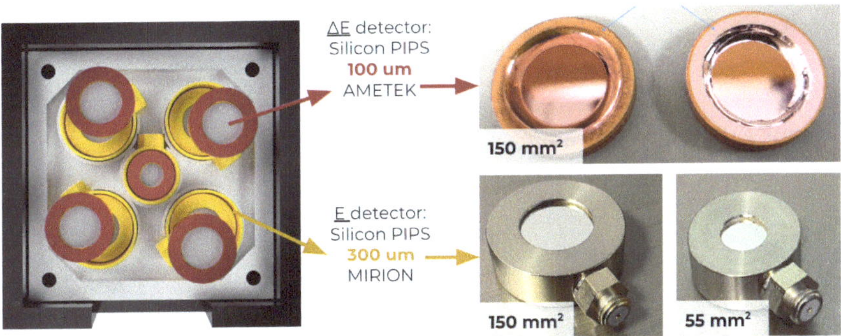

Figure 9. Mounting arrangement of the 5 pairs of silicon detectors within the LEM. On the right-hand side of the picture, some pictures of the PIPS detectors manufactured by AMETEK/ORTEC and by MIRION/CANBERRA are reported.

The 100 μm PIPS detectors are covered by aluminum and gold layers on both sides. Each layer has a grammage of 50 μg/cm^2 and 40 μg/cm^2, respectively. These layers are important for making the PIPS detectors light-tight and "ruggedized". On the other hand, the five bottom sensors, with a thickness of 300 μm, will be produced by Canberra/MIRION [28]. The 300 μm thick PIPS detectors have aluminum layers with a grammage of approximately 70 μg/cm^2 and 250 μg/cm^2 on their two sides. This treatment ensures that the detectors are also light-tight.

An experimental setup to assess the performance of a PIPS detector was developed at the INFN-TIFPA laboratory. During the characterizations, the PIPS sensor's depletion voltage was set at 60 V. An initial measurement of the power consumption of the employed Charge-Sensitive Preamplifier (CSA) was below 100 mW per channel. We characterized the sensor acquiring various particle types in a telescopic configuration, including atmospheric muons, γ-rays from a ^{176}Lu source, as well as alpha particles and γ-rays from a ^{241}Am source. During these tests, a good linearity of the energy scale was achieved. Furthermore, the PIPS detector's response to particles with significantly different specific ionization levels (such as muons, recoiling electrons, and alpha particles) was determined to be consistent within a few percentage points, as anticipated.

In Figure 10 are reported some preliminary results from the characterization of the DAQ developed by Nuclear Instruments SRL. For the test, the silicon detector AP-CAM25 (manufactured by Mirion with an embedded CSA) was used. The detector was exposed to different radioactive sources: ^{137}Cs and ^{241}Am. ^{137}Cs, with a branching ratio of 94% decays in an excited state of ^{137}Ba through a beta decay. The excited state relaxes releasing a 662 keV photon. Since Silicon is a material with a low atomic number, the photoelectric peak is suppressed. Nevertheless, it is possible to record the Compton edge of the spectrum with an end-point at $\approx$478 keV. Moreover, the detector was exposed to the 59.5-keV gamma line emission by the ^{241}Am source.

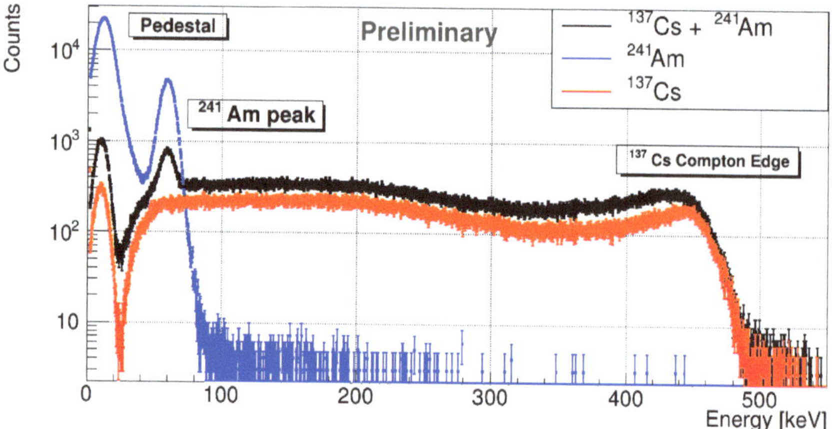

Figure 10. Some preliminary measurements were performed using the DAQ manufactured by Nuclear Instruments SRL. For these measurements, a fully depleted silicon detector with an embedded CSA was used (AP-CAM25 manufactured by MIRION). The spectra were acquired by exposing the detector to different radioactive sources. As indicated in the legend, the red curve represents the Compton edge of the 662 keV gamma-ray from the decay of ^{137}Cs. The blue spectrum displays the peak at 59.5 keV from the decay of the ^{241}Am radioactive source. The black spectrum was obtained by exposing the silicon detector to both radioactive sources, ^{241}Am and ^{137}Cs.

There are two important requirements for the readout chain of the LEM: a relatively low energy threshold and a rapid response. The measurements (Figure 10) obtained confirmed the practicability of the 40 keV energy threshold utilized in LEM simulations and validate the achievement of a $\approx$5-keV energy resolution. Moreover, another crucial aspect of the LEM concerns the characterization of the signal decay time (defined by the time constant of the CSA), which is connected to the detector occupancy—a possible concern in the harsh conditions of the South Atlantic Anomaly (SAA). The measured signal decay time of approximately 200 ns effectively mitigates the risk of signal overlap (pile-up) from various particles in the SAA, thus addressing this potential issue.

5. Conclusions

The Low-Energy Module (LEM), a small particle analyzer that will be part of Zirè on-board the NUSES mission, is currently being built and tested. The INFN-TIFPA laboratory tested prototypes of the PIPS detector readout, confirming an energy resolution of around 5 keV and a signal decay time of around 200 ns. The LEM is intended to monitor the energy, direction, and composition of relatively low-energy charged particles, with a kinetic energy limit of 0.1 MeV. Its goals include measuring particle fluxes in the South Atlantic Anomaly (SAA), studying the interaction of the lithosphere and magnetosphere, and monitoring Space Weather.

Supplementary Materials: The following supporting information can be downloaded at: https://www.mdpi.com/article/10.3390/instruments7040040/s1, for detailed author affiliations please see File S1. Full NUSES Author List: R. Aloisio, C. Altomare, F.C.T. Barbato, R. Battiston, M. Bertaina, E. Bissaldi, D. Boncioli, L. Burmistrov, I. Cagnoli, M. Casolino, A.L. Cummings, N. D'Ambrosio, I. De Mitri, G. De Robertis, C. De Santis, A. Di Giovanni, A. Di Salvo, M. Di Santo, L. Di Venere, J. Eser, M. Fernandez Alonso, G. Fontanella, P. Fusco, S. Garbolino, F. Gargano, R.A. Giampaolo, M. Giliberti, F. Guarino, M. Hellero, R. Iuppa, J.F. Krizmanic, A. Lega, F. Licciulli, F. Loparco, L. Lorusso, M. Mariotti, O.M.N. Mazziotta, M. Mese, H. Miyamoto, T. Montarulio, A. Nagaio, R. Nicolaidis, F. Nozzoli, A.V. Olinto, D. Orlandi, G. Osteria, P.A. Palmieri, B. Panico, G. Panzarini, A. Parenti, L. Perrone, P. Picozza, R. Pillera, R. Rando, O.M. Rinaldi, A. Rivetti, V. Rizi, F. Salamida, E. Santero Mormile, V. Scherini, V. Scotti, D. Serini, I. Siddique, L. Silveri, A. Smirnov, R. Sparvoli, S. Tedesco, C. Trimarellio, L. Wu, P. Zuccon, S.C. Zugravel.

Author Contributions: Conceptualization, F.N. and R.N.; methodology, F.N. and R.N.; software, R.N.; validation, F.N. and R.N.; formal analysis, R.N.; investigation, F.N. and R.N.; resources, F.N., R.N., G.P. and on behalf of the NUSES coll.; data curation, R.N. and F.N.; writing—original draft preparation, R.N. and F.N.; writing—review & editing, R.N., F.N., G.P. and on behalf of NUSES coll.; visualization, R.N., F.N. and G.P.; supervision, F.N. and G.P.; project administration, F.N., G.P. and NUSES coll.; and funding acquisition, F.N., G.P. and on behalf the of NUSES coll. All authors have read and agreed to the published version of the manuscript.

Funding: NUSES is funded by the Italian Government (CIPE n. 20/2019), by the Italian Minister of Economic Development (MISE reg. CC n. 769/2020), by the Italian Space Agency (CDA ASI n. 15/2022), by the Swiss National Foundation (SNF grant n. 178918) and by the European Union-NextGenerationEU under the MUR National Innovation Ecosystem grant ECS00000041-VITALITY-CUP D13C21000430001.

Institutional Review Board Statement: Not applicable.

Informed Consent Statement: Not applicable.

Data Availability Statement: The data presented in this study are available on request from the corresponding author.

Conflicts of Interest: The authors declare no conflict of interest.

Abbreviations

The following abbreviations are used in this manuscript:

ACD	Anti-Coincidence Detector
CSA	Charge Sensitive Amplifier
EAS	Extensive Air Showers
FOV	Field Of View
GDML	Geometry Description Markup Language
GEANT4	GEometry ANd Tracking 4
GRB	Gamma-Ray Burst
IRENE	International Radiation Environment Near Earth
LAIM	Lithosphere Atmosphere Ionosphere Magnetosphere
LEM	Low-Energy Module
LEO	Low Earth Orbit
LYSO	Lutetium–Yttrium OxyorthoSilicate
MILC	Magnetosphere ionosphere lithosphere coupling

MIP	Minimum Ionizing Particle
MPV	Most Probable Value
NUSES	NeUtrino and Seismic Electromagnetic Signals
PID	Particle Identification
PIPS	Passivated Implanted Planar Silicon
SAA	South Atlantic Anomaly
TGF	Terrestrial Gamma-ray Flash
UAS	Upgoing Air Showers
VAB	Van Allen Belt

References

1. Burmistrov, L. Terzina on board NUSES: A pathfinder for EAS Cherenkov Light Detection from space. *Proc. Eur. Phys. J. Web Conf.* **2023**, *283*, 06006. [CrossRef]
2. Aloisio, R.; Burmistrov, L.; Heller, M.; Montaruli, T.; Trimarelli, C.. The Terzina instrument onboard the NUSES space mission. *Proc. Sci.* **2023**, *444*, 391. [CrossRef]
3. De Mitri, I.; Di Santo, M. The NUSES space mission. In *Proceedings of the Journal of Physics: Conference Series*; IOP Publishing: Bristol, UK, 2023; Volume 2429, p. 012007. [CrossRef]
4. Di Giovanni, A.; Di Santo, M. The NUSES space mission. *Proc. Sci.* **2022**, *414*, 354. [CrossRef]
5. Di Santo, M.; Di Giovanni, A. The NUSES space mission. *Proc. Sci.* **2023**, *423*, 143. [CrossRef]
6. Mazziotta, M.N.; Altomare, C.; Bissaldi, E.; De Gaetano, S.; De Robertis, G.; Dipinto, P.; Di Venere, L.; Franco, M.; Fusco, P.; Gargano, F.; et al. A light tracker based on scintillating fibers with SiPM readout. *Nucl. Instrum. Methods Phys. Res. Sect. A Accel. Spectrom. Detect. Assoc. Equip.* **2022**, *1039*, 167040. [CrossRef]
7. Mazziotta, M.N.; Pillera, R. The light tracker based on scintillating fibers with SiPM readout of the Zire instrument on board the NUSES space mission. In Proceedings of 38th International Cosmic Ray Conference— PoS(ICRC2023), Nagoya, Japan, 26 July–3 August 2023. [CrossRef]
8. Olinto, A.V.; Krizmanic, J.; Adams, J.; Aloisio, R.; Anchordoqui, L.; Anzalone, A.; Bagheri, M.; Barghini, D.; Battisti, M.; Bergman, D.; et al. The POEMMA (probe of extreme multi-messenger astrophysics) observatory. *J. Cosmol. Astropart. Phys.* **2021**, *2021*, 7. [CrossRef]
9. Olinto, A.; Adams, J.; Aloisio, R.; Anchordoqui, L.; Bergman, D.; Bertaina, M.; Bertone, P.; Christl, M.; Csorna, S.; Eser, J.; et al. POEMMA: Probe Of Extreme Multi-Messenger Astrophysics. In Proceedings of the 35th International Cosmic Ray Conference (ICRC2017), Busan, Republic of Korea, 12–20 July 2017; Volume 301, p. 542. [CrossRef]
10. Denton, P.B.; Kini, Y. Ultrahigh-energy tau neutrino cross sections with GRAND and POEMMA. *Phys. Rev. D* **2020**, *102*, 123019. [CrossRef]
11. Krizmanic, J. POEMMA: Probe of extreme multi-messenger astrophysics. In *Proceedings of the EPJ Web of Conferences*; EDP Sciences: 2019, Les Ulis, France; Volume 210, p. 06008. [CrossRef]
12. Kahler, S. Solar flares and coronal mass ejections. *Annu. Rev. Astron. Astrophys.* **1992**, *30*, 113–141. [CrossRef]
13. Fishman, G.J.; Meegan, C.A. Gamma-ray bursts. *Annu. Rev. Astron. Astrophys.* **1995**, *33*, 415–458. [CrossRef]
14. Piran, T. The physics of gamma-ray bursts. *Rev. Mod. Phys.* **2005**, *76*, 1143. [CrossRef]
15. D'Angelo, G.; Piersanti, M.; Battiston, R.; Bertello, I.; Carbone, V.; Cicone, A.; Diego, P.; Papini, E.; Parmentier, A.; Picozza, P.; et al. Haiti Earthquake (Mw 7.2): Magnetospheric–Ionospheric–Lithospheric Coupling during and after the Main Shock on 14 August 2021. *Remote Sens.* **2022**, *14*, 5340. [CrossRef]
16. Battiston, R.; Vitale, V. First evidence for correlations between electron fluxes measured by NOAA-POES satellites and large seismic events. *Nucl. Phys. B-Proc. Suppl.* **2013**, *243*, 249–257. [CrossRef]
17. Li, X.; Xu, Y.; An, Z.; Liang, X.; Wang, P.; Zhao, X.; Wang, H.; Lu, H.; Ma, Y.; Shen, X.; et al. The high-energy particle package onboard CSES. *Radiat. Detect. Technol. Methods* **2019**, *3*, 1–11. [CrossRef]
18. Scarduelli, V.; Gasques, L.; Chamon, L.; Lépine-Szily, A. A method to optimize mass discrimination of particles identified in ΔE-E silicon surface barrier detector systems. *Eur. Phys. J. A* **2020**, *56*, 1–7. [CrossRef]
19. Evensen, L.; Westgaard, T.; Avdeichikov, V.; Carlen, L.; Jakobsson, B.; Murin, Y.; Martensson, J.; Oskarsson, A.; Siwek, A.; Whitlow, H.; et al. Thin detectors for the CHICSi/spl Delta/EE telescope. *IEEE Trans. Nucl. Sci.* **1997**, *44*, 629–634. [CrossRef]
20. Carboni, S.; Barlini, S.; Bardelli, L.; Le Neindre, N.; Bini, M.; Borderie, B.; Bougault, R.; Casini, G.; Edelbruck, P.; Olmi, A.; et al. Particle identification using the ΔE-E technique and pulse shape discrimination with the silicon detectors of the FAZIA project. *Nucl. Instrum. Methods Phys. Res. Sect. A Accel. Spectrom. Detect. Assoc. Equip.* **2012**, *664*, 251–263. [CrossRef]
21. Alonso, M.F.; Mitri, I.D. The Zire experiment on board the NUSES space mission. In Proceedings of 38th International Cosmic Ray Conference—PoS(ICRC2023), Nagoya, Japan, 26 July–3 August 2023. [CrossRef]
22. Agostinelli, S.; Allison, J.; Amako, K.a.; Apostolakis, J.; Araujo, H.; Arce, P.; Asai, M.; Axen, D.; Banerjee, S.; Barrand, G.; et al. GEANT4—A simulation toolkit. *Nucl. Instrum. Methods Phys. Res. Sect. A Accel. Spectr. Detect. Assoc. Equip.* **2003**, *506*, 250–303. [CrossRef]

23. Chytracek, R.; McCormick, J.; Pokorski, W.; Santin, G. Geometry description markup language for physics simulation and analysis applications. *IEEE Trans. Nucl. Sci.* **2006**, *53*, 2892–2896. [CrossRef]
24. Sloan, K.; Hindi, M.; luzpaz.; lambdam94.; Lei, Z.; JakubDmochowski. *KeithSloan/GDML: New Release for FreeCAD*, v. 0.20. 2022; Zenodo. [CrossRef]
25. Sullivan, J. Geometric factor and directional response of single and multi-element particle telescopes. *Nucl. Instr. Methods* **1971**, *95*, 5–11. [CrossRef]
26. Ginet, G.P.; O'Brien, T.P.; Huston, S.L. et al. AE9, AP9 and SPM: New Models for Specifying the Trapped Energetic Particle and Space Plasma Environment. *Space Sci. Rev.* **2013**, *179*, 579–615.
27. ORTEC/AMETEK. Si Charged Particle Radiation Detectors for Research Applications. Available online: https://www.ortec-online.com/products/radiation-detectors/silicon-charged-particle-radiation-detectors/si-charged-particle-radiation-detectors-for-research-applications (accessed on 11 October 2023).
28. Technologies, M. PIPS Detectors. Available online: https://mirionprodstorage.blob.core.windows.net/prod-20220822/cms4_mirion/files/pdf/spec-sheets/c39313_passivated_pips_super_spec_1.pdf (accessed on 11 October 2023).

instruments

MDPI

Article

The Imaging X-ray Polarimetry Explorer (IXPE) and New Directions for the Future

Paolo Soffitta

Istituto di Astrofisica e Planetologia Spaziali (INAF-IAPS), Via Fosso del Cavaliere 100, 00133 Rome, Italy; paolo.soffitta@inaf.it

Abstract: An observatory dedicated to X-ray polarimetry has been operational since 9 December 2021. The Imaging X-ray Polarimetry Explorer (IXPE), a collaboration between NASA and ASI, features three X-ray telescopes equipped with detectors sensitive to linear polarization set to 120°. This marks the first instance of a three-telescope SMEX mission. Upon reaching orbit, an extending boom was deployed, extending the optics and detector to a focal length of 4 m. IXPE targets each celestial source through dithering observations. This method is essential for supporting on-ground calibrations by averaging the detector's response across a section of its sensitive plane. The spacecraft supplies power, enables attitude determination for subsequent on-ground attitude reconstruction, and issues control commands. After two years of observation, IXPE has detected significant linear polarization from nearly all classes of celestial sources emitting X-rays. This paper outlines the IXPE mission's achievements after two years of operation in orbit. In addition, we report developments for future high-throughput X-ray optics that will have much smaller dead-times by using a new generation of Applied Specific Integrated Circuits (ASIC), and may provide 3D reconstruction of photo-electron tracks.

Keywords: astrophysics; X-rays; polarimetry; gas detectors; X-ray optics; ASICs

Citation: Soffitta, P. The Imaging X-ray Polarimetry Explorer (IXPE) and New Directions for the Future. *Instruments* **2024**, *8*, 25. https://doi.org/10.3390/instruments8020025

Academic Editors: Matteo Duranti and Valerio Vagelli

Received: 4 January 2024
Revised: 12 March 2024
Accepted: 15 March 2024
Published: 25 March 2024

1. Introduction

Cyclotron emission, synchrotron emission, and non-thermal bremsstrahlung [1–3] are the most common emission processes in X-ray astronomy providing polarized radiation. Even if emitted as intrinsically non-polarized thermal radiation, radiation can become polarized via scattering in accretion disks, blobs, and accreting columns, which are structures commonly found in astrophysical sources [4,5].

Moreover, X-ray polarimetry can probe isolated neutron stars such as magnetars, as well as neutron stars in binary systems, uncovering the long-sought quantum electrodynamics effect of vacuum birefringence [6–8]. Despite theorists' expectations for the reasons mentioned above, until very recently the only notable detection was the measurement of polarization from the Crab Nebula [9]. At the time this was a significant measurement, as it confirmed for the first time the extension of synchrotron emission to X-rays in this source.

In fact, a new generation of X-ray detectors [10–12], the Gas Pixel Detectors (GPDs), has allowed polarization to be measured by means of the photoelectric effect in gas. Using this device, we designed a space mission providing sensitive measurement in the classical energy band of X-ray astronomy. Although some Chinese colleagues had previously launched a CubeSat mission equipped with a single Gas Pixel Detector (GPD) and a collimator before IXPE, achieving low-significance results on bright galactic sources over months-long observing times [13–17], it has become evident that sensitive polarimetry requires a substantial number of detected photons. This level of sensitivity can be achieved through the use of X-ray mirrors.

Imaging polarimetry's advancements and the launch of the Imaging X-ray Polarimetry Explorer (IXPE) [18,19] have made X-ray polarimetry a standard tool in astrophysics, akin to its use at other wavelengths for the first time. IXPE's data are publicly accessible, allowing every scientist to utilize this newly available resource.

While in the future we aim to conduct experiments with optics having a large throughput, the present ASIC of IXPE suffers from high dead time. While a five-fold improved ASIC has already been obtained by INFN, for very large optics a much larger step forward remains necessary. Below, we describe how a new generation of digital ASICs with a parallel readout allows for a drastic reduction in dead time, accompanied by the possibility of 3D imaging of the photoelectron track. Such ASICs promise to devise a photoelectric X-ray polarimeter with a dead time compliant with future high-throughput X-ray missions.

2. The IXPE Mission in Summary

IXPE, as the 14th Small Explorer (SMEX) NASA mission in partnership with ASI, was built under the supervision of NASA-MSFC (the PI institution; Philip E. Kaaret serves as PI, with Martin Weisskopf as emeritus). INAF, INFN, and industrial partner OHB-Italia devised, built, tested, and calibrated the three Detector Units, plus one spare unit,containing the GPD, the filter and onboard calibration system and the payload computer named the Detector Service Unit (DSU).

The IXPE mission, along with its optics and instrumentation [18–20], is shown operating in orbit in Figure 1.

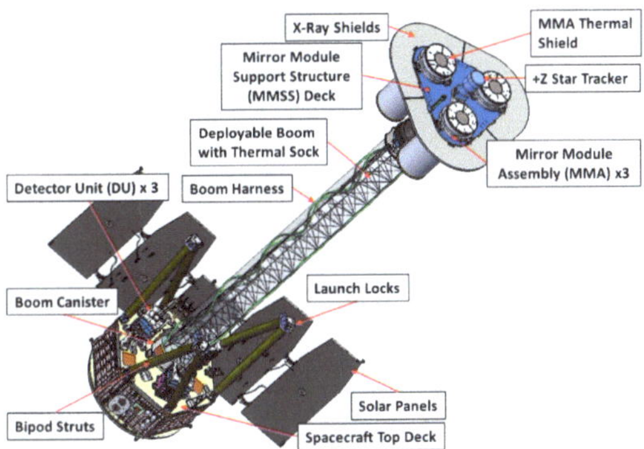

Figure 1. Elements of the deployed IXPE mission [18].

NASA-MSFC fabricated and calibrated three mirror modules [20] with the contribution of Nagoya University (thermal shields) along with one spare unit. An instrument located in the service module at the mirror focal plane, provided by ASI and composed of three detector units [19,21], is separated from the mirror by a focal length of 4 m.

The IXPE spacecraft has a global positioning system (GPS), allowing for the timing of the events with μs accuracy. Two other star trackers (rear and front) are employed to correct images after dithering by using photon-by-photon ground transmission. An X-ray shield, in conjunction with stray-light collimators on top of each Detector Unit (hearafter DU), absorbs cosmic background X-ray photons originating from outside the field of view. An ion–UV filter is located on top of each DU [22].

The DU calibration system [23] is composed of commercial ^{55}Fe isotopes (see Figure 2) with a K_α line at 5.89 keV and a K_β line at 6.5 keV. Polarized radiation at 3 keV (by means of a silver target) and 5.9 keV is produced through 45° Bragg reflection off a graphite mosaic crystal (Cal-A). Unpolarized 5.9 keV and 6.5 keV X-rays (spot ~3 mm and flood ~15 × 15 mm) are source Cal-B and source Cal-C, respectively. Finally, source Cal-D uses a silicon target that produces a wide beam at 1.7 keV (Si K_α) thanks to a ^{55}Fe. Cal sources are used during flight operations and Earth occultation. Cal-C and Cal-D provide the final gain correction for energy determination.

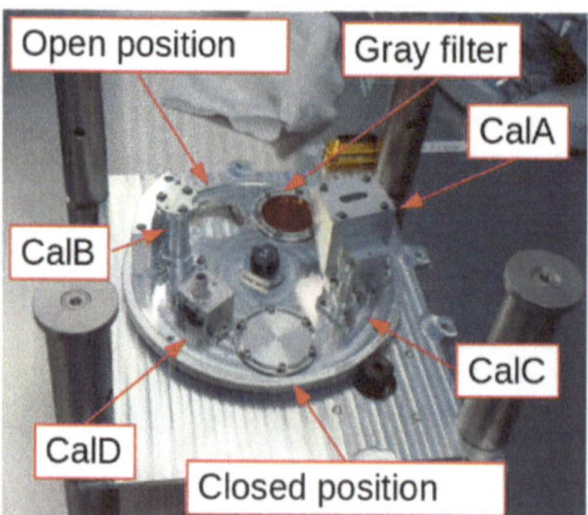

Figure 2. The filter and calibration wheel (FCW) inside each detector unit for onboard calibration. In addition to the calibration system, the FCW hosts a filter made of kapton for high-flux sources and an aluminum cap used for gathering the background.

A residual [24] miscalibration of a few (2–3) tens of eV is irreducible; this is possibly caused by the gas gain decrement due to ions and secondary electrons attaching to the exposed dielectric surface of the Gas Electron Multiplier (GEM) (charging). Because this effect is rate- and energy-dependent, it may differ during flight calibration and during observation of celestial sources [24]. An extensible boom covered with a thermal sock and thermal shields for the mirrors completes the payload system.

The IXPE mirrors (see Figure 3a) were fabricated using the classical technique of replica of electro-formed nickel–cobalt shells. The main design of the IXPE mission was based on Pegasus-XL fairing; thus, the very small thickness of the mirror shell allows for both light weight and the necessary effective area (see Figure 3b). Eventually, the Falcon-9 launcher was adopted after a competitive tender. The Falcon 9 rocket is shown in Figure 4.

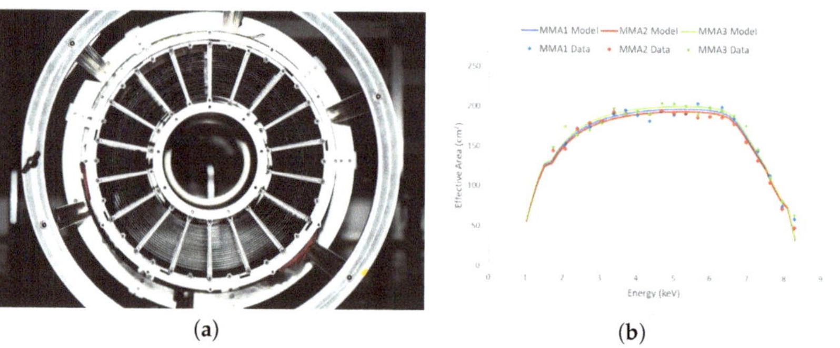

(**a**) (**b**)

Figure 3. (**a**) Top view of a mirror fabricated by NASA-MSFC and (**b**) effective area of each flight mirror [20].

Figure 4. The Falcon-9 rocket and its its firing before being attached at the launch pad.

The IXPE DUs performed as expected after on-ground calibration using both polarized and unpolarized monochromatic X-ray sources (see [24]). After extensive ground calibration at INAF-IAPS [25], the three flight DUs were electrically integrated into the flight Detector Service Unit at the same laboratories (see Figure 5) on the optical bench. The instrumentation underwent extensive laboratory testing, including all of the available payload operation modes.

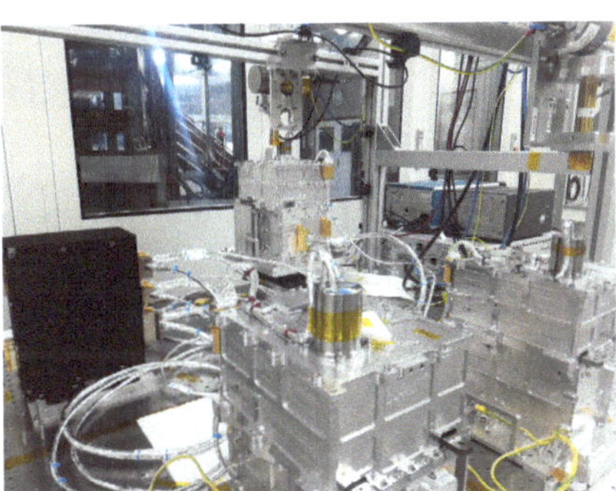

Figure 5. The three detector units integrated into the Detector Service Unit on the optical bench.

The initial analysis of IXPE data revealed that the source counting rate measured by Detector Unit 1 was somewhat higher compared to those measured by Detector Units 2 and 3. The initial response matrices did not accurately account for the differing pressures of the gas mixture inside the detectors. Consequently, this led to a variance in the detected photon flux at a given energy (notably at 1 keV, referred to as the normalization) when observing celestial sources with DU2 and DU3. This issue has since been addressed with updated response matrices that more accurately reflect the time-dependent absorption of dimethyl ether by components within the Gas Pixel Detector (GPD).

Indeed, the efficiency of the three detectors slightly diminishes with time because of the absorption of dimethyl ether by the epoxy used for sealing the detector body (Supreme 10HT by Masterbond) and possibly by the beryllium window support structure. The internal gas pressure is asymptotic, with a slow time constant of 2–3 years and a fast time constant of 1 month, as shown in [24].

However, the modulation factor is slightly better due to the increased track length, meaning that the decrease in sensitivity is not dramatic. The introduction of weights (the asymmetric tracks weigh more) [26] provides 13% better sensitivity with respect to the unweighted analysis. HEASARC analysis tools allow weighted analysis to be available to the general user. In addition, a neural network weighted analysis approach [27–29] was developed, with an improvement of about 8% with respect to the standard weighted moment analysis [30].

IXPE was designed to fit in a Pegasus-XL launcher. After the launch, we discovered a boom motion due to sunlight-to-night thermal expansion (see Figure 6). We used the portion of the orbit with active star trackers (front or rear) and the temperature sensors on the payload to model the (~1 arcmin) shift. Eventually, this very accurate modeling was included in the flight pipeline to make it transparent to the general user.

In contrast to the first two years of operation, when the IXPE collaboration was carried out based on the observation plan, general observers with a competitive tender managed by HEASARC now decide on the new observations. The IXPE collaboration consisted of about 190 scientists, including about 90 participants from about thirteen countries worldwide.

Table 1 summarizes the sources observed during the first two years. The largest group is for binary neutron star and blazar science. The magnetars and SNR group required the largest observing time. Bright source observations are followed by dim sources due to the small size of the onboard memory and the constraints of the S band used at the ASI Ground

Station for receiving data, located at Malindi, Kenya. A gray filter is used to cope with the very high flux. We successfully used the gray filter during the observation of Sco X-1 and the target of opportunity source Swift J1727.8-1613.

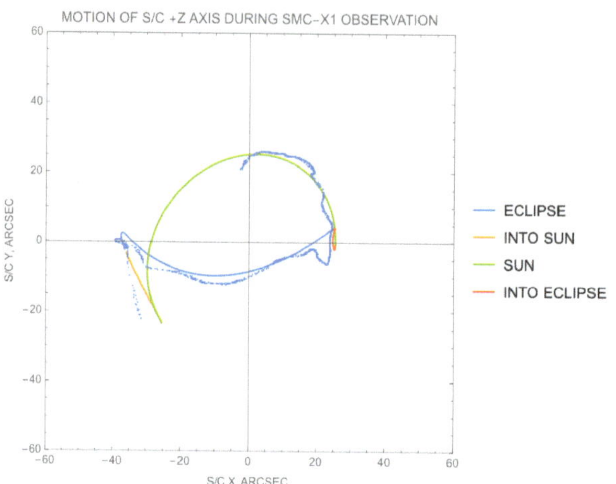

Figure 6. Boom motion due to thermal–elastic expansion along the orbit, as accurately modeled and corrected thanks to post facto reconstruction to remove the dithering [31].

Table 1. Celestial sources observed by IXPE during the first two years of operations.

WGs	Sources Observed
PWN and Pulsars	Crab Nebula and pulsar, Vela PWN, MSH 15-52, PSR B0540-69
SNR	Cas A, Tycho SNR, SN 1006 NE, RCW86, RX J1713.7-3, G21.5-0.9, Vela Jr.
BH-BN	Cyg X-1, 4U 1630-47, LMC X-1, Cyg X-3, 4U 1957 + 115, LMC X-3, Swift J1727.8-1613
NS-BN	Cen X-3, Her X-1, GS 1826-238, Vela X-1, Cyg X-2, GX301-2, X Persei, XTE J1701-462, GX9 + 9, Swift J0243.6 + 6124, IC 4329A, GRO J1008-57,EXO 2030 + 37,LS V + 44 17GX 5-1, GX 13 + 1, SMC X-1
Magnetars	4U 0142 + 61, 1RXS J170849.0, SGR 1806-20, 1E 2259 + 586
RQ-AGN	Sgr A* Complex, MCG-05-23-16, Circinus galaxy, NGC 4151, NGC1068
RL-AGN and Blazars	Mrk 501, S5 0716 + 714, 1ES 1959 + 650, Mrk 421, BL Lac, 3C 454.3, PG1553 + 113, 3C 273, 3C 279, 1ES 0229 + 20 S4 0954 + 65, 3C 454.3, PKS2155-304

Table 2 shows the celestial sources for which a significance larger than 6σ was arrived at from quick-look analysis of their polarization. This is a very limited list, as this analysis does not resolve polarimetry in terms of the angle, energy, or time, and as no background rejection [32] or subtraction was applied. Indeed, we detected significant polarization for a much larger number of sources (about 70%) by exploiting this capability once the full capabilities of IXPE were utilized.

Table 2. Quick-look analysis results providing polarimetry with a significance larger than 6σ. ⋆ Cas A, Tycho SNR, and SN 1006 show significant polarization when angularly resolved. † NGC4151 and Circinus galaxy show significant polarization when the background and energy selection are correctly taken into account.

WGs	Celestial Sources
WG1	Crab Nebula and pulsar, Vela PWN, MSH 15-52, G21.5-0.9
WG2	none ⋆
WG3	Cyg X-1, 4U 1630-47, Cyg X-3, LMC-X3, Swift J1727.8-1613
WG4	Cen X-3, Her X-1, GX301-2, X Persei XTE J1701-462, GX9 + 9, Swift J0243.6 + 6124 GRO J1008-57,LSV 44-17, GX 5-1, Swift J0243.6 + 6124, Sco X-1, GX 13 + 1
WG5	4U 0142 + 61, 1RXS J170849.0, SGR 1806-20, 1E 2259 + 586
WG6	none †
WG7	Mrk 501, Mrk 421, 1ES 0229 + 20, 3C 454.3, 1ES 1959 + 650

The Main Limitation of IXPE

Although a significant success (see Appendix A), the achievements of IXPE in X-ray polarimetry suggest potential for further improvement. The limited effective area of the mirrors restricted the ability to conduct comprehensive 'population studies'. In practice, only the brightest X-ray sources from each category were within reach. Future designs aim for much larger mirror areas, as envisioned for eXTP and Athena. Although Athena does not include a polarimeter, its design goal is to achieve a square meter of effective area. However, such large telescopes cannot utilize the current ASIC technology of IXPE, even considering recent advancements, as noted in [33] (see Section 3).

Additionally, IXPE's results indicate that a promising direction for future missions would involve wide-band X-ray polarimetry extending beyond 8 keV. This approach would enhance the analysis of celestial sources where reflection (from disks, tori, winds, molecular clouds, etc.) plays a significant role in the spectrum, a facet that IXPE is barely able to examine. The employment of large multi-layer optics could enable study of the transportation of radiation in magnetized plasma at cyclotron line energies in binary pulsars or of the dynamics between power-law and hard energy tails characteristic of magnetars. Importantly, improved capacity to handle high flux could significantly reduce calibration times, which for IXPE required 40 days per detector operating continuously. This efficiency is crucial, as many missions must limit calibration time to adhere to schedules.

3. A Possible Path to the Future of X-Ray Polarimetry

One of the main drawbacks of the GPD currently flying onboard the IXPE is the large dead time [19,21], though this is mitigated by a new version of the ASIC [33]. As a matter of fact, a drastic reduction in dead time is already possible thanks to the new generation of ASICs allowing parallel readout with digital information on the pulse amplitude. These ASICs, developed by an international collaboration and are derived from the MEDIPIX family, are the TimePIX3 [34] and the most recent TimePIX4 [35]. Their design allows for data-driven operation with dead time-free operation up to 40 Mpixels s^{-1} cm^{-2} for TimePIX3 and up to 3.5 Mpixels s^{-1} mm^{-2} for TimePIX4. These ASICs (see Figure 7a) allow for a sparse readout as well as simultaneous per-pixel measurement of the time of arrival (with a resolution of 1.56 ns for TimePIX3 and <200 ps for TimePIX4) and time over threshold, with the latter being proportional to the charge content for each pixel. TimePIX3 features 65,536 pixels in a square pattern, with a pixel pitch of 55 μm and a noise of 60 e_{rms}. A practical implementation of TimePIX3 as the front-end for a gas detector is the GridPix [36]) configuration, where the multiplication stage is obtained by applying precise photolithographic techniques to make a metallic Micromega grid above the sensitive ASIC plane at a distance of a few tens of μm (see Figure 7b, [37]).

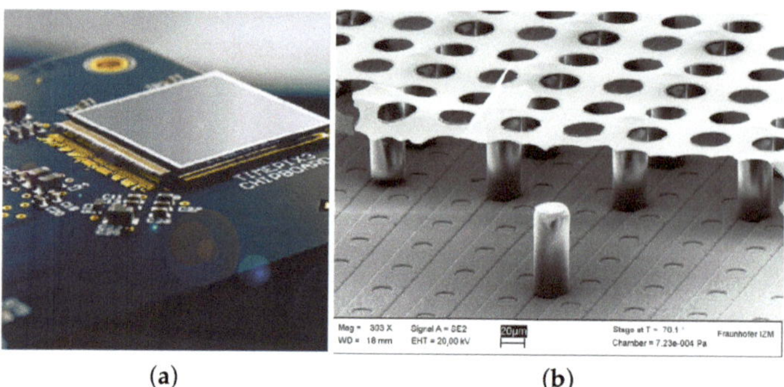

Figure 7. (**a**). TimePIX3 three-side buttable (https://kt.cern/technologies/timepix3 (accessed on 17 March 2024)) and (**b**) Ingrid solution for the multiplication region of GridPIX [37]. The pillars are 50 μm long.

In principle, this design, allows for full 3D photoelectron track reconstruction. We previously proved the suitability of this approach for increased polarization sensitivity [38]. Before this practical implementation becomes mature enough for a space experiment, it is first necessary to: (1) prove the performance in terms of the modulation factor and lack of spurious modulation; (2) determine the energy resolution; (3) prove the resistance against heavy ion interaction with the gas;and (4) build a sealed detector body. and will be carried out in the near future.

4. Conclusions

IXPE is now a real flown polarimetry mission, and is discovering and explaining new physical phenomena in previously known X-ray sources. In addition, it is helping to disentangle geometry from physics, thereby maintaining what scientists have been promising for decades since the first rocket launches and the discoveries of OSO-8. Thanks to the perseverance of many scientists, we now are in possession of a rapidly developing observational tool to better understand a wide variety of X-ray sources and their environments. The expectations of theory can be tested with the help of accurate X-ray polarimetry. The same scientists are studying a new detector based on a modern ASIC, which promises to overcome the main limitations of the current ASIC employed onboard the IXPE.

Funding: The Imaging X-ray Polarimetry Explorer (IXPE) is a joint US and Italian mission. The US contribution is supported by the National Aeronautics and Space Administration (NASA) and led and managed by its Marshall Space Flight Center (MSFC), with industry partner Ball Aerospace (contract NNM15AA18C). The Italian contribution is supported by the Italian Space Agency (Agenzia Spaziale Italiana, ASI) through the contract ASI-OHBI-2022-13-I.0, the agreements ASI-INAF-2022-19-HH.0 and ASI-INFN-2017.13-H0, and its Space Science Data Center (SSDC) with agreements ASI-INAF-2022-14-HH.0 and ASI-INFN 2021-43-HH.0, as well as by the Istituto Nazionale di Astrofisica (INAF) and the Istituto Nazionale di Fisica Nucleare (INFN) in Italy. This research used data products provided by the IXPE Team (MSFC, SSDC, INAF, and INFN) and distributed with additional software tools by the High-Energy Astrophysics Science Archive Research Center (HEASARC) at the NASA Goddard Space Flight Center (GSFC). The author acknowledges funding from the Ministry of Universities and Research (PRIN 2020MZ884C) "Hype-X: High Yield Polarimetry Experiment in X-rays", the INAF MiniGrant "New generation of 3D detectors for X-ray polarimetry: simulation of performances" and the MAECI Contribution 2023–2025 Grant "RES-PUBLICA Research Endeavour for Science Polarimetry with the University of Bonn in Liason with INAF for Cosmic Applications".

Data Availability Statement: IXPE data are publicly available at https://heasarc.gsfc.nasa.gov/docs/heasarc/missions/ixpe.html (accessed on 17 March 2024).

Acknowledgments: The author acknowledges the IXPE Science collaboration, of which he is a part, and discussions with Klaus Desch, Jochen Kaminsky, Markus Gruber, and Vladislavs Plesanovs from the University of Bonn regarding GridPIX detectors.

Conflicts of Interest: The author declares no conflicts of interest.

Appendix A. Selected Scientific Results of IXPE

Appendix A.1. Pulsar Wind Nebulae and Radio Pulsars

Pulsar wind nebulae (PWNe) shine in X-rays emitted via the synchrotron process. Bubbles of plasma accelerated up to 10–100 TeV and magnetic fields produced by a spinning neutron star interact with the interstellar medium. Thees are responsible for the complex morphologies seen in X-rays. The Crab Nebula was the only source for which OSO-8 detected polarized radiation in the 1970s [9] thanks to its collimated Bragg diffraction polarimeter, and has been more recently re-detected by Polarlight [13,14]. The angularly resolved polarimetry from IXPE observations have already been published for Vela PWN [39] (see Figure A1a), the Crab Nebula and its pulsars [40] (see Figure A1b), and MSH 15-52 and its pulsars [41] (see Figure A1c). The polarization map obtained by IXPE for these two PWNs are shown in Figure A1.

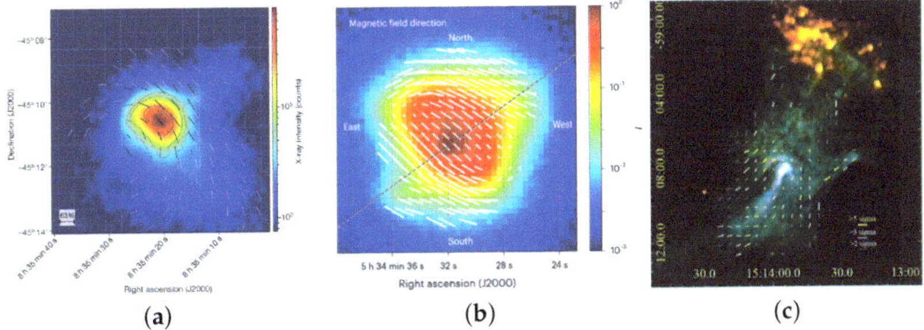

Figure A1. (**a**) IXPE polarization map for the Vela PWN [39]; (**b**) IXPE polarization map for the Crab Nebula [40]; (**c**) IXPE polarization map for the MSH15−52 Nebula [41].

The high level of polarization in Vela (up to 67–72%), Crab PWNe (up to 45–50%), and MSH 15-52 (up to 70%), along with the direction of the magnetic field, show that the turbulence is much less effective than expected. The IXPE image of Vela PWN shows that the polarization structure is symmetric about the projected pulsar spin axis, which corresponds to its proper direction of motion. For Crab PWN, the integrated polarization degree is 20% and the polarization angle is about $145°$. While the polarization degrees are consistent between IXPE and OSO-8, the polarization angle has a small but statistically significant difference from the $154°$ measured [9] by OSO-8. Such a difference could be due to a change in the morphology of the inner structure of the Crab Nebula. In MSH 15-52, the magnetic field follows the thumb, fingers, and other linear structures. The polarization reaches about 70% at the end of the jet, while the magnetic field is less ordered at the base of the inner jet.

IXPE further investigated the polarization properties of the Crab and MSH 15-52 pulsars, facilitated by its imaging capabilities. For the Crab Pulsar, after subtracting the residual nebular component under the pulsar point spread function (PSF), the phase-resolved polarization properties shows significant detection only at the center of the main (P1) pulse, which is 15% with a polarization angle of about $105°$. The phase-integrated polarimetry of the Crab Pulsar is $2.6^{+2.7}_{-2.6}$. Such small polarization is in contrast with most of the existing PSR models [42,43]. For MSH-1552, a single significant polarization bin at the maximum of the phase-resolved lightcurve is interpreted as a possible extension of its radio emission.

Appendix A.2. Supernova Remnants

At the time of writing, IXPE had observed five supernova remnants so far: Cas A (see Figure A2a), Tycho SNR (see Figure A2b), SN 1006 north-east rim (see Figure A2c), RCW86, and RX J1713; however, only the first three have been published [44–46] In order to measure the polarization of Cas A [44] and Tycho SNR [45], we first selected an energy range between the calcium/argon line and the iron line, where the thermal emission is expected to be at a minimum. On the contrary, no lines are present in the SN 1006 NE limb [46], and the energy range that maximized the source-to-background ratio was selected. We then performed analysis on a pixel-by-pixel basis (see Figure A2). The results for Tycho and Cas A were inconclusive; thus, we adopted a different technique. Assuming a circular symmetry for the polarization direction, we recalculated the Stokes parameter for each event [47] by calculating a new zero for the direction of the photoelectrons and its position angle with respect to the rotated celestial coordinates, taking the center of both supernovae as the origin. This procedure resulted in new values for the Stokes parameters, providing an overall signal for the signal in all regions corresponding to the tangential and radial Q and U Stokes parameters. For every annular or circular region selected, we found that the polarization was tangential. Because synchrotron emissions require a magnetic field perpendicular to the polarization angle, we discovered that for Cas A and Tycho SNR, just as in the radio wavelength, the magnetic field has a radial global orientation. X-rays are actually emitted close to the accelerating shock fronts, and the 10–100 TeV electrons responsible for this emission have a short lifetime due to cooling. Further, interstellar magnetic fields in the outer shock (and in the reverse shock in Cas A) are eventually compressed tangentially, meaning that the instability mechanism should act quickly to realign the magnetic field in the radial direction. The tangential polarization degree for the whole Cas A emission is 1.8% ± 0.3%, which is smaller than in the radio band. The corresponding average polarization degrees for the sole synchrotron emission, considering the external shock rim, are 2.5% and 5%. For Tycho SNR, the global tangential polarization degree is 3.5% ± 0.7%, corresponding to 9.1% ± 2.0% for the synchrotron component, while for the external rim it is 11.9% ± 2.2%. For Tycho, the levels of polarization are larger than those in the radio band. It is worth noting [45] that in Tycho SNR the west non-circular region containing the stripes shows a significant expected polarization (∼23%), possibly indicating the presence of nonlinear diffusive shock acceleration [48].

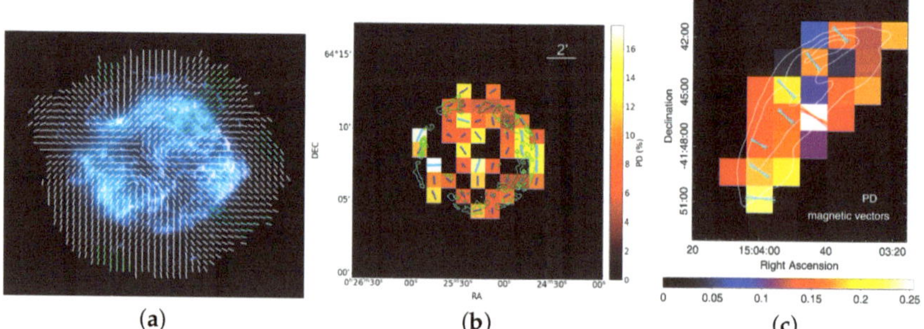

Figure A2. (**a**) Magnetic−field map for the Cas A SNR (Image credit: NASA/CXC/SAO/NASA/MSFC/Vink et al.). The region in green has a higher-confidence measurement. The magnetic field is mostly radial. (**b**) IXPE polarization map for Tycho SNR [45]. The polarization directions show a mostly radial magnetic field. (**c**) Polarization map for the SN 1006 NE limb [46]. The polarization directions show a mostly magnetic field perpendicular to the limb.

SN 1006 shows larger polarization than radio emissions, with an average value of about 20% for the whole shell. As in the other SNRs, the direction of the magnetic field is perpendicular to the rim. As a matter of fact, all of the SNRs show a smaller polarization degree with respect to the maximum obtainable by synchrotron emission ($\approx$80%).

Appendix A.3. Accreting Stellar-Mass Black Holes

The first black hole binary system observed by IXPE was Cyg X-1 [49]. During this first point, Cyg X-1 was in a low and hard state, and the polarization found in the IXPE energy band, at $\sim$4%, was much larger than expected based only on the orbital inclination. This suggests a disk with its most internal part observed more edge-on than expected—a sort of warped disk. A hint of an increase in polarization with energy was found in the data as well (see Figure A3a). The other important result is that the polarization angle was found to be parallel to the radio jet (see Figure A3b). Because most of the emitted X-rays are due to the corona in the low and hard state, the polarization direction excludes a lamppost geometry (see Figure A3c). In such a geometry, the polarization angle should be perpendicular to the radio jet. The corona geometry must be sandwiched against the disk, while the polarization can be either parallel or perpendicular to the disk but the jet cannot be parallel to the disk. Thus, this is the first time that the inner flow toward the black hole has been observed to be perpendicular to the jet direction.

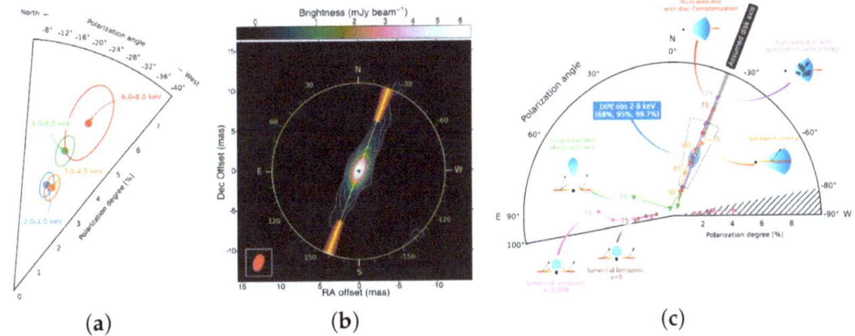

(**a**)   (**b**)   (**c**)

Figure A3. (**a**) The polarization degree shows a possible increase with energy. (**b**) The polarization angle is parallel to the radio jet. This discovery (1) establishes that the disk axis is parallel to the jet and (2) that the corona geometry cannot be a lamppost. (**c**) The different expected polarization degrees and angles for different corona models. All figures are from [49].

A sandwich corona excludes the aborted jet origin, and points to plasma instabilities across the surface. Other black holes were observed, as indicated in Table 1. The most puzzling are 4U1630-47 [50,51] and Cyg X-3 [52]. 4U1630-47 was observed at two different levels of luminosity in a high soft state where the disk emission dominates. Its complex behavior challenges a simple geometrically thin and optically thick disk model. Cyg X-3 shows polarization perpendicular to the radio ejection, thought to be due to reflection from the circumnuclear material and a polarization degree as high as $\sim$25%.

Appendix A.4. Accreting White Dwarfs and Neutron Stars

During the first two years of IXPE operation, we observed both low-magnetized neutron star binaries (LMNSB) and X-ray binary pulsars, with the latter being more polarized than the former. This was not unexpected, as the magnetic field is much larger for pulsars (few 10^{12} Gauss) and the photon opacity is anisotropic with respect to the magnetic field direction. In LMXRB, instead, residual polarization may derive from the scattering of primary radiation either on the accretion disk that extends down to the neutron star's surface, from the spreading layer (the layer of material accreting onto the neutron star's surface, which is approximately perpendicular to the accretion disk), or from the boundary layer, which is the parallel layer between the truncated disk and the neutron star surface. The sources observed thus far are listed in Table 1. Among these, Cyg X-2 [53], XTE J1702-

462 [54], GX5-1 [55], and Sco X-1 [56] are called "Z sources" because of the characteristic "Z" shape in the color–color diagram. For "Atoll", the observed sources were GS 1826-238 [57], GX 9 + 9 [58], and 4U1830-303 [59].

X-ray pulsars show a much smaller polarization degree ($\sim$10–15%) than was expected ($\sim$60–80%) [5,60,61]. The reason for this may be that the reprocessing geometry [62] is much more complex than the simple "fan" or "pencil" model, which involves only simple columns and hot spots at the poles.

The low polarization degree found in the archetypal wind-accreting high-mass X-ray binary system Vela X-1 [63] as in the other X-Ray pulsars, could be related to the inverse temperature structure of the neutron star atmosphere, the same as for the other XRPs.The low polarization degree found in Vela X-1 may also be due to the evolution of the polarization degree with the energy (a 90° rotation in the IXPE band) and pulse phase.

Despite the smaller than expected observed polarization, thanks to IXPE it was possible to disentangle the physics from the geometry by applying the rotating vector model derived from radio-polarimetry. For the first time, we measured the magnetic obliquity (the angle between the magnetic dipole axis and spin axis and the projection of the spin axis to the plane of the sky). Interestingly, an orthogonal rotator with magnetic obliquity close to $\sim$90° was found by IXPE [64].

Appendix A.5. Magnetars

Magnetars are isolated neutron stars powered by an extreme magnetic field far larger than what is available on Earth, ranging from 10^{14} to 10^{15} Gauss. These very useful phenomena allow for studying photon propagation in highly magnetized atmospheres and magnetospheres. IXPE has published results from four magnetars (see Table 1), exploring their energy and phase-resolved polarization and finding very different behavior between 4U0142 + 61 [65] (see Figure A4a) and 1RXS J170849.0 [66] (see Figure A4b) in terms of their energy-resolved polarization (see Figure A4).

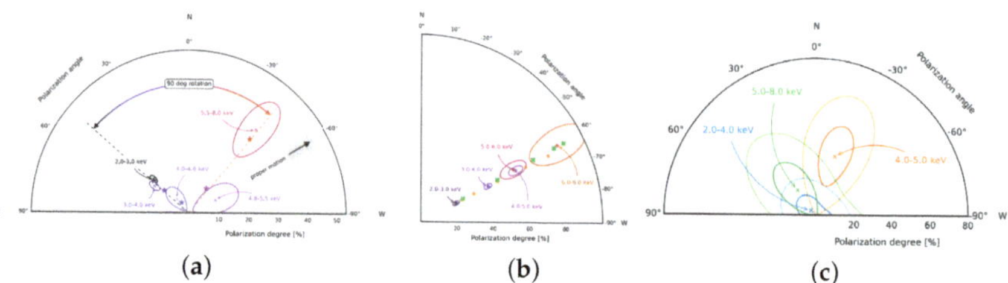

(a)   (b)   (c)

Figure A4. (**a**) Spectro−polarimetry for 4U0142 + 61. Crosses indicate the measured values and stars indicate the model (the equatorial belt–condensed surface RCS models are indicated by the stars). Contours enclose the 68.3% confidence level. The gray shaded area and the black arrow indicate the direction of the proper motion and its uncertainty [65]. (**b**) Spectro-polarimetry of 1RXS J1708, showing the 50% confidence regions for joint measurement of the polarization degree and angle. Green crosses and orange stars show the prediction of the two different possible emission regions' structures [66]. (**c**) Spectro-polarimetry of SGR1806. The crosses indicate the measures. The model is frozen from the one determined by XMM (black body plus power law). The contours are 68.3% and 99% [67].

This difference is explained by the different kinds of emitting regions on the surface (i.e., geometry and physical status). Although vacuum birefringence is considered in the modeling while evaluating polarimetry expectations, the size of the emitting region is not yet sufficiently extended to require unambiguously this QED effect. A large extended region, as determined by a small pulsed fraction and a high polarization degree, are necessary for securing the vacuum birefringence at work in these systems. A third magnetar, SGR 1806-20 (see Figure A4c), was observed to be similar to 4U0142 + 61, albeit with a much smaller significance [67], and was modeled with two hot spots placed near the magnetic equator of the bare neutron star's surface. The fourth and last magnetar observed by IXPE was 1E 2259 + 5586 [68]. For this source, as for SGR 1806, the IXPE results were interpreted in light of the spectral analysis derived from simultaneous observations with XMM. The presence of a condensed surface and a plasma loop that scatters the radiation in the magnetosphere is considered the conclusive model for this source.

In fact, the four magnetars do not show the unambiguous presence of vacuum polarization and birefringence; thus, we need to wait for observations of additional sources with a much wider emitting surface region and high polarization in order to definitively unveil them.

*Appendix A.6. Radio-Quiet AGNs and Sgr A**

Accretion disks in AGNs emit mostly in the UV–optical energy band, and the primary X-ray emission is thought to be due to inverse Compton radiation in a hot corona embedding the colder accretion disk [69]. Such a geometry can produce polarized radiation [70], and from the degree of polarization it is possible to derive information on the geometry of the corona. An aborted jet origin is derived from a lamppost corona while the presence of instabilities is derived from a corona sandwiching of the accretion disk. An angle of polarization parallel to the disk axis, detected as the direction of the commonly present weakly emitting extended radio emission, is the signature of a corona sandwiching the disk. This is the case for NGC4151; indeed, the measured polarization $(4.9 \pm 1.1)\%$ is thought to be entirely due to the reflection from the accretion disk. Only the upper limits [71] were found for NGC-5-23-16. Interestingly, for IXPE observation of the Circinus galaxy, a Compton-thick AGN which is observed almost edge-on with respect to its symmetry axis, confirms the presence of a thick obscuring torus as a neutral reflector due to polarization [58] $(28 \pm 7\%)$. In fact, for this AGN the polarization direction is normal with respect to to the weak radio jet (see Figure A5).

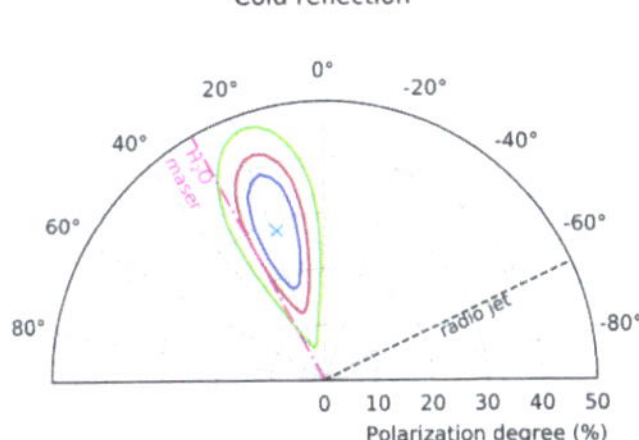

Figure A5. The polarization angle of the Circinus Galaxy is directed along the accretion disk traced by the H_2O maser shown in the figure; Together with the presence of reflection spectrum from cold matter, this is the signature of an obscuring torus responsible for the observed polarization. The polarization degree and angles contours represent 68%, 90% and 99% confidence level. Based on a comparison of the simulation and the observed polarization, the aperture of the torus is 45–55° [58].

Much closer to us, our galactic supermassive black hole is a very dim X-ray source with occasional fast flares. Cold molecular clouds shining in X-rays may reflect [72] photons emitted in the past from Sgr A*. Thus, the reflected and observed radiation should be polarized [73], with the polarization vector indicating the origin of the radiation and, eventually, the Sgr A*. IXPE has established this for certain (see Figure A6) [74].

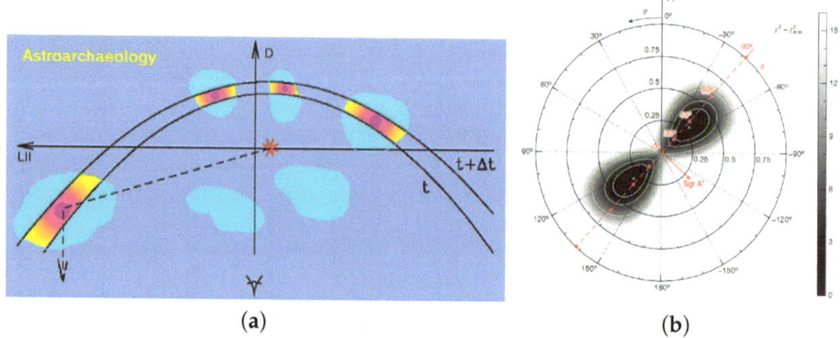

(**a**) (**b**)

Figure A6. (**a**) Polarimetry map of the molecular clouds in the vicinity of the galactic center. This mapping allows the past X-ray flares of the galactic center to be reconstructed based on their polarization degree and angle. (**b**) Measurements taken by IXPE show that Sgr A* was 10^6 times brighter in the X-ray wavelength some 200 years ago. The mapping of different molecular clouds could allow for the determination of whether a single flare or a multiple flares occurred in the past [74].

Appendix A.7. Blazars and Radio Galaxies

The IXPE energy band is particularly suitable for analyzing blazars' polarimetry. Blazars with X-rays either in the synchrotron peak (high-synchrotron peaked HSP) or the Inverse Compton (IC) peak (low-synchrotron peaked LSP) can be probed using polarimetry. Based on the sensitivity of IXPE, only HSP blazars were found to be polarized [75–77], while LSP blazars such as BL-Lac were found to be unpolarized [78]. The upper limits remain too high to discriminate hadronic versus leptonic models as the origin of the hIC peak [79], which was not totally unexpected given their lower fluxes. Interestingly, an observation of BL Lac (LSP) during a flare showed significant polarization, with X-rays moved into the synchrotron peak [80].

Restricting ourselves to HSP blazars such as Mrk 501 and Mrk 421, we note that for Mrk 501 IXPE observation [75] showed a polarization degree of ∼10%, which is twice as much as in the optical band, with the polarization angle directed along the jet. Together with a modest, if not null, polarization variability, these characteristic features are considered the signature of an energy-stratified shock acceleration process.

The first IXPE observation of Mrk 421 showed a polarization vector that was not coincident with the jet direction [76] but rather with a polarization degree of $(15 \pm 2)\%$, ∼3 times larger than that observed in the optical-infrared-mm region. Another later observation surprisingly showed a polarization angle that was rotating quickly with time [77] (see Figure A7a). This rotation indicates the presence of a helical magnetic field (see Figure A7b) in addition to energy-stratified shock acceleration.

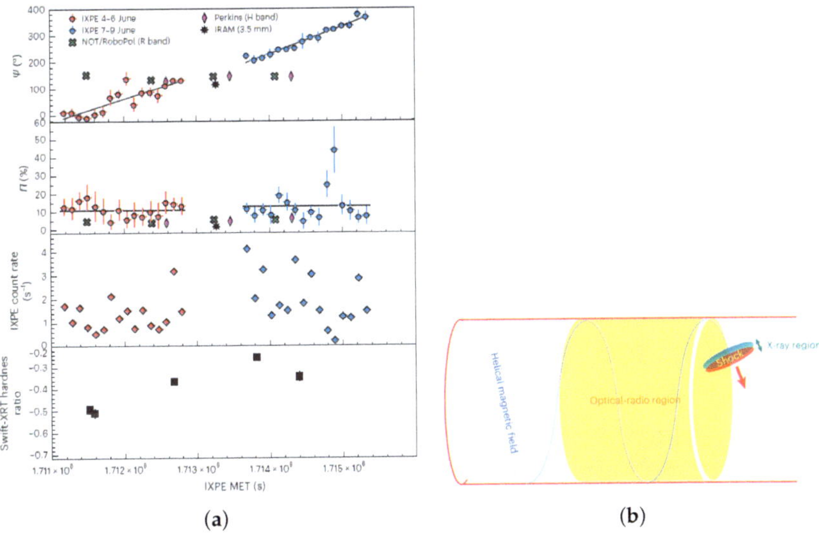

Figure A7. (**a**) The rotation of the polarization angle in Mrk 421 measured in X-rays is much faster (80°–90°/day) than that previously measured in the optical band [77] (8°–9°/day) for this source. (**b**) Energy-stratified shock acceleration is active in an environment embedded with a helicoidal magnetic field [77].

References

1. Westfold, K.C. The Polarization of Synchrotron Radiation. *Astrophys. J.* **1959**, *130*, 241. [CrossRef]
2. Gnedin, I.N.; Sunyaev, R.A. Polarization of optical and X-radiation from compact thermal sources with magnetic field. *Astron. Astrophys.* **1974**, *36*, 379–394.
3. Rees, M.J. Expected polarization properties of binary X-ray sources. *Mon. Not. R. Astron. Soc.* **1975**, *171*, 457. [CrossRef]
4. Sunyaev, R.A.; Titarchuk, L.G. Comptonization of low-frequency radiation in accretion disks Angular distribution and polarization of hard radiation. *Astron. Astrophys.* **1985**, *143*, 374–388. [CrossRef]
5. Mèszàros, P.; Novick, R.; Szentgyorgyi, A.; Chanan, G.A.; Weisskopf, M.C. Astrophysical implications and observational prospects of X-ray polarimetry. *Astrophys. J.* **1988**, *324*, 1056. [CrossRef]
6. Gnedin, I.N.; Pavlov, G.G.; Shibanov, I.A. The effect of vacuum birefringence in a magnetic field on the polarization and beaming of X-ray pulsars. *Sov. Astron. Lett.* **1978**, *4*, 214.
7. Ventura, J.; Nagel, W.; Meszaros, P. Possible vacuum signature in the spectra of X-ray pulsars. *Astrophys. J.* **1979**, *233*, L125. [CrossRef]
8. Mèszàros, P.; Ventura, J. Vacuum polarization effects on radiative opacities in a strong magnetic field. *Phys. Rev. D* **1979**, *19*, 3565. [CrossRef]
9. Weisskopf, M.C.; Silver, E.H.; Kestenbaum, H.L.; Long, K.S.; Novick, R. A precision measurement of the X-ray polarization of the Crab Nebula without pulsar contamination. *Astrophys. J.* **1978**, *220*, L117. [CrossRef]
10. Costa, E.; Soffitta, P.; Bellazzini, R.; Brez, A.; Lumb, N.; Spandre, G. An efficient photoelectric X-ray polarimeter for the study of black holes and neutron stars. *Nature* **2001**, *411*, 662. [CrossRef]
11. Bellazzini, R.; Spandre, G.; Minuti, M.; Baldini, L.; Brez, A.; Cavalca, F.; Latronico, L.; Omodei, N.; Massai, M.M.; Sgro', C.; et al. Direct reading of charge multipliers with a self-triggering CMOS analog chip with 105 k pixels at 50 μm pitch. *Nucl. Instrum. Methods Phys. Res. A* **2006**, *566*, 552. [CrossRef]
12. Bellazzini, R.; Spandre, G.; Minuti, M.; Baldini, L.; Brez, A.; Latronico, L.; Omodei, N.; Razzano, M.; Massai, M.M.; Pesce-Rollins, M.; et al. A sealed Gas Pixel Detector for X-ray astronomy. *Nucl. Instrum. Methods Phys. Res. A* **2007**, *579*, 853. [CrossRef]
13. Feng, H.; Jiang, W.; Minuti, M.; Wu, Q.; Jung, A.; Yang, D.; Citraro, S.; Nasimi, H.; Yu, J.; Jin, G.; et al. PolarLight: A CubeSat X-ray polarimeter based on the gas pixel detector. *Exp. Astron.* **2019**, *47*, 225–243. [CrossRef]
14. Feng, H.; Li, H.; Long, X.; Bellazzini, R.; Costa, E.; Wu, Q.; Huang, J.; Jiang, W.; Minuti, M.; Wang, W.; et al. Author Correction: Re-detection and a possible time variation of soft X-ray polarization from the Crab. *Nat. Astron.* **2020**, *4*, 808. [CrossRef]

15. Long, X.; Feng, H.; Li, H.; Zhu, J.; Wu, Q.; Huang, J.; Minuti, M.; Jiang, W.; Wang, W.; Xu, R.; et al. X-ray Polarimetry of the Crab Nebula with PolarLight: Polarization Recovery after the Glitch and a Secular Position Angle Variation. *Astrophys. J. Lett.* **2021**, *912*, L28. [CrossRef]

16. Long, X.; Feng, H.; Li, H.; Zhu, J.; Wu, Q.; Huang, J.; Minuti, M.; Jiang, W.; Yang, D.; Citraro, S.; et al. A Significant Detection of X-ray Polarization in Sco X-1 with PolarLight and Constraints on the Corona Geometry. *Astrophys. J.* **2022**, *924*, L13. [CrossRef]

17. Long, X.; Feng, H.; Li, H.; Kong, L.D.; Heyl, J.; Ji, L.; Tao, L.; Muleri, F.; Wu, Q.; Zhu, J.; et al. X-ray Polarimetry of the Accreting Pulsar 1A 0535+262 in the Supercritical State with PolarLight. *Astrophys. J.* **2023**, *950*, 76. [CrossRef]

18. Weisskopf, M.C.; Soffitta, P.; Baldini, L.; Ramsey, B.D.; O'Dell, S.L.; Romani, R.W.; Matt, G.; Deininger, W.D.; Baumgartner, W.H.; Bellazzini, R.; et al. The Imaging X-ray Polarimetry Explorer (IXPE): Pre-Launch. *J. Astron. Telesc. Instrum. Syst.* **2022**, *8*, 026002. [CrossRef]

19. Soffitta, P.; Baldini, L.; Bellazzini, R.; Costa, E.; Latronico, L.; Muleri, F.; Del Monte, E.; Fabiani, S.; Minuti, M.; Pinchera, M.; et al. The Instrument of the Imaging X-Ray Polarimetry Explorer. *Astron. J.* **2021**, *162*, 208. [CrossRef]

20. Ramsey, B.D.; Bongiorno, S.D.; Kolodziejczak, J.J.; Kilaru, K.; Alexander, C.; Baumgartner, W.H.; Breeding, S.; Elsner, R.F.; Le Roy, S.; McCracken, J.; et al. Optics for the imaging X-ray polarimetry explorer. *J. Astron. Telesc. Instrum. Syst.* **2022**, *8*, 024003. [CrossRef]

21. Baldini, L.; Barbanera, M.; Bellazzini, R.; Bonino, R.; Borotto, F.; Brez, A.; Caporale, C.; Cardelli, C.; Castellano, S.; Ceccanti, M.; et al. Design, construction, and test of the Gas Pixel Detectors for the IXPE mission. *Astropart. Phys.* **2021**, *133*, 102628. [CrossRef]

22. La Monaca, F.; Fabiani, S.; Lefevre, C.; Morbidini, A.; Piazzolla, R.; Amici, F.; Attina', P.; Brienza, D.; Costa, E.; Di Cosimo, S.; et al. IXPE DU-FM ions-UV filters characterization. In *Proceedings of the Space Telescopes and Instrumentation 2020: Ultraviolet to Gamma Ray*; den Herder, J.W.A., Nikzad, S., Nakazawa, K., Eds.; SPIE: Bellingham, WA, USA, 2021; Volume 11444, p. 1144463. [CrossRef]

23. Ferrazzoli, R.; Muleri, F.; Lefevre, C.; Morbidini, A.; Amici, F.; Brienza, D.; Costa, E.; Del Monte, E.; Di Marco, A.; Di Persio, G.; et al. In-flight calibration system of imaging x-ray polarimetry explorer. *J. Astron. Telesc. Instrum. Syst.* **2020**, *6*, 048002. [CrossRef]

24. Rankin, J.; Muleri, F.; Ferrazzoli, R.; Baldini, L.; Baumgartner, W.; Costa, E.; Del Monte, E.; Di Marco, A.; Dietz, K.; Ehlert, S.; et al. In-flight monitoring and calibration of the x-ray polarimeters on board IXPE. In *Proceedings of the UV, X-ray, and Gamma-ray Space Instrumentation for Astronomy XXIII, San Diego, CA, USA, 1–5 August 2023*; Society of Photo-Optical Instrumentation Engineers (SPIE) Conference Series; SPIE: Bellingham, WA, USA, 2023.

25. Muleri, F.; Piazzolla, R.; Di Marco, A.; Fabiani, S.; La Monaca, F.; Lefevre, C.; Morbidini, A.; Rankin, J.; Soffitta, P.; Tobia, A.; et al. The IXPE instrument calibration equipment. *Astropart. Phys.* **2022**, *136*, 102658. [CrossRef]

26. Di Marco, A.; Costa, E.; Muleri, F.; Soffitta, P.; Fabiani, S.; La Monaca, F.; Rankin, J.; Xie, F.; Bachetti, M.; Baldini, L.; et al. A Weighted Analysis to Improve the X-Ray Polarization Sensitivity of the Imaging X-ray Polarimetry Explorer. *Astron. J.* **2022**, *163*, 170. [CrossRef]

27. Peirson, A.; Romani, R.; Marshall, H.; Steiner, J.; Baldini, L. Deep ensemble analysis for Imaging X-ray Polarimetry. *Nucl. Instrum. Methods Phys. Res. Sect. A Accel. Spectrometers Detect. Assoc. Equip.* **2021**, *986*, 164740. [CrossRef]

28. Peirson, A.L.; Romani, R.W. Toward Optimal Signal Extraction for Imaging X-ray Polarimetry. *Astrophys. J.* **2021**, *920*, 40. [CrossRef]

29. Cibrario, N.; Negro, M.; Moriakov, N.; Bonino, R.; Baldini, L.; Di Lalla, N.; Latronico, L.; Maldera, S.; Manfreda, A.; Omodei, N.; et al. Joint machine learning and analytic track reconstruction for X-ray polarimetry with gas pixel detectors. *Astron. Astrophys.* **2023**, *674*, A107. [CrossRef]

30. Di Marco, A.; Tennant, A.; La Monaca, F.; Muleri, F.; Rankin, J.; Rushing, J.; Soffitta, P.; Baglioni, G.; Baldini, L.; Costa, E.; et al. Validation of neural network software by using IXPE ground calibration data. In *Proceedings of the SPIE Astronomical Telescopes + Instrumentation, Montréal, QC, Canada, 31 August 2022*; den Herder, J.W.A., Nikzad, S., Nakazawa, K., Eds.; Society of Photo-Optical Instrumentation Engineers (SPIE): Bellingham, WA, USA, 2022; Volume 12181, p. 1218157. [CrossRef]

31. Soffitta, P.; Baldini, L.; Baumgartner, W.; Bellazzini, R.; Bongiorno, S.D.; Bucciantini, N.; Costa, E.; Dovčiak, M.; Ehlert, S.; Kaaret, P.E.; et al. The Imaging X-ray polarimetry explorer (IXPE) at last! In *UV, X-ray, and Gamma-ray Space Instrumentation for Astronomy XXIII*; Siegmund, O.H., Hoadley, K., Eds.; Society of Photo-Optical Instrumentation Engineers (SPIE) Conference Series; SPIE: Bellingham, WA, USA, 2023; Volume 12678, p. 1267803. [CrossRef]

32. Di Marco, A.; Soffitta, P.; Costa, E.; Ferrazzoli, R.; La Monaca, F.; Rankin, J.; Ratheesh, A.; Xie, F.; Baldini, L.; Del Monte, E.; et al. Handling the Background in IXPE Polarimetric Data. *Astron. J.* **2023**, *165*, 143. [CrossRef]

33. Minuti, M.; Baldini, L.; Bellazzini, R.; Brez, A.; Ceccanti, M.; Krummenacher, F.; Latronico, L.; Lucchesi, L.; Manfreda, A.; Orsini, L.; et al. XPOL-III: A new-generation VLSI CMOS ASIC for high-throughput X-ray polarimetry. *Nucl. Instrum. Methods Phys. Res. A* **2023**, *1046*, 167674. [CrossRef]

34. Poikela, T.; Plosila, J.; Westerlund, T.; Campbell, M.; De Gaspari, M.; Llopart, X.; Gromov, V.; Kluit, R.; van Beuzekom, M.; Zappon, F.; et al. Timepix3: A 65K channel hybrid pixel readout chip with simultaneous ToA/ToT and sparse readout. *J. Instrum.* **2014**, *9*, C05013. [CrossRef]

35. Llopart, X.; Alozy, J.; Ballabriga, R.; Campbell, M.; Casanova, R.; Gromov, V.; Heijne, E.H.M.; Poikela, T.; Santin, E.; Sriskaran, V.; et al. Timepix4, a large area pixel detector readout chip which can be tiled on 4 sides providing sub-200 ps timestamp binning. *J. Instrum.* **2022**, *17*, C01044. [CrossRef]

36. Ligtenberg, C.; Heijhoff, K.; Bilevych, Y.; Desch, K.; van der Graaf, H.; Hartjes, F.; Kaminski, J.; Kluit, P.M.; Raven, G.; Schiffer, T.; et al. Performance of a GridPix detector based on the Timepix3 chip. *Nucl. Instrum. Methods Phys. Res. A* **2018**, *908*, 18–23. [CrossRef]
37. Lupberger, M. The Pixel-TPC: First results from an 8-InGrid module. *J. Instrum.* **2014**, *9*, C01033. [CrossRef]
38. Kim, D.E.; Di Marco, A.; Soffitta, P.; Costa, E.; Fabiani, S.; Muleri, F.; Ratheesh, A.; La Monaca, F.; Rankin, J.; Del Monte, E.; et al. The future of X-ray polarimetry towards the 3-Dimensional photoelectron track reconstruction. *J. Instrum.* **2024**, *19*, C02028. [CrossRef]
39. Xie, F.; Di Marco, A.; La Monaca, F.; Liu, K.; Muleri, F.; Bucciantini, N.; Romani, R.W.; Costa, E.; Rankin, J.; Soffitta, P.; et al. Vela pulsar wind nebula X-rays are polarized to near the synchrotron limit. *Nature* **2022**, *612*, 658–660. [CrossRef] [PubMed]
40. Bucciantini, N.; Ferrazzoli, R.; Bachetti, M.; Rankin, J.; Di Lalla, N.; Sgrò, C.; Omodei, N.; Kitaguchi, T.; Mizuno, T.; Gunji, S.; et al. Simultaneous space and phase resolved X-ray polarimetry of the Crab pulsar and nebula. *Nat. Astron.* **2023**, *7*, 602–610. [CrossRef]
41. Romani, R.W.; Wong, J.; Di Lalla, N.; Omodei, N.; Xie, F.; Ng, C.Y.; Ferrazzoli, R.; Di Marco, A.; Bucciantini, N.; Pilia, M.; et al. The Polarized Cosmic Hand: IXPE Observations of PSR B1509-58/MSH 15-5^2. *Astrophys. J.* **2023**, *957*, 23. [CrossRef]
42. Dyks, J.; Harding, A.K.; Rudak, B. Relativistic Effects and Polarization in Three High-Energy Pulsar Models. *Astrophys. J.* **2004**, *606*, 1125–1142. [CrossRef]
43. Pétri, J. Phase-resolved polarization properties of the pulsar striped wind synchrotron emission. *Mon. Not. R. Astron. Soc.* **2013**, *434*, 2636–2644. [CrossRef]
44. Vink, J.; Prokhorov, D.; Ferrazzoli, R.; Slane, P.; Zhou, P.; Asakura, K.; Baldini, L.; Bucciantini, N.; Costa, E.; Di Marco, A.; et al. X-Ray Polarization Detection of Cassiopeia A with IXPE. *Astrophys. J.* **2022**, *938*, 40. [CrossRef]
45. Ferrazzoli, R.; Slane, P.; Prokhorov, D.; Zhou, P.; Vink, J.; Bucciantini, N.; Costa, E.; Di Lalla, N.; Di Marco, A.; Soffitta, P.; et al. X-Ray Polarimetry Reveals the Magnetic-field Topology on Sub-parsec Scales in Tycho's Supernova Remnant. *Astrophys. J.* **2023**, *945*, 52. [CrossRef]
46. Zhou, P.; Prokhorov, D.; Ferrazzoli, R.; Yang, Y.J.; Slane, P.; Vink, J.; Silvestri, S.; Bucciantini, N.; Reynoso, E.; Moffett, D.; et al. Magnetic Structures and Turbulence in SN 1006 Revealed with Imaging X-ray Polarimetry. *Astrophys. J.* **2023**, *957*, 55. [CrossRef]
47. Kislat, F.; Clark, B.; Beilicke, M.; Krawczynski, H. Analyzing the data from X-ray polarimeters with Stokes parameters. *Astropart. Phys.* **2015**, *68*, 45–51. [CrossRef]
48. Bykov, A.M.; Ellison, D.C.; Osipov, S.M.; Pavlov, G.G.; Uvarov, Y.A. X-ray Stripes in Tycho's Supernova Remnant: Synchrotron Footprints of a Nonlinear Cosmic-ray-driven Instability. *Astrophys. J. Lett.* **2011**, *735*, L40. [CrossRef]
49. Krawczynski, H.; Muleri, F.; Dovčiak, M.; Veledina, A.; Rodriguez Cavero, N.; Svoboda, J.; Ingram, A.; Matt, G.; Garcia, J.A.; Loktev, V.; et al. Polarized x-rays constrain the disk-jet geometry in the black hole X-ray binary Cygnus X-1. *Science* **2022**, *378*, 650–654. [CrossRef]
50. Ratheesh, A.; Dovčiak, M.; Krawczynski, H.; Podgorný, J.; Marra, L.; Veledina, A.; Suleimanov, V.; Rodriguez Cavero, N.; Steiner, J.; Svoboda, J.; et al. The high polarisation of the X-rays from the Black Hole X-ray Binary 4U 1630-47 challenges standard thin accretion disc scenario. *arXiv* **2023**, arXiv:2304.12752. [CrossRef].
51. Rodriguez Cavero, N.; Marra, L.; Krawczynski, H.; Dovčiak, M.; Bianchi, S.; Steiner, J.F.; Svoboda, J.; Capitanio, F.; Matt, G.; Negro, M.; et al. The First X-ray Polarization Observation of the Black Hole X-ray Binary 4U 1630-47 in the Steep Power Law State. *arXiv* **2023**, arXiv:2305.10630. [CrossRef].
52. Veledina, A.; Muleri, F.; Poutanen, J.; Podgorný, J.; Dovčiak, M.; Capitanio, F.; Churazov, E.; De Rosa, A.; Di Marco, A.; Forsblom, S.; et al. Astronomical puzzle Cyg X-3 is a hidden Galactic ultraluminous X-ray source. *arXiv* **2023**, arXiv:2303.01174. [CrossRef].
53. Farinelli, R.; Fabiani, S.; Poutanen, J.; Ursini, F.; Ferrigno, C.; Bianchi, S.; Cocchi, M.; Capitanio, F.; De Rosa, A.; Gnarini, A.; et al. Accretion geometry of the neutron star low mass X-ray binary Cyg X-2 from X-ray polarization measurements. *Mon. Not. R. Astron. Soc.* **2023**, *519*, 3681–3690. [CrossRef]
54. Cocchi, M.; Gnarini, A.; Fabiani, S.; Ursini, F.; Poutanen, J.; Capitanio, F.; Bobrikova, A.; Farinelli, R.; Paizis, A.; Sidoli, L.; et al. Discovery of strongly variable X-ray polarization in the neutron star low-mass X-ray binary transient XTE J1701$-$462. *Astron. Astrophys.* **2023**, *674*, L10. [CrossRef]
55. Fabiani, S.; Capitanio, F.; Iaria, R.; Poutanen, J.; Gnarini, A.; Ursini, F.; Farinelli, R.; Bobrikova, A.; Steiner, J.F.; Svoboda, J.; et al. Discovery of a variable energy-dependent X-ray polarization in the accreting neutron star GX 5-1. *arXiv* **2023**, arXiv:2310.06788. [CrossRef].
56. La Monaca, F.; Di Marco, A.; Poutanen, J.; Bachetti, M.; Motta, S.E.; Papitto, A.; Pilia, M.; Xie, F.; Bianchi, S.; Bobrikova, A.; et al. Highly Significant Detection of X-Ray Polarization from the Brightest Accreting Neutron Star Sco X-1. *arXiv* **2023**, arXiv:2311.06359. [CrossRef].
57. Capitanio, F.; Fabiani, S.; Gnarini, A.; Ursini, F.; Ferrigno, C.; Matt, G.; Poutanen, J.; Cocchi, M.; Mikusincova, R.; Farinelli, R.; et al. Polarization Properties of the Weakly Magnetized Neutron Star X-ray Binary GS 1826-238 in the High Soft State. *Astrophys. J.* **2023**, *943*, 129. [CrossRef]
58. Ursini, F.; Marinucci, A.; Matt, G.; Bianchi, S.; Marin, F.; Marshall, H.L.; Middei, R.; Poutanen, J.; Rogantini, D.; De Rosa, A.; et al. Mapping the circumnuclear regions of the Circinus galaxy with the Imaging X-ray Polarimetry Explorer. *Mon. Not. R. Astron. Soc.* **2023**, *519*, 50–58. [CrossRef]

59. Di Marco, A.; La Monaca, F.; Poutanen, J.; Russell, T.D.; Anitra, A.; Farinelli, R.; Mastroserio, G.; Muleri, F.; Xie, F.; Bachetti, M.; et al. First Detection of X-Ray Polarization from the Accreting Neutron Star 4U 1820-303. *Astrophys. J. Lett.* **2023**, *953*, L22. [CrossRef]
60. Caiazzo, I.; Heyl, J. Polarization of accreting X-ray pulsars. I. A new model. *Mon. Not. R. Astron. Soc.* **2021**, *501*, 109–128. [CrossRef]
61. Caiazzo, I.; Heyl, J. Polarization of accreting X-ray pulsars—II. Hercules X-1. *Mon. Not. R. Astron. Soc.* **2021**, *501*, 129–136. [CrossRef]
62. Tsygankov, S.S.; Doroshenko, V.; Poutanen, J.; Heyl, J.; Mushtukov, A.A.; Caiazzo, I.; Di Marco, A.; Forsblom, S.V.; González-Caniulef, D.; Klawin, M.; et al. The X-Ray Polarimetry View of the Accreting Pulsar Cen X-3. *Astrophys. J. Lett.* **2022**, *941*, L14. [CrossRef]
63. Forsblom, S.V.; Poutanen, J.; Tsygankov, S.S.; Bachetti, M.; Di Marco, A.; Doroshenko, V.; Heyl, J.; La Monaca, F.; Malacaria, C.; Marshall, H.L.; et al. IXPE Observations of the Quintessential Wind-accreting X-ray Pulsar Vela X-1. *Astrophys. J. Lett.* **2023**, *947*, L20. [CrossRef]
64. Tsygankov, S.S.; Doroshenko, V.; Mushtukov, A.A.; Poutanen, J.; Di Marco, A.; Heyl, J.; La Monaca, F.; Forsblom, S.V.; Malacaria, C.; Marshall, H.L.; et al. X-ray pulsar GRO J1008−57 as an orthogonal rotator. *Astron. Astrophys.* **2023**, *675*, A48. [CrossRef]
65. Taverna, R.; Turolla, R.; Muleri, F.; Heyl, J.; Zane, S.; Baldini, L.; González-Caniulef, D.; Bachetti, M.; Rankin, J.; Caiazzo, I.; et al. Polarized x-rays from a magnetar. *Science* **2022**, *378*, 646–650. [CrossRef] [PubMed]
66. Zane, S.; Taverna, R.; González-Caniulef, D.; Muleri, F.; Turolla, R.; Heyl, J.; Uchiyama, K.; Ng, M.; Tamagawa, T.; Caiazzo, I.; et al. A Strong X-Ray Polarization Signal from the Magnetar 1RXS J170849.0-400910. *Astrophys. J. Lett.* **2023**, *944*, L27. [CrossRef]
67. Turolla, R.; Taverna, R.; Israel, G.L.; Muleri, F.; Zane, S.; Bachetti, M.; Heyl, J.; Di Marco, A.; Gau, E.; Krawczynski, H.; et al. IXPE and XMM-Newton observations of the Soft Gamma Repeater SGR 1806-20. *arXiv* **2023**, arXiv:2308.01238. [CrossRef].
68. Heyl, J.; Taverna, R.; Turolla, R.; Israel, G.L.; Ng, M.; Kırmızıbayrak, D.; González-Caniulef, D.; Caiazzo, I.; Zane, S.; Ehlert, S.R.; et al. The detection of polarized x-ray emission from the magnetar 1E 2259+586. *Mon. Not. R. Astron. Soc.* **2023**, *527*, 12219–12231. [CrossRef]
69. Haardt, F.; Maraschi, L. X-Ray Spectra from Two-Phase Accretion Disks. *Astrophys. J.* **1993**, *413*, 507. [CrossRef]
70. Haardt, F.; Matt, G. X-ray polarization in the two-phase model for AGN and X-ray binaries. *Mon. Not. R. Astron. Soc.* **1993**, *261*, 346–352. [CrossRef]
71. Marinucci, A.; Muleri, F.; Dovciak, M.; Bianchi, S.; Marin, F.; Matt, G.; Ursini, F.; Middei, R.; Marshall, H.L.; Baldini, L.; et al. Polarization constraints on the X-ray corona in Seyfert Galaxies: MCG-05-23-16. *Mon. Not. R. Astron. Soc.* **2022**, *516*, 5907–5913. [CrossRef]
72. Sunyaev, R.A.; Markevitch, M.; Pavlinsky, M. The center of the Galaxy in the recent past—A view from GRANAT. *Astrophys. J.* **1993**, *407*, 606–610. [CrossRef]
73. Churazov, E.; Sunyaev, R.; Sazonov, S. Polarization of X-ray emission from the Sgr B2 cloud. *Mon. Not. R. Astron. Soc.* **2002**, *330*, 817–820. [CrossRef]
74. Marin, F.; Churazov, E.; Khabibullin, I.; Ferrazzoli, R.; Di Gesu, L.; Barnouin, T.; Di Marco, A.; Middei, R.; Vikhlinin, A.; Costa, E.; et al. X-ray polarization evidence for a 200-year-old flare of Sgr A*. *Nature* **2023**, *619*, 41–45. [CrossRef]
75. Liodakis, I.; Marscher, A.P.; Agudo, I.; Berdyugin, A.V.; Bernardos, M.I.; Bonnoli, G.; Borman, G.A.; Casadio, C.; Casanova, V.; Cavazzuti, E.; et al. Polarized blazar X-rays imply particle acceleration in shocks. *Nature* **2022**, *611*, 677–681. [CrossRef]
76. Di Gesu, L.; Donnarumma, I.; Tavecchio, F.; Agudo, I.; Barnounin, T.; Cibrario, N.; Di Lalla, N.; Di Marco, A.; Escudero, J.; Errando, M.; et al. The X-ray Polarization View of Mrk 421 in an Average Flux State as Observed by the Imaging X-ray Polarimetry Explorer. *Astrophys. J. Lett.* **2022**, *938*, L7. [CrossRef]
77. Di Gesu, L.; Marshall, H.L.; Ehlert, S.R.; Kim, D.E.; Donnarumma, I.; Tavecchio, F.; Liodakis, I.; Kiehlmann, S.; Agudo, I.; Jorstad, S.G.; et al. Discovery of X-ray polarization angle rotation in the jet from blazar Mrk 421. *Nat. Astron.* **2023**, *7*, 1245–1258. [CrossRef]
78. Middei, R.; Liodakis, I.; Perri, M.; Puccetti, S.; Cavazzuti, E.; Di Gesu, L.; Ehlert, S.R.; Madejski, G.; Marscher, A.P.; Marshall, H.L.; et al. X-ray Polarization Observations of BL Lacertae. *Astrophys. J. Lett.* **2023**, *942*, L10. [CrossRef]
79. Zhang, H.; Böttcher, M. X-Ray and Gamma-Ray Polarization in Leptonic and Hadronic Jet Models of Blazars. *Astrophys. J.* **2013**, *774*, 18. [CrossRef]
80. Peirson, A.L.; Negro, M.; Liodakis, I.; Middei, R.; Kim, D.E.; Marscher, A.P.; Marshall, H.L.; Pacciani, L.; Romani, R.W.; Wu, K.; et al. X-ray Polarization of BL Lacertae in Outburst. *Astrophys. J. Lett.* **2023**, *948*, L25. [CrossRef]

instruments

Review

Scattering Polarimetry in the Hard X-ray Range

Enrico Costa

Istituto di Astrofisica e Planetologia Spaziali-INAF, Via del Fosso del Cavaliere 100, 00133 Roma, Italy; enrico.costa@inaf.it

Abstract: In one and a half years, the Imaging X-ray Polarimetry Explorer has demonstrated the role and the potentiality of Polarimetry in X-ray Astronomy. The next steps include extension to higher energies. There is margin for an extension of the photoelectric approach up to 20–25 keV, but above that energy the only technique is Compton Scattering. Grazing incidence optics can focus photons up to 80 keV, not excluding a marginal extension to 150–200 keV. Given the physical constraints involved, the passage from photoelectric to scattering approach can make less effective the use of optics because of the high background. I discuss the choices in terms of detector design to mitigate the problem and the guidelines for future technological developments.

Keywords: X-ray astronomy; X-ray polarimetry; space instrumentation

Citation: Costa, E. Scattering Polarimetry in the Hard X-ray Range. *Instruments* **2024**, *8*, 20. https://doi.org/10.3390/instruments8010020

Academic Editors: Matteo Duranti and Valerio Vagelli

Received: 7 January 2024
Revised: 11 February 2024
Accepted: 13 February 2024
Published: 2 March 2024

1. Introduction

Results of the Imaging Polarimetry Explorer eventually demonstrated, after 60 years of predictions, that X-ray polarimetry can be a powerful diagnostic for most classes of sources in the domain of High Energy Astrophysics. A short history of this subject can be found in [1]. The breakthrough performance of IXPE is due to a detector exploiting the photoelectric process. Measuring both the interaction point and the angle, the Gas Pixel Detector is suitable to be used as a focal plane detector [2]. For the future, we can predict a more extensive use of polarimetry techniques in X-ray Astronomy. This can include:

- a better exploitation in the IXPE band, with a larger area, as in the enhanced X-ray Timing and Polarimetry Mission [3], better angular resolution and faster operations, and
- the design of wide field instruments.

But both theoretic predictions and IXPE data suggest that an important step forward is the opening of the band above 10 keV. The photoelectric technique can be extended up to 20–25 keV [4,5], but most of instruments are based on scattering. Extensive reviews of scattering polarimetry can be found in Chattopadhyay (2021), Del Monte (2023) [6,7] and, for Gamma-Ray Bursts, in McConnell (2017) [8]. In this paper, I discuss how and when a polarimeter based on Compton scattering was and can be conceived, which implementations have been realized so far and which technical developments are needed in view of another future breakthrough. Conceptually every Compton Telescope, namely every instrument conceived to derive the direction of a photon from the kinematics of Compton scattering between two detecting units of the instrument, is by definition also a polarimeter and is out of this presentation. I only discuss those instruments that can be considered as an extension of the IXPE band, so I neglect the instruments operating only above 100–150 keV.

2. Plenty of Configurations

Any polarimeter is based on an analyzer, namely a material subject to a physical process that depends on polarization, and all the needed equipment to define the direction of the input radiation to detect the output radiation and to record, somehow, the angle selected by the interaction. In the optical domain, this is typically a rotating filter interposed in the path from the optics to the detector. The modulation of the rate with angle, typically following a (cos^2) law, is the basis of the measurement of polarization. This is named a *dispersive* polarimeter in the sense that each angle is sampled at one time and the measurement needs one (or possibly several) complete rotation to provide a result. Also, in the optical band, there may be filters or polarizing prisms at fixed angles. In this case, the polarimeter is sampling three or four angles of the modulation curve, and this is sufficient to measure the polarization at every moment. This is a *not dispersive* polarimeter. In general, a polarimeter based on scattering is composed of the following components:

- A *scatterer*, namely a block of material toward which the input radiation is conveyed by an optics or a collimator.
- An *absorber*, namely a detector capable to detect the scattered photon and possibly measure angles and energy. Depending on the experiment concept, the scatterer/absorber configuration can be single or multiple, namely replicated to achieve a large area.

A scattering polarimeter is *not dispersive* in the sense that most of the angles are sampled simultaneously. A major difference with respect to photoelectric polarimeters is that some angles are forbidden or covered non-uniformly for several reasons, including mechanical mounting, or the different self-absorption within the scatterer, and the geometry of detecting arrays. Consequently, the coverage of angles is different. This is the source of serious complications that can be faced in different ways but, in any case, make significantly cumbersome the analysis of data. On the basis of the physics of the interactions, we can also identify two groups:

- *One phase*: Same material for the analyzer (scatterer) and the absorber.
- *Two phases*: Different materials for the analyzer and the absorber.

  Another way of subdividing is as follows:

- *Active scatterer*, when the scatterer is a detector to be put in coincidence with the absorber.
- *Passive Scatterer*, when the scatterer is an inert material.

  A further division is the following:

- *Wide field* to monitor wide regions of the sky and detect sources from unpredicted directions, such as Gamma-Ray Burst.
- *Narrow field* to study a source at a time. These can include large area detectors with a *collimator* or instruments for the *focal plane* of a telescope.

  Two last divisions are not strictly technical. A polarimeter can be one of the two:

- *Dedicated*: designed and built to perform polarimetry.
- *Byproduct*: designed and built for some other purpose also performing some polarimetry.

An instrument not designed for polarimetry can also offer some information on scattering events, so, in principle, can perform some polarimetry. Historically, some use of this type was proposed. These instruments as polarimeters are much less sensitive and/or reliable than a dedicated polarimeter but, of course, have more chances to arrive in the orbit. I name them *byproduct*. Lastly, a polarimeter can be the following:

- *Stand-alone*, namely aboard a dedicated satellite.
- Part of a *multi-instrument* payload.

The problem of systematics and of uneven coverage of angles is usually solved with the rotation of the instrument around the observation axis. Of course, this is not feasible with instruments devoted to Gamma-Ray Bursts, given that the direction is unknown. Also, *byproduct* polarimetry based on imagers cannot benefit from rotation. All these configurations have been proposed or studied. A few have been implemented. Very few have arrived to be real experiment. I mainly review these configurations and propose my personal view for the future.

3. Basic Statistics and Physics

3.1. The Basic Statistics

To discuss the various configurations, I recall the basic statistics of detection of polarization in a regime of Poisson distribution, that can be found in several publications as in Weisskopf (2010), Strohmayer (2013) or Muleri (2022) [9–11]. The parameter driving the observing strategy and quantifying the scientific performance is the Minimum Detectable Polarization, namely the polarization to be exceeded to keep the probability of statistical fluctuation below a certain value. The general convention is to offer the *MDP* at 99%,

$$MDP = \frac{4.29}{\mu \varepsilon S} \times \sqrt{\frac{\varepsilon S + B}{T}} \qquad (1)$$

where ε is the efficiency of the instrument, S the flux of the source, B is the background rate, T is observing time. μ is the modulation factor, the parameter measuring the response of the instrument to a 100% polarized source. $\mu = 1$ for an ideal analyzer. Except the time, all the parameters in the equation are energy dependent and the proper convolution integrals should be used instead of the variable, but for the purpose of this discussion, I use this simplified formalism. Also, in the literature, as in the papers presenting the IXPE results, data are analyzed and results are shown with the formalism of Stokes Parameters, coherently with the use in other wavelengths. This has many advantages in performing the analysis and showing the results [11], but would be a useless complication here. So I will carry on the discussion in terms of Polarization Degree and Angle. Starting from the interaction cross-sections, I discuss the value that can be achieved for these parameters with the various above-mentioned configurations of scattering polarimeters.

3.2. The Basic Physics

I follow the approach of Fabiani (2014) [12]. From the Compton formula, the energy of the incoming photon E and the energy of the scattered photon E' are connected through the polar scattering angle θ.

$$E' = \frac{E}{1 + \frac{E}{mc^2}(1 - \cos\theta)}. \qquad (2)$$

The difference E-E' of the energy of the photons is given to an electron of the scatterer, which is stopped with a range much shorter than the interaction length of the X-ray photon. In practice, for the sake of discussion, with a reasonable approximation, we can assume a local energy loss for the electron the angular distributions for scattering on free electrons for the emerging photon.

The polarization of the incoming photon determines the azimuth distribution. The Klein Nishina formula,

$$\frac{d\sigma}{d\Omega} = \frac{r_0^2}{2} \frac{E'^2}{E^2} \left[\frac{E}{E'} + \frac{E'}{E} - 2\sin^2\theta\cos^2\varphi \right] \qquad (3)$$

gives the angular distribution of the scattered photons. φ is the azimuth scattering angle. The distribution in θ is independent from polarization, while the distribution in φ is dependent and has the maximum for azimuth angle defining a plane of scattering perpendicular to the polarization of the photon. A complete treatment can be found in [13]. From these equations, the two distributions can be derived which are the most relevant for our discussion. One is the modulation (around azimuth angle φ) as a function of the polar scattering angle and of the energy as shown in Figure 1.

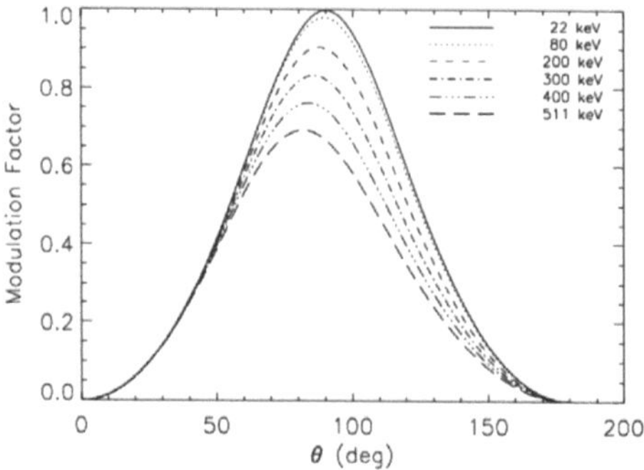

Figure 1. Compton scattering has a high modulation around 90° decreasing to 0 for forward and backward scattering. From Fabiani (2013) [12].

From the figure, it is clear that, since μ is the parameter with the maximum impact on sensitivity in Equation (1), the photons scattered around 90° are the most useful ones. On the other side, the photons which are not collected do not contribute to efficiency ϵ. Every scattering polarimeter limits the accepted paths for the scattered photons, trying to optimize the MDP. The geometric configuration determines the scattering angles accepted and fixes the trade-off between the two parameters. Given that both μ and ϵ depend on energy, the trade-off configuration is energy dependent and the design of the experiment is based on a hypothetical optimization of the total scientific throughput of the mission. With a more ambitious approach, viable with nowadays technology, when the point of scattering and the point of absorption can be measured, this information can be used by assigning to each event a weight (substantially proportional to μ), but this is not easy at all.

Given that the polarimeter is designed to accept mostly photons scattered at angles around 90°, it is interesting to see the energies involved. The energy given to an electron in the scatterer is

$$T_e = E - E' = E\frac{\gamma(1 - \cos\theta)}{1 + \gamma(1 - \cos\theta)} \tag{4}$$

This is the second distribution driving the design. In all cases of interest for the discussion, this energy given to the electron is at maximum of a few tens of keV. With solid detectors, it can be assumed that it is converted in ionization or excitation within a few microns.

In Figure 2, this energy is plotted for angles around 90°. I notice that for photons of energy < 20 keV, the energy lost in the scatterer is less than 1 keV. A detailed description of this point can be found in Chattopadhyay (2021) [6].

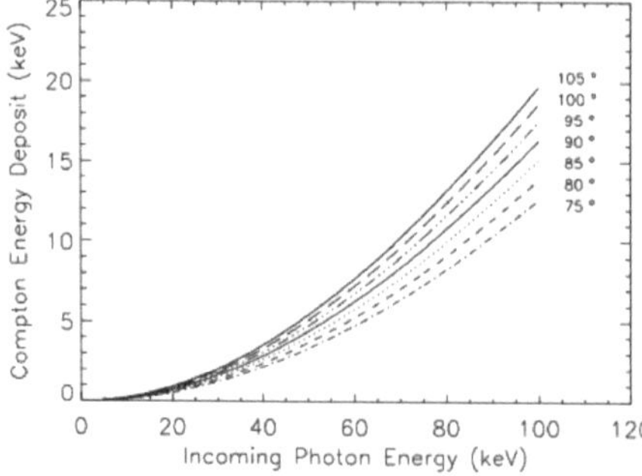

Figure 2. The angles of scattering around 90° are those more interesting for polarimetry. Following Equation (4), the energy transferred to the scatterer can be computed as a function of the energy of the photon and of the scattering angle. From Fabiani (2013) [12].

4. Practical Implementations

In this paper, I only discuss scattering polarimeters sensitive in the hard X-ray band, namely at energies > 15 keV, where the scattering is no more overwhelmed by photoabsorption and <150 keV, not to enter in the γ-ray range. From the equation, this corresponds to a few tens of keV at the high-energy side and to a few keV (or even a fraction of keV) at the low-energy side. I cannot conduct a systematic discussion of all possible configurations or their combination. Therefore, I select some examples of actually implemented instruments or of instruments with an adequate level of study.

The Materials Involved

In a two-phase polarimeter, the scatterer is always a material of low atomic number and of reasonable density (no gas). Lithium and Beryllium are used for passive scattering polarimeters, as well as a plastic scintillator for active scattering polarimeters. Lithium Hydride is, in theory, the best material (lowest Z, denser than Lithium), but it has some instabilities that discourage its use. In any case, both Lithium and Lithium Hydride are hygroscopic and must be encased in a thin Beryllium container. Therefore, some Beryllium is present in any case. In Table 1, the materials used in practice are shown.

Table 1. Materials used as a scatterer in a two-phase polarimeter. The third column displays the energy where the scattering equalizes absorption and in practice where the technique is fully operative. LiH, Li and Be are the favorites for passive scatterer configurations, while a plastic scintillator (or other organic scintillators) is the baseline as an active scatterer.

Material	ρ (g $\cdot$ cm^{-3})	E (keV) Scatt = Absorption
Lithium	0.53	8.7
LiH	0.82	8.2
Beryllium	1.85	14
Plastc Scintillator (PVT)	1.03	20

One-phase polarimeters are (of course) only active. The material must be suitable as a detector. The two basic design include arrays of plastic scintillator and arrays of medium-atomic-number scintillators, such as Tallium activated Cesium Iodide (CsI) or Cerium activated Gadolinium Aluminium Gallium Garnet (GAGG).

5. Without Optics

5.1. Passive Scatterer

Following an historical sequence, the first implementation was conducted by the Columbia Team lead by Robert Novick. The payload was a set of Lithium blocks surrounded by proportional counters. The blocks were aligned with the rocket spinning axis pointed to the source [14,15]. This was, in fact, the very first attempt to perform X-ray polarimetry, and the first of a long sequence of upper limits. The second time, a stage based on Bragg crystals was added to the rocket payload, and from this combination, the first positive detection arrived [15]. The experience of rockets showed that the scattering technique in this implementation was much less sensitive than Bragg, so it was abandoned in the rocket age and in the early satellite age.

Many years later, a little block of Beryllium was set in between some Germanium detectors of the RHESSI mission. The band arrived to low energies (20 keV) due to the use of Be as a scatterer (typical two-phase polarimeter). A certain protection from the high background of direct unscattered photons was achieved with the capability to identify photons absorbed in the lower part of Germanium detectors, so that the upper part acted in practice as a shield [16].

But the results for RHESSI as a polarimeter were modest, basically upper limits, and still confirmed the mismatching of sensitivity of a polarimeter with the other instruments and the consequent poor throughput from what I named a byproduct polarimeter.

Only a dedicated satellite can effectively apply this technique. The POLIX instrument includes a collimator, aligned with the spin axis, a Beryllium scatterer in a well, of four proportional counters, heritage of ASTROSAT. POLIX is hosted aboard the XPoSAT mission by ISRO that was launched on 1 January 2024. The nominal range is 8–30 keV. POLIX [17] is mainly aimed to study bright sources on the basis of pointing of the order of a few weeks, possible with a dedicated satellite.

5.2. Active Scatterer

A polarimeter can be conceived as a combination of detectors. For known sources, a collimator limits the direction of primary photons, while only a large field delimiter is used for bursts. The temporal coincidence identifies the path of the scattered photon from the scatterer to the final absorber. The sum of the two detected energies is the total energy. For tens of years (as in [18,19]), this was merely conceptual. The straightforward implementation is with a low-atomic-number detector as a scatterer and a higher-atomic-number detector as an absorber. Typical pairs are a plastic scintillator and an CsI. Many different configurations have been proposed. The New Hampshire University Team has a long record of proposed and prototypized payloads. The Gamma-Ray polarimeter experiment (GRAPE) based on an array of bars of plastic scintillators surrounded by bars of CsI, read with a multi-anode photomultiplier, was tested aboard balloon flights. With a collimator, it can be used to measure known sources; without a collimator, it can study GRBs. A more recent version uses Si PMTs instead of MAPMTs. A version with seven moduli named a Large Area Burst Polarimeter (LEAP) should be the first such polarimeter to proceed on orbit aboard the ISS [20].

A small experiment for Gamma-Ray Bursts was IKAROS-GAP [21]. It was a single block of a plastic scintillator, surrounded with 12 CsI detectors with individual photo-multipliers acting as absorbers. This was a raw and effective design but robust and well calibrated.

A mission in progress based on the concept of an active scatterer is the CUbesat Solar Polarimeter (CUSP) [22] aimed to develop a constellation of two CubeSats to measure the linear polarisation of solar flares in the hard X-ray band, in progress at IAPS-INAF under the management of Italian Space Agency. The payload is based on an array of bars of a plastic scintillator, surrounded by bars of GAGG, which is faster than CsI. Plastic scintillators are read with four multi-anode photomultipliers, R7600, while the GAGG bars are read with avalanche photo-diodes as shown in Figure 3.

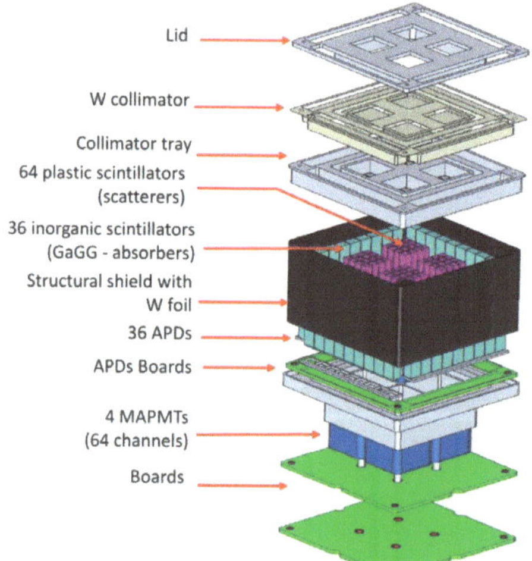

Figure 3. Exploded view of the CUSP payload. Photons from solar flares are scattered on the plastic scintillators and absorbed by GAGG scintillators. From Fabiani (2022) [22].

One-phase active scattering polarimeters using the same material for both functions are conceptually less performing. A good efficiency would be achieved with a high probability of scattering in the first detector and a high probability of absorption in the second. Since the two processes compete, this is not possible by definition. At low energies where the absorption is mainly photoelectric and thence fast depending on the energy, and where the energies of the two processes are very different (as clear from Figure 2), the scattering/absorption is very ineffective. Yet, there is the possibility of a first Compton interaction in a detector and a second Compton interaction in another detector. The probability of this second interaction, after scattering at angles around 90°, can be maximized with a an array of thin wire-like detectors of a large area. The process is also modulated with polarization. The difficulty is that the sum of the two energies lost is lesser than the energy of the incoming photon. The modulation factor depends on the energy, so if the energy of the photon is not known, the conversion from the modulation to the polarization is very ambiguous. In any case, these experiments by simulations and calibrations can produce the polarization on a broad band that is absolutely correct if the spectrum is available from an independent instrument or from another mission.

The best implementation of this concept, based on plastic–plastic scattering, are the balloon payloads of the POGO family [23]. POGO is conceived to observe known sources with a narrow field of view. This is achieved with a tight passive/active collimator and with a heavy anticoincidence shield. In fact, POGO is the only one achieving results on discrete sources in the hard X-ray range [24].

A strong argument in favor of the plastic–plastic configuration is the possibility of having large arrays of wire-like scintillators with fine subdivision using the same photonic device reading all the sensing units. This allows for a better use of space and makes everything simpler, from the alignment to the optical contacts to the readout electronics. The best implementation of this approach for a wide field instrument was POLAR [25,26], an array of plastic scintillator wires, read with a multi-anode photo multiplier. Flown aboard the Chinese space lab, POLAR was a very successful mission, the best for GRB polarimetry, but in a typical γ-ray band, marginal to our range of interest. But, POLAR-2, a new version in an advanced stage [27], will increase the area and use Silicon photo multipliers as in Figure 4. The lower energy threshold in POLAR-2 is somewhere between 20 and 30 keV, as shown in Figure 5, an interesting extension of the technique toward the X-ray band.

POLAR-2

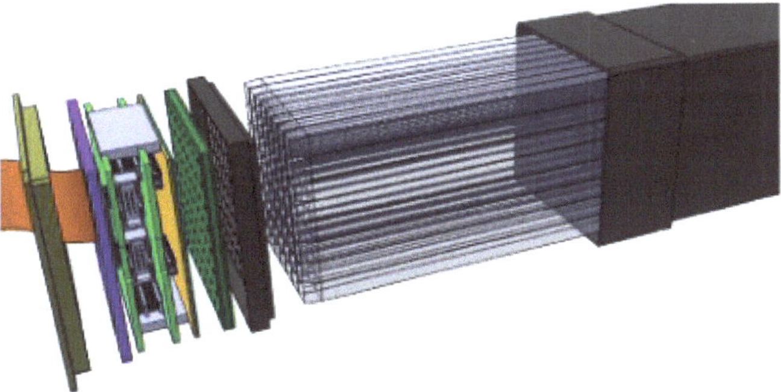

Figure 4. POLAR-2 is an assembly of moduli similar to POLAR but 4 times larger. The main difference is the use of Silicon Photomultipliers, allowing for a significant decrease in the threshold on the first interaction and, as a consequence, a very effective decrease in the low energy threshold of the whole. From Kole (2019) [27].

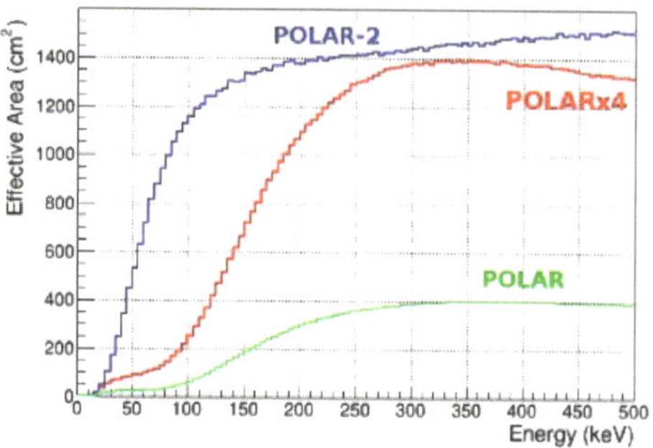

Figure 5. POLAR-2 with respect of POLAR has a 4-fold larger area and a significantly lower threshold. From Kole (2019) [27].

5.3. Byproduct Polarimetry

By byproduct polarimetry I mean instruments designed and built for some other purpose also conducting some polarimetry. Structured instruments sometimes include intermediate data that contain information on linear polarization. If these data are transmitted (by original design or by late additions), the instrument can be used as a polarimeter.

Given that polarization is more difficult to detect than spectra, images or timing of the technique usually apply to a very limited subset of the brightest sources.

Moreover, polarimetry requires an extreme (almost maniacal) care in the prevention of systematics that is absent in candidate byproduct polarimeters. In most cases, it does not work at all. The only substantial exception is ASTROSAT [28]. The Cadmium–Zinc–Telluride Imager (CZTI) is a hard X-ray coded mask camera working in the band of 10–100 keV. Pixels of CZT, 5 mm thick, have a reasonable fraction of Compton interactions at higher energies. Some of these scattered photons are absorbed by other pixels. Laboratory tests showed that the corrected angular distribution is modulated by polarization. One problem of such an approach is that the distribution is sensitive to the interaction point, and this can be very critical in a focal plane instrument. But, in the case of ASTROSAT, this is substantially mitigated with a parallel beam. Moreover, even though the instrument was calibrated as a polarimeter before the launch only on the axis [28,29], all the simulated response, including the dependence of this modulation on the offset angle, was verified with measurements performed on ground on a representative physical model [30,31]. So, in this case, we have the needed reliability, but, of course, the point that it only works with very strong sources holds.

6. In the Focal Plane

6.1. Optics in X-ray Astronomy and Optics in X-ray Polarimetry

The introduction of optics was the turning point in X-ray Astronomy as proposed by Riccardo Giacconi soon after the first discovery. With the mission *Einstein* in 1978, X-ray Astronomy achieved the capability to image extended sources [32]. But the major breakthrough was the capability to detect very weak sources because, with an imaging detector in the focus, the flux of the source is compared with the fluctuations of the background in the point spread function and not on the whole detector or one half of that (as in experiments with collimators or coded masks). The conventional polarimeters, based on Bragg diffraction, were totally mismatched in sensitivity with imagers and found no more place in multi-instrument missions. Since then, the the path to the polarimetry of known sources has been the quest for an imaging detector. This is based on photoelectric effect at low energies. But the technology of X-ray optics extends to hard X-rays due to multi-layer technology, and in this energy range, the viable process is scattering.

IXPE was possible because the Gas Pixel Detector allows for the reconstruction of the impact point of the photon and the angle of ejection of the photoelectron [2,33]. This means that to the counts from a point-like source background, counts are added from a surface of the order of 0.5 mm^2, namely of less than 20 µCrab. In practice, the background has no impact on sensitivity for any point-like source when integrating on the 2–8 keV band. While IXPE has demonstrated that focal plane photoelectric polarimetry is viable, the equivalent for scattering has many criticalities, most of all the poor localization of the first interaction. Here, I discuss how these affect the concept and how they can be mitigated or overcome. Any focal plane scattering polarimeter is a scatterer centered on the focus or near it surrounded with detectors. The optimal design is a cylindrical scatterer long enough to provide a reasonable efficiency and large enough to include all the divergent beams from the telescope and any possible misalignment. In practice, the scatterer needs to have a length of several cm and a diameter of <1 cm. An ideal detector should be cylindrical itself as shown in Figure 6, but in practical implementations, major or minor deviations from this geometry were and are needed.

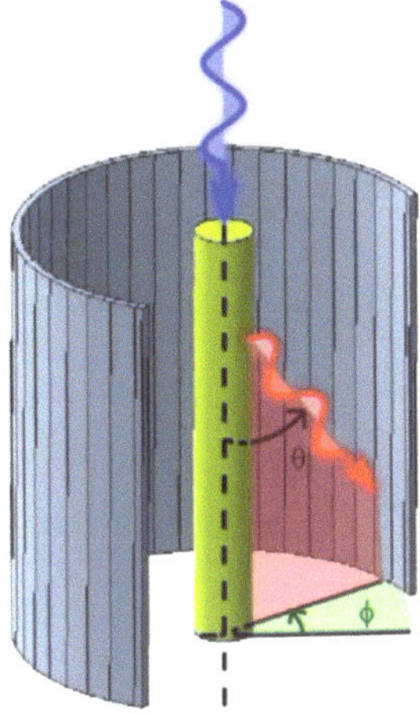

Figure 6. A focal plane scattering polarimeter is always a cylindrical scatterer, centered on the axis and in the focal plane, surrounded with a well of detectors, ideally of cylindrical geometry. The scatterer can be a detector itself. In this case, it is named an Active Scatterer Focal Polarimeter. From Fabiani (2012) [5].

6.2. Passive Scatterer in Focus

The ambitious Spectrum X-Gamma (SRG) of the Soviet Union hosted two large telescopes manufactured in Denmark [34]. The focal plane of one of them hosted the Stellar X-ray Polarimetry (SXRP) lead by Robert Novick [35], with a contribution of Italian teams. At the focus there was a scatterer of Lithium, encased in Beryllium, surrounded with a well of four proportional counters. In order to compensate possible misalignment [36], the detectors were positioned relatively far from the scatterer, and this was larger than the convergence of the beam. SXRP was the first exploitation of the optics in polarimetry. Starting from an area of the optics of around 1000 cm^2, the effective area of the polarimeter was around 50 cm^2, still a considerable value. But the background rate, due to the large area of the detectors, ranged between one-fourth and one half of the rate from the Crab. Therefore, the advantage of being in the focal plane was effective only for a few bright sources; a step forward with respect to OSO-8 but not yet a breakthrough.

SXRP was built and tested until acceptance, but the SRG satellite was never completed and flown. The work of calibration and simulation [37], however, performed for SXRP was a good basis for the future proposals of X-ray polarimetry.

A straightforward consequence was that the system should be more compact than SXRP. But this would not be sufficient. In the interplay between efficiency and background, the first problem is the thickness of the scatterer. A lithium scatterer, to have reasonable efficiency, must have a length of more or less 10 cm. On the other hand, a system that is too long accepts photons scattered at large polar angles, which are poorly modulated. A scatterer of Beryllium could be long, about one-third but not less, given that at 30 keV, 3 cm of Be are transparent to 45% of photons. On the contrary, most of the decisions on modulation vs efficiency vs background trade-off, gives a larger value. The well of detectors must be as long as the scatterer. Also, a design tighter then SXRP has a radius of centimeters, and thence a surface of tens of square centimeters, nothing to do with p.s.f. of a fraction of a square millimeter of photoelectric low-energy detectors. The ratio of S/B in Equation (1) is not reduced at the same level. This is a simple truth, directly derived from cross-sections. A design achieving an optimal trade-off between the efficiency and the surface of the absorber will never escape to this. Moreover, in any detector, the instrumental background increases with energy. The realistic limit to the sample of targets available for these instruments is the flux for which the counts from the source are equal to those of the background. This limit can be lowered with a compact design and with techniques of background reduction. The best implementation of this concept, 20 years or more after SXRP, is the X-Calibur [38] mission and its evolution XL-Calibur [39], a scattering polarimeter in the focus of a multi-layer telescope onboard a stratospheric balloon, clearly also conceived as a pathfinder for a future satellite mission [40]. The X-Calibur telescope has a focal length of 8 m and an effective area of 93 cm^2 at 20 keV. In the focus, a stick of Beryllium is the scatterer surrounded from a square well of CZT detectors acting as absorbers. The whole is surrounded with a CsI anticoincidence.

A flight from Antartica in 2018, with the observation of the bright source, GX301-2 [39], demonstrated the functionality of the whole but also showed the difficulty to achieve the real breakthrough with the introduction of optics only. The high background rate limited the sample of sources on which the measurement would be significant. This was mainly due to the limited efficiency (also due to the high zenith angle of bright sources at near-polar latitudes) and the high background (also maximum in polar regions). The analysis of this first flight drove the design of the evolved version of the experiment named XL-Calibur [41,42], also with the inclusion of the POGO team. The new focal plane set up is shown in Figure 7. The major improvements are as follows:

- A telescope with an increased collecting area (of 300 cm^2 at 20 keV), also with a longer focal length of 12 m.
- An anticoincidence shield of BGO instead of CsI.
- Thinner detectors to reduce background.

The massive anticoincidence is somehow unavoidable since it is well known that in the hard X-ray range only active shielding with inorganic scintillators can drastically reduce the background.

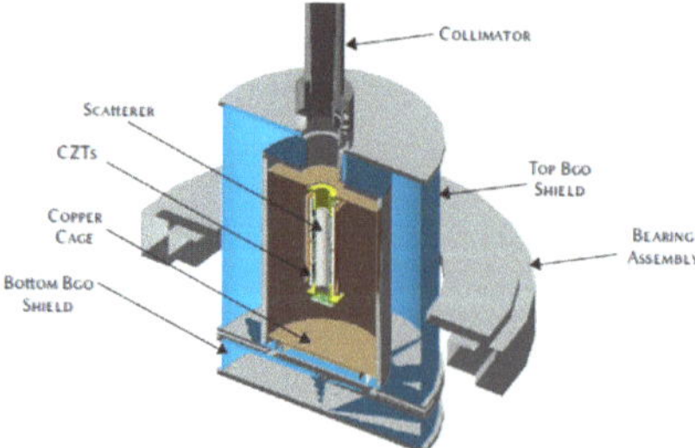

Figure 7. The present configuration of the XL-Calibur focal plane instrument. The Beryllium scatterer is surrounded with four strings of CZT dtectors. All around, a thick BGO shield reduces the background. From Iyer (2023) [42].

With these improvements, the background rate in an arctic balloon should be of the order of 100 mCrab. Of course, a similar configuration aboard a satellite should be more sensitive because of higher efficiency, especially at lower energies.

6.3. Active Scatterer in the Focus

A way to overcome the problem of large background is to have an active scatterer, namely a detector in coincidence with the absorber. The rate of coincidences between the scatterer and the absorber should be much lower than the rate on the absorber only. But this has consequences in terms of efficiency. In order to understand whether this can be, in some cases, a viable solution, the materials involved should be discussed starting from Table 1. No detector exists based on Lithium or Beryllium. So the lowest (in practice, the only) useful materials are organic scintillators, where the scattering element is basically Carbonium. In terms of efficiency and background rate,

- The passive scatterer with Li has the lowest energy energy of transition from photo-absorption to scattering. With Be, the energy of transition is higher; higher still with an organic scintillator. The passive scatterer is more efficient also because of materials, and Lithium is better than Beryllium.
- With the active scatterer, the count rate from the source is lower than that with the passive scatterer with the same materials, because not every event of energy loss provides a signal suited to trigger the readout electronics and switch the coincidence.

In fact, the solution with the active scatterer was the original design of X-Calibur [40]. The scatterer was a stick of a plastic scintillator, 12 cm long, read with a photomultiplier. After a test flight, various measurements and simulations, it was found [38] that the reduction in the background of one order of magnitude was not adequate to compensate for the drastic drop of efficiency due, beside the aforementioned effect of materials, to the low coincidence trigger efficiency. So, eventually, X-Calibur and XL-Calibur went back to the passive scatterer design, which also benefitted from the larger density of Beryllium vs that of the plastic scintillator.

This choice was likely the best in the specific conditions but not necessarily the best for any case. In the literature, the performances of a scatterer of a plastic scintillator were studied at least twice [5,43], also with a certain number of tests. It is not clear how the background can be reduced, but it is evident that the crucial parameters are the overall efficiency and the trigger efficiency. The latter is a matter of energy but also of light collection. A passive scatterer can be made as long as possible, achieving an efficiency not far from one. In an active scatterer, the length is a trade-off to maximize the product of the interaction efficiency by the trigger efficiency and this for sure leads to a shorter scatterer. Both studies mentioned above show that the triggering efficiency increases with the energy of the incoming photon. With an increase in light collection efficiency, an active scattering solution should be more sensitive than a passive scatter one, at least above a certain energy that could range from 20 to 30 keV. Where exactly this occurs is not trivial. With a spectral slope of E^{-3}, more than one half of the photons has 20 keV < E < 30 keV. With a passive scatterer, some photons of <20 keV can also be detected, but, for instance, in a balloon-borne instrument, the atmosphere absorbs most of photons of <30 keV.

Much depends on other factors. In an experiment like XL-Calibur, the design can be optimized on the basis of the instrument by itself. But if the polarimeter is combined with some other instrument peaked on a nearby band, the optimization will be performed for the combination of both, and the choice for the scattering stage can be different. In my opinion, if the photoelectric technique can be extended up to 20–25 keV, for the scattering stage, the active option becomes the best. This implies that the scatterer can be designed to maximize the sensitivity above 25 keV where the active scattering solution will be more effective.

7. The Path to the Future

From the IXPE experience and from the theoretical analysis, X-ray polarimetry is a technique of high scientific impact. This is true for all the three ranges and related detectors where Polarimetry is affordable, namely the low energy in the range of IXPE (2–10 keV), the medium energy still based on GPD with pressurized Argon filling (5–25 keV) [4,5,44] and the high energy based on scattering (20–80 keV). Moreover, the possibility to perform broad-band polarimetry is even more attractive.

A mission with three optics in parallel is possible; the possibility to stack two or more instruments is very useful, also allowing the presence of a telescope devoted to polarimetry and other telescopes pointing the same source to perform spectra and timing. In the past, instead of GPD, a Time Projection Chamber has been proposed for use for the low energies with a rear window of Beryllium, transparent to higher energy photons [44]. After this window, a scattering polarimeter active at high energies has also been hypothesized . The proposal is interesting and for sure can produce good measurements, but this configuration is not imaging in both stages and, after IXPE, giving up the possibility to study SuperNova Remnants, Pulsar Wind Nebulae, reflection clouds and Jets is difficult to accept. On the other hand, the GPD configuration with the drift field on the pointing direction unavoidably has the ASIC chip obstructing the path of higher-energy photons. One possibility with a potentially dramatic impact is to make the ASIC as thin as possible to leave a reasonable transparency to photons above 20 keV. In these chips, a thickness of 100 μm seems feasible and would guarantee the possibility to stack a LEP or a MEP with a scattering polarimeter in the rear.

Also, in a multi-telescope configuration, a combination of stacked instruments can be imaged to maximize the broad band throughput. In such a configuration, combination TPC/LEP - MEP - HEP stacked on the same telescope can be excellent for any point-like source, while the extended sources (Supernova remnants, Pulsar Wind Neblae, jets) of high interest but of limited number could be resolved with all the MEPs and with a single LEP in the imaging configuration. In my opinion, also in an active scattering configuration, a good anticoincidence is also useful for the MEP given that IXPE data also show that the background of photoelectric detectors seriously increases with energy. Last but not least, after achieving the necessary confidence, a Soft X-ray Telescope based on diffraction on a multi-layer with a laterally graded multi-layer component can enlarge the band to 0.15–0.30 keV as proposed, after more than 20 years of study, and described in Marshall (2018) [45].

Incidentally, I notice that a thin silicon device of ≤100 μm such as a Silicon Drift Detector (or, less likely given the high noise a Silicon Photomultiplier), substantially transparent to hard X-rays) can be used to read a cylinder of a scintillator from two sides with the two detectors in coincidence, drastically reducing the threshold and so increasing the trigger efficiency. It is a matter of fact that most of the thickness of these semiconductors is not hosting any electric component, but is needed as mechanical support or to make connections easier.

To conclude, a certain number of Research and Development activities can be the path to future missions of X-ray polarimetry extended to the hard X-ray band by the inclusion of one or more scattering stages.

1. Feasibility of thinner ASIC pixel chips by testing the capability to support mechanical troubles and to allow connections.
2. Feasibility of thinner photonic sensors to use two (or more) of them for the scatterer.
3. Improving the radiation hardness of Si PMTs.
4. Testing the windows of Silicon Nitride.
5. Testing the triggering threshold of long plastic scintillators.
6. Comparing the yield and the transparency of alternative organic scintillators (Anthracene, Stilbene, etc.).
7. Following the progress of single-crystal diamond detectors that have been recently proposed as potential scatterers [46], although they are far from the needed performances.
8. Simulation of the background in satellite orbits and potential anticoincidence materials.

The result of these studies could allow performance of experiments of polarimetry with the goal to achieve a more balanced sensitivity in terms of mCrab with a balanced combination of detectors in different bands.

Funding: This research received no external funding.

Data Availability Statement: No new data were created for this research.

Acknowledgments: This research could benefit of useful discussions with colleagues Ettore Del Monte, Sergio Fabiani, Fabio Muleri and Paolo Soffitta.

Conflicts of Interest: The author declares no conflict of interest.

References

1. Costa, E. General History of X-ray Polarimetry in Astrophysics. In *Handbook of X-ray and Gamma-ray Astrophysics*; Springer-Nature: Berlin/Heidelberg, Germany, 2022; p. 36. [CrossRef]
2. Costa, E.; Soffitta, P.; Bellazzini, R.; Brez, A.; Lumb, N.; Spandre, G. An efficient photoelectric X-ray polarimeter for the study of black holes and neutron stars. *Nature* **2001**, *411*, 662–665. [CrossRef]
3. Zhang, S.; Santangelo, A.; Feroci, M.; Xu, Y.; Lu, F.; Chen, Y.; Feng, H.; Zhang, S.; Brandt, S.; Hernanz, M.; et al. The enhanced X-ray Timing and Polarimetry mission—eXTP. *Sci. China Phys. Mech. Astron.* **2019**, *62*, 29502. [CrossRef]
4. Muleri, F.; Bellazzini, R.; Costa, E.; Soffitta, P.; Lazzarotto, F.; Feroci, M.; Pacciani, L.; Rubini, A.; Morelli, E.; Baldini, L.; et al. An X-ray polarimeter for hard X-ray optics. *Proc. SPIE* **2006**, *6266*, 62662X. [CrossRef]
5. Fabiani, S.; Bellazzini, R.; Berrilli, F.; Brez, A.; Costa, E.; Minuti, M.; Muleri, F.; Pinchera, M.; Rubini, A.; Soffitta, P.; et al. Performance of an Ar-DME imaging photoelectric polarimeter. *Proc. SPIE* **2012**, *8443*, 84431C . [CrossRef]
6. Chattopadhyay, T. Hard X-ray polarimetry—An overview of the method, science drivers, and recent findings. *J. Astrophys. Astron.* **2021**, *42*, 106. [CrossRef]
7. Del Monte, E.; Fabiani, S.; Pearce, M. Compton Polarimetry. In *Handbook of X-ray and Gamma-ray Astrophysics*; Bambi, C., Santangelo, A., Eds.; Springer-Nature: Berlin/Heidelberg, Germany, 2023; p. 127. [CrossRef]
8. McConnell, M.L. High energy polarimetry of prompt GRB emission. *New Astron. Rev.* **2017**, *76*, 1–21. [CrossRef]
9. Weisskopf, M.C.; Elsner, R.F.; O'Dell, S.L. On understanding the figures of merit for detection and measurement of X-ray polarization. *Proc. SPIE* **2010**, *7732*, 77320E. [CrossRef]
10. Strohmayer, T.E.; Kallman, T.R. On the Statistical Analysis of X-ray Polarization Measurements. *Astrophys. J.* **2013**, *773*, 103. [CrossRef]
11. Muleri, F. Analysis of the Data from Photoelectric Gas Polarimeters. In *Handbook of X-ray and Gamma-ray Astrophysics*; Springer-Nature: Berline/Heidelberg, Germany, 2022; p. 6. [CrossRef]
12. Fabiani, S.; Campana, R.; Costa, E.; Del Monte, E.; Muleri, F.; Rubini, A.; Soffitta, P. Characterization of scatterers for an active focal plane Compton polarimeter. *Astropart. Phys.* **2013**, *44*, 91–101. [CrossRef]
13. McMaster, W.H. Matrix Representation of Polarization. *Rev. Mod. Phys.* **1961**, *33*, 8–27. [CrossRef]
14. Angel, J.R.; Novick, R.; vanden Bout, P.; Wolff, R. Search for X-ray Polarization in Sco X-1. *Phys. Rev. Lett.* **1969**, *22*, 861–865. [CrossRef]
15. Novick, R.; Weisskopf, M.C.; Berthelsdorf, R.; Linke, R.; Wolff, R.S. Detection of X-ray Polarization of the Crab Nebula. *Astrophys. J.* **1972**, *174*, L1. [CrossRef]
16. McConnell, M.L.; Ryan, J.M.; Smith, D.M.; Lin, R.P.; Emslie, A.G. RHESSI as a Hard X-ray Polarimeter. *Sol. Phys.* **2002**, *210*, 125–142. .:1022413708738 [CrossRef]
17. Paul, B.; Gopala Krishna, M.R.; Puthiya Veetil, R. POLIX: A Thomson X-ray polarimeter for a small satellite mission. In Proceedings of the 41st COSPAR Scientific Assembly, Istanbul, Turkey, 30 July–7 August 2016; Volume 41, Abstract id E1.15-8-16.
18. McConnell, M.; Forrest, D.; Levenson, K.; Vestrand, W.T. The design of a gamma-ray burst polarimeter. In *AIP Conference Proceedings, Proceedings of the Compton Gamma-ray Observatory, St. Louis, MO, USA, 15–17 October 1992*; American Institute of Physics Conference Series; Friedlander, M., Gehrels, N., Macomb, D.J., Eds.; AIP: Kentwood, MI, USA, 1993; Volume 280, pp. 1142–1146. [CrossRef]
19. Costa, E.; Cinti, M.N.; Feroci, M.; Matt, G.; Rapisarda, M. Design of a scattering polarimeter for hard X-ray astronomy. *Nucl. Instrum. Methods Phys. Res. A* **1995**, *366*, 161–172. [CrossRef]
20. McConnell, M.L.; Baring, M.; Bloser, P.; Briggs, M.S.; Ertley, C.; Fletcher, G.; Gaskin, J.; Gelmis, K.; Goldstein, A.; Grove, E.; et al. The LargE Area burst Polarimeter (LEAP) a NASA mission of opportunity for the ISS. *Proc. SPIE* **2021**, *11821*, 118210P. [CrossRef]
21. Yonetoku, D.; Murakami, T.; Gunji, S.; Mihara, T.; Sakashita, T.; Morihara, Y.; Kikuchi, Y.; Takahashi, T.; Fujimoto, H.; Toukairin, N.; et al. Gamma-Ray Burst Polarimeter (GAP) aboard the Small Solar Power Sail Demonstrator IKAROS. *Publ. Astron. Soc. Jpn.* **2011**, *63*, 625–638. . [CrossRef]

22. Fabiani, S.; Baffo, I.; Bonomo, S.; Contini, G.; Costa, E.; Cucinella, G.; De Cesare, G.; Del Monte, E.; Del Re, A.; Di Cosimo, S.; et al. CUSP: A two CubeSats constellation for space weather and solar flares X-ray polarimetry. *Proc. SPIE* **2022**, *12181*, 121810J. [CrossRef]

23. Friis, M.; Kiss, M.; Mikhalev, V.; Pearce, M.; Takahashi, H. The PoGO+ Balloon-Borne Hard X-ray Polarimetry Mission. *Galaxies* **2018**, *6*, 30. [CrossRef]

24. Chauvin, M.; Florén, H.G.; Friis, M.; Jackson, M.; Kamae, T.; Kataoka, J.; Kawano, T.; Kiss, M.; Mikhalev, V.; Mizuno, T.; et al. Shedding new light on the Crab with polarized X-rays. *Sci. Rep.* **2017**, *7*, 7816. [CrossRef]

25. Produit, N.; Bao, T.W.; Batsch, T.; Bernasconi, T.; Britvich, I.; Cadoux, F.; Cernuda, I.; Chai, J.Y.; Dong, Y.W.; Gauvin, N.; et al. Design and construction of the POLAR detector. *Nucl. Instrum. Methods Phys. Res. A* **2018**, *877*, 259–268. [CrossRef]

26. Suarez-Garcia, E.; Haas, D.; Hajdas, W.; Lamanna, G.; Lechanoine-Leluc, C.; Marcinkowski, R.; Mtchedlishvili, A.; Orsi, S.; Pohl, M.; Produit, N.; et al. A method to localize gamma-ray bursts using POLAR. *Nucl. Instrum. Methods Phys. Res. A* **2010**, *624*, 624–634. [CrossRef]

27. Kole, M. POLAR-2: The First Large Scale Gamma-ray Polarimeter. In Proceedings of the 36th International Cosmic Ray Conference (ICRC2019), Madison, WI, USA, 24 July–1 August 2019; Volume 36, p. 572. [CrossRef]

28. Vadawale, S.V.; Chattopadhyay, T.; Rao, A.R.; Bhattacharya, D.; Bhalerao, V.B.; Vagshette, N.; Pawar, P.; Sreekumar, S. Hard X-ray polarimetry with Astrosat-CZTI. *Astron. Astrophys.* **2015**, *578*, A73. [CrossRef]

29. Chattopadhyay, T.; Vadawale, S.V.; Rao, A.R.; Sreekumar, S.; Bhattacharya, D. Prospects of hard X-ray polarimetry with Astrosat-CZTI. *Exp. Astron.* **2014**, *37*, 555–577. [CrossRef]

30. Chattopadhyay, T.; Gupta, S.; Iyyani, S.; Saraogi, D.; Sharma, V.; Tsvetkova, A.; Ratheesh, A.; Gupta, R.; Mithun, N.P.S.; Vaishnava, C.S.; et al. Hard X-ray Polarization Catalog for a Five-year Sample of Gamma-Ray Bursts Using AstroSat CZT Imager. *Astrophys. J.* **2022**, *936*, 12. [CrossRef]

31. Vaishnava, C.S.; Mithun, N.P.S.; Vadawale, S.V.; Aarthy, E.; Patel, A.R.; Adalja, H.L.; Kumar Tiwari, N.; Ladiya, T.; Navale, N.; Chattopadhyay, T.; et al. Experimental verification of off-axis polarimetry with cadmium zinc telluride detectors of AstroSat-CZT Imager. *J. Astron. Telesc. Instrum. Syst.* **2022**, *8*, 038005. [CrossRef]

32. Giacconi, R.; Branduardi, G.; Briel, U.; Epstein, A.; Fabricant, D.; Feigelson, E.; Forman, W.; Gorenstein, P.; Grindlay, J.; Gursky, H.; et al. The Einstein (HEAO 2) X-ray Observatory. *Astrophys. J.* **1979**, *230*, 540–550. [CrossRef]

33. Baldini, L.; Barbanera, M.; Bellazzini, R.; Bonino, R.; Borotto, F.; Brez, A.; Caporale, C.; Cardelli, C.; Castellano, S.; Ceccanti, M.; et al. Design, construction, and test of the Gas Pixel Detectors for the IXPE mission. *Astropart. Phys.* **2021**, *133*, 102628. [CrossRef]

34. Schnopper, H.W. SODART telescopes on the Spectrum X-Gamma (SRG) and their complement of instruments. *Proc. SPIE* **1994**, *2279*, 412–423. [CrossRef]

35. Kaaret, P.; Novick, R.; Martin, C.; Hamilton, T.; Sunyaev, R.; Lapshov, I.; Silver, E.; Weisskopf, M.; Elsner, R.; Chanan, G.; et al. SXRP. A focal plane stellar X-ray polarimeter for the SPECTRUM-X-Gamma mission. *Proc. SPIE* **1989**, *1160*, 587–597.

36. Soffitta, P.; Costa, E.; Kaaret, P.; Dwyer, J.; Ford, E.; Tomsick, J.; Novick, R.; Nenonen, S. Proportional counters for the stellar X-ray polarimeter with a wedge and strip cathode pattern readout system. *Nucl. Instrum. Methods Phys. Res. A* **1998**, *414*, 218–232. [CrossRef]

37. Kaaret, P.E.; Schwartz, J.; Soffitta, P.; Dwyer, J.; Shaw, P.S.; Hanany, S.; Novick, R.; Sunyaev, R.; Lapshov, I.Y.; Silver, E.H.; et al. Status of the stellar X-ray polarimeter for the Spectrum-X-Gamma mission. *Proc. SPIE* **1994**, *2010*, 22–27.

38. Kislat, F.; Abarr, Q.; Beheshtipour, B.; De Geronimo, G.; Dowkontt, P.; Tang, J.; Krawczynski, H. Optimization of the design of X-Calibur for a long-duration balloon flight and results from a one-day test flight. *J. Astron. Telesc. Instrum. Syst.* **2018**, *4*, 011004. [CrossRef]

39. Abarr, Q.; Baring, M.; Beheshtipour, B.; Beilicke, M.; de Geronimo, G.; Dowkontt, P.; Errando, M.; Guarino, V.; Iyer, N.; Kislat, F.; et al. Observations of a GX 301-2 Apastron Flare with the X-Calibur Hard X-ray Polarimeter Supported by NICER, the Swift XRT and BAT, and Fermi GBM. *Astrophys. J.* **2020**, *891*, 70. [CrossRef]

40. Beilicke, M.; Baring, M.G.; Barthelmy, S.; Binns, W.R.; Buckley, J.; Cowsik, R.; Dowkontt, P.; Garson, A.; Guo, Q.; Haba, Y.; et al. Design and tests of the hard X-ray polarimeter X-Calibur. *Nucl. Instrum. Methods Phys. Res. A* **2012**, *692*, 283–284. [CrossRef]

41. Abarr, Q.; Awaki, H.; Baring, M.G.; Bose, R.; De Geronimo, G.; Dowkontt, P.; Errando, M.; Guarino, V.; Hattori, K.; Hayashida, K.; et al. XL-Calibur—A second-generation balloon-borne hard X-ray polarimetry mission. *Astropart. Phys.* **2021**, *126*, 102529. [CrossRef]

42. Iyer, N.K.; Kiss, M.; Pearce, M.; Stana, T.A.; Awaki, H.; Bose, R.G.; Dasgupta, A.; De Geronimo, G.; Gau, E.; Hakamata, T.; et al. The design and performance of the XL-Calibur anticoincidence shield. *Nucl. Instrum. Methods Phys. Res. A* **2023**, *1048*, 167975. [CrossRef]

43. Chattopadhyay, T.; Vadawale, S.V.; Goyal, S.K.; Mithun, N.P.S.; Patel, A.R.; Shukla, R.; Ladiya, T.; Shanmugam, M.; Patel, V.R.; Ubale, G.P. Development of a hard X-ray focal plane compton polarimeter: A compact polarimetric configuration with scintillators and Si photomultipliers. *Exp. Astron.* **2016**, *41*, 197–214. [CrossRef]

44. Jahoda, K.; Krawczynski, H.; Kislat, F.; Marshall, H.; Okajima, T. The X-ray Polarization Probe mission concept. *Bull. Am. Astron. Soc.* **2019**, *51*, 181.

45. Marshall, H.L.; Günther, H.M.; Heilmann, R.K.; Schulz, N.S.; Egan, M.; Hellickson, T.; Heine, S.N.T.; Windt, D.L.; Gullikson, E.M.; Ramsey, B.; et al. Design of a broadband soft X-ray polarimeter. *J. Astron. Telesc. Instrum. Syst.* **2018**, *4*, 011005. [CrossRef]
46. Poulson, D.; Bloser, P.F.; Ogasawara, K.; Lemire, C.R.; Trevino, J.A.; Legere, J.S.; Ryan, J.M.; McConnell, M.L. Using single-crystal diamond detectors as a scattering medium in Compton telescopes. *Proc. SPIE* **2022**, *12181*, 121812I. [CrossRef]

 *instruments*

Review

Mini-EUSO on Board the International Space Station: Mission Status and Results

Laura Marcelli on behalf of the JEM-EUSO Collaboration

INFN, Sezione di Roma Tor Vergata, Via della Ricerca Scientifica 1, 00173 Rome, Italy; laura.marcelli@roma2.infn.it

Abstract: The telescope Mini-EUSO has been observing, since 2019, the Earth in the ultraviolet band (290–430 nm) through a nadir-facing UV-transparent window in the Russian Zvezda module of the International Space Station. The instrument has a square field of view of 44°, a spatial resolution on the Earth surface of 6.3 km and a temporal sampling rate of 2.5 microseconds. The optics is composed of two 25 cm diameter Fresnel lenses and a focal surface consisting of 36 multi-anode photomultiplier tubes, 64 pixels each, for a total of 2304 channels. In addition to the main camera, Mini-EUSO also contains two cameras in the near infrared and visible ranges, a series of silicon photomultiplier sensors and UV sensors to manage night-day transitions. Its triggering and on-board processing allow the telescope to detect UV emissions of cosmic, atmospheric and terrestrial origin on different time scales, from a few microseconds up to tens of milliseconds. This makes it possible to investigate a wide variety of events: the study of atmospheric phenomena (lightning, transient luminous events (TLEs) such as ELVES and sprites), meteors and meteoroids; the search for nuclearites and strange quark matter; and the observation of artificial satellites and space debris. Mini-EUSO is also potentially capable of observing extensive air showers generated by ultra-high-energy cosmic rays with an energy above 10^{21} eV and can detect artificial flashing events and showers generated with lasers from the ground. The instrument was integrated and qualified in 2019 in Rome, with additional tests in Moscow and final, pre-launch tests in Baikonur. Operations involve periodic installation in the Zvezda module of the station with observations during the crew night time, with periodic downlink of data samples, and the full dataset being sent to the ground via pouches containing the data disks. In this work, the mission status and the main scientific results obtained so far are presented, in light of future observations with similar instruments.

Keywords: UV telescope; space telescope; UV emissions; ISS; ultra-high energy cosmic rays (UHECRs); meteors; strange quark matter; transient luminous events

Citation: Marcelli, L., on behalf of the JEM-EUSO Collaboration. Mini-EUSO on Board the International Space Station: Mission Status and Results. *Instruments* **2024**, *8*, 6. https://doi.org/10.3390/instruments8010006

Academic Editor: Antonio Ereditato

Received: 25 December 2023
Revised: 12 January 2024
Accepted: 14 January 2024
Published: 24 January 2024

1. Introduction

The Mini-EUSO (Multiwavelength Imaging New Instrument for the Extreme Universe Space Observatory) experiment [1] is part of the program carried out by the JEM-EUSO (Joint Exploratory Missions for an Extreme Universe Space Observatory) collaboration [2]. The goal of the collaboration is to construct a large space telescope to detect ultra-high-energy cosmic rays (UHECRs) from space for the first time.

Sixty years after the first detection of a particle with an energy of 10^{20} eV [3], the origin and nature of UHECRs are still unknown. This is mainly due to the extremely low flux of these particles—about 1 particle/(km^2 × millennium). Currently, two ground-based observatories are observing the sky searching for UHECRs: the Pierre Auger Observatory [4], from the Southern Hemisphere, and Telescope Array [5], from the Northern one. In the future, the observation of UHECRs from space-based experiments will be complementary to that from ground observatories and will offer significant advantages such as an extremely large instantaneous observational area and the capability of observing both Earth's hemispheres with a single instrument, reducing possible systematic uncertainties.

UHECR observation from space is based on the measurement of the fluorescence and Cherenkov light produced in an extensive air shower (EAS). A UHECR that hits the atmosphere produces secondary particles which, on their turn, collide with atoms in the air, producing a shower dominated by electrons and positrons. As they pass through the atmosphere, these particles excite atmospheric molecules, particularly nitrogen, which emit isotropically the characteristic fluorescence light in the ultraviolet (UV) band during de-excitation; the EAS therefore produces a streak of fluorescent light along its path through the atmosphere, depending on the energy and zenith angle of the primary particle. Another detectable component is the Cherenkov light emitted in the direction of travel by the charged, relativistic particles of the EAS and reflected into space from the ground or clouds. Thus, by looking at the Earth's atmosphere from space, a purpose-built telescope can detect these fluorescence and Cherenkov light contributions and study UHECRs.

1.1. The JEM-EUSO Experiments

During the last ten years, the JEM-EUSO collaboration has accomplished many successful missions (see Figure 1) by operating on ground (EUSO-TA [6] (2013–Current)), on stratospheric balloons (EUSO-Balloon [7,8] (2014), EUSO-SPB1 [9] (2017), EUSO-SPB2 [10,11] (2023)) and in space (TUS [12,13] (2016), Mini-EUSO [1] (2019)). Other missions are foreseen for the coming years: K-EUSO [14] and POEMMA [15].

Figure 1. A summary of the projects accomplished by the JEM-EUSO Collaboration: on ground (EUSO-TA), on stratospheric balloons (EUSO-Balloon, EUSO-SPB1, EUSO-SPB2), and in space (TUS, Mini-EUSO).

1.1.1. EUSO-TA (2013–Current)

EUSO-TA [6] is a ground telescope installed at the Telescope Array (TA) site in Utah, USA, in front of the TA fluorescence detector station at Black Rock Mesa. EUSO-TA optical system is composed by two Fresnel lenses and a focal surface with 6×6 Multi-Anode PhotoMultiplier Tubes (MAPMTs) with 64 channels each, for a total of 2304 channels. The overall field of view is $\simeq 10.6° \times 10.6°$. The telescope detects cosmic ray events with high spatial resolution of $\simeq 0.2°$ and a temporal resolution of 2.5 µs. About ten UHECR events have been observed to date. In 2022, the detector was upgraded by replacing the focal surface and acquisition system and implementing one similar to that of Mini-EUSO. Moreover, since 2013, the TA site and the EUSO-TA telescope have also been used as an auxiliary experiment for the calibration and testing of the other JEM-EUSO detectors.

1.1.2. EUSO-Balloon (2014)

EUSO-Balloon [7,8], a balloon flight of the CNES (French Space Agency), was a one-night mission with several key innovative features such as Fresnel optics, dedicated ASIC (first-generation SPACIROC, Spatial Photomultiplier Array Counting Integrated Read-OutChip) for the front-end electronics and efficient data processing. It worked nominally and recorded the Earth's night-time UV emissions, showing an anticorrelation of UV-IR brightness. Tracks of laser light from a helicopter that flew below the balloon and Xenon flashers from the ground have also been recorded.

1.1.3. TUS (2016)

The TUS detector [12,13] was the first space-based mission aimed for UHECRs detection. TUS was launched on board the Russian Lomonosov satellite in April 2016 and operated till December 2017. Almost 90,000 events were recorded during the mission, among them lightning discharges, meteors, transient luminous events (TLEs), polar lights and anthropogenic signals. No event has been classified as UHECR candidate.

1.1.4. EUSO-SPB1 (2017)

EUSO-SPB1 (Extreme Universe Space Observatory on a Super Pressure Balloon) [9] was launched on board a NASA long-duration super pressure balloon (SPB) from the NASA balloon facility in Wanaka, New Zealand, in April 2017. Even though the telescope functioned nominally, the flight was shortened to 12 days due to a balloon leak, preventing the detection of a real cosmic ray shower and the detector recovery. Despite this, 25.1 h of data were downloaded allowing the measurement of Earth night-time UV emissions, which represent the background for the detection of EASs, over different kinds of surfaces such as land, ocean and clouds. The EUSO-SPB1 focal surface was improved compared to that of EUSO-Balloon, including SPACIROC-3 ASIC [16], more compact focal surface elements, an enhanced optics performance and an autonomous trigger for events.

1.1.5. EUSO-SPB2 (2023)

On 13 May 2023, EUSO-SPB2 [10,11], a second NASA super pressure balloon, was launched from Wanaka but, again due to a leak in the balloon, the flight lasted only about 32 h, and sank in the Pacific Ocean. It aimed to search for UHECRs (E > EeV) and very high-energy neutrinos (E > PeV) using ultraviolet fluorescence and Cherenkov radiation, respectively. For these purposes, the mission comprised two independent optical telescopes: a fluorescence telescope (FT) having 108 MAPMTs (three EUSO-SPB1 focal surfaces side by side) at the focal point of a Schmidt telescope with a diameter of one meter and a Cherenkov Telescope (CT) using a Silicon Photomultiplier camera. In addition, an infrared camera (IR) was installed for cloud monitoring. Although the flight was short, all the telescopes performed as planned, confirming their expected functionality through extensive simulations, laboratory and field tests. During the flight period, a large amount of data was downloaded, about 56 GB, consisting of more than 120,000 FT triggers and over 32,000 CT events. The collaboration is currently analyzing this data. No cosmic ray candidates have so far been identified in FT events (in line with the low expected rate of one cosmic ray event every 15 h). The CT data include several triggers from below the limb, providing valuable insights into potential neutrino observations for future missions, and several near-horizontal above-limb Cherenkov signals from EAS caused by cosmic rays: this result not only convalidates the developed triggering procedure, but also proves the feasibility of the technique of detection itself, which was another main goal of the mission.

In what follows in this paper, the in-flight performance of the Mini-EUSO instrument during the first four years of data collection and its first scientific results are described.

2. The Mini-EUSO Instrument

Mini-EUSO [1] is a UV telescope (range 290–430 nm) operating in the International Space Station (ISS) from the UV-transparent window facing the nadir located in the Russian

Zvezda module. Its dimensions ($37 \times 37 \times 62$ cm^3) are therefore determined by the window size and the requirements associated with the Soyuz spacecraft. In addition, its design takes into account safety requirements, such as the absence of sharp edges, a low surface temperature and general robustness to ensure the crew's well-being. Installation on the window is via a mechanical adapter flange and the only connection to the ISS is via a 28 V power supply and grounding cable. The telescope power consumption is $\simeq$60 W, and its weight is $\simeq$35 kg (5 kg flange included). The instrument field of view is squared with a side of 44°, and its spatial resolution on the Earth's surface is 6.3 km^2 (depending on the altitude of the ISS). Mini-EUSO has also single-photon-counting capabilities.

The optical system is composed of two 25 cm diameter Poly(Methyl Methacrylate)—PMMA Fresnel lenses, which focus light on a focal surface, or a photon detector module (PDM), consisting of a matrix of 6×6 MAPMTs (Hamamatsu R11265-M64, Hamamatsu Photonics K.K., Shizuoka, Japan), 64 pixels each, for a total of 2304 pixels (see Figure 2, left side). Each MAPMT has a BG3 UV bandpass filter on the input window and is powered by a Cockroft–Walton power supply board, and its front-end electronics consists of a SPACIROC3 board. Data from the entire PDM are then processed by a Xilinx Zynq-based FPGA board that runs a multilevel trigger [17], allowing for the measurement of triggered UV transients for 128 frames on time scales of both 2.5 µs (defined as 1 gate time unit, GTU) and 320 µs. Moreover, a non-triggered acquisition mode with 40.96 ms frames allows for continuous data acquisition. Data collection and storage on 512 GB USB Solid State Disk (SSD) cards, inserted into the telescope side by the cosmonaut before the session, are handled by a PCIe/104 form factor CPU. No direct telecommunication with Earth is present.

Mini-EUSO is also provided with two auxiliary cameras to integrate near-infrared and visible UV measurements [18], three single-pixel UV sensors (a linear photodiode (Analog Devices AD8304ARUZ, Analog Devices, Inc., Norwood, MA, USA) a logarithmic photodiode (Lapis Semiconductor ML8511, LAPIS Semiconductor Co., Yokohama, Japan) and a single-pixel silicon photomultiplier (Hamamatsu C13365, Hamamatsu Photonics K.K., Shizuoka, Japan) used to handle day/night transitions during the data taking, and an 8×8 Silicon PhotoMultiplier (SiPM) imaging array (Hamamatsu C14047-3050EA08, Hamamatsu Photonics K.K., Shizuoka, Japan) [19].

Figure 2. Some photographs shot during the integration. (**Left**): the Mini-EUSO focal surface (FS); UV sensors (below the FS) and the photomultiplier array (above the FS) are visible through the lens frames. Image taken from [1] (© reproduced with permission from AAS). (**Right**): the instrument completely assembled but with the mechanics box opened. Image taken from [20] (© reproduced with permission from SNCSC).

Two models of the detector were produced: the engineering model (EM) and the flight model (FM). The two copies are identical, with the exception that in the PDM of the EM, only the central four MAPMTs are installed, while the remaining components are substituted by inert mass dummies. Additionally, in the EM, the Fresnel lenses have been replaced by flat PMMA elements of equivalent weight.

2.1. Integration and Tests

The instruments were integrated at the INFN Laboratories located in Frascati and Rome Tor Vergata. Figure 2 provides some photographs shot during the integration of the FM. For a more detailed description about the telescope and its first observations, refer to [1].

A series of qualification tests [20] were performed on both the two models. These tests were performed to assure the instruments could safely withstand transport to the launch site, the launch itself, and subsequent operations on-board the ISS. Tests included vibration and shock, electromagnetic interference and compatibility (EMI and EMC, respectively), and thermal-vacuum and environmental tests. Additionally, the Mini-EUSO FM was subjected to field tests in dark sky conditions.

Following qualification and field tests, the FM successfully underwent also several acceptance tests, first in Rome, then in Moscow, and at last at the Baikonur cosmodrome. Acceptance tests consist of a sequence of final checking procedures aimed at certifying that the detector is able to endure the environmental launch conditions and the operations in space, while also assessing conformity with safety requirements.

Currently, the EM is being used as a training model for the various crews responsible for operating Mini-EUSO on the ISS.

2.2. Launch and In-Flight Operations

Mini-EUSO was launched to the International Space Station (ISS) on 22 August 2019, from the Baikonur Cosmodrome in Kazakhstan on board the unmanned Soyuz MS-14 capsule. The instrument was switched on for the very first time on 7 October 2019, after the trained cosmonaut responsible for its operation arrived on the ISS (see Figure 3). Since then, the telescope has been systematically collecting data at regular intervals, with a total of 96 operational sessions operated up to now over four years.

Figure 3. Mini-EUSO installed by two cosmonauts on the UV-transparent window of the Zvezda module.

Mini-EUSO is switched on approximately every two weeks and put in acquisition mode for about 12 h, during the local night of the ISS. At the beginning of each observation session the instrument is recovered from storage, the lens cover removed and the detector mounted on the UV-transparent window of the Zvezda module. The power and ground cables are then connected, a USB SSD card inserted on the side of the instrument and the power switched on. Timing is managed internally with a real-time clock, since no external connections are available to the ISS (the clock's daily drift was measured on the ground and is periodically cross-checked with data collected on board).

At start-up, the initialization program checks whether on the SSD card there are software and/or firmware updates or updated operating parameters intended to override existing ones, and applies them if so. This very flexible approach allows the collaboration to continuously improve operations. At the end of each session, the detector and the SSD card are securely stowed, and the log file and a subset of the acquired data files (typically about 10%, corresponding to the beginning and end of the session) are transmitted to the ground via the ISS telemetry channel to verify the proper functioning of the system.

The pouches containing 25 SSDs are returned to ground every 6–12 months, and at a similar interval, a new pouch containing new SSD cards is dispatched to the ISS.

During these four years of operations, the JEM-EUSO collaboration also performed various in-flight calibration campaigns for studying the instrument's response when exposed to a light source of known intensity. This was carried out by sending pulses of LED light from the ground toward the sky during the ISS passage. Further test campaigns will be conducted in the future.

3. Scientific Objectives and Selected Results

Mini-EUSO was developed, within the JEM-EUSO program, with the main objective of proving the feasibility of studying ultra-high-energy cosmic rays (UHECRs) from space. This primarily consists of demonstrating that a space telescope has a high enough duty cycle, defined as the fraction of time during which atmospheric or man-made light sources do not make it impossible to observe UHECRs from space. The objective is also to establish the potential of detecting short light transients (SLTs) that present similarities in terms of light intensity or pulse duration to what is expected from an extensive air shower (EAS) cascade in the atmosphere. Nonetheless, it is important to note that the lens size of Mini-EUSO (25 cm of diameter) results in a minimum energy threshold for the detection of UHECRs well-above 10^{21} eV, an energy range in which no events have been detected so far. However, the collaboration intends to set an upper limit for the particle flux at these energies, since Mini-EUSO has yet accumulated an exposure comparable to that of ground-based hybrid experiments to date [21,22].

Moreover, through the observation of terrestrial and atmospheric UV emissions from space, Mini-EUSO can achieve many other scientific goals: studying atmospheric processes such as lightning and transient luminous events (TLEs), which include ELVES; observing meteors and meteoroids; searching for interstellar meteors and strange quark matter (SQM); proving the practical feasibility of detecting and tracking space debris from space; and constructing the map of terrestrial night-time UV emissions, both natural and anthropogenic. An overview of Mini-EUSO primary scientific objective is provided in Figure 4 for reference.

In Figure 5, the total signal detected by the focal surface in function of time is shown for signals of different time scales, from the fastest sampling of 2.5 µs (D1 acquisition mode), to the average of 128 D1 frames for D2 mode (320 µs) and to the average of 128×128 2.5 µs frames for D3 acquisitions (40.96 ms). In D3 acquisition mode, the gradual increases are due to passing over regions covered by clouds, while the sharp spikes correspond to lightning.

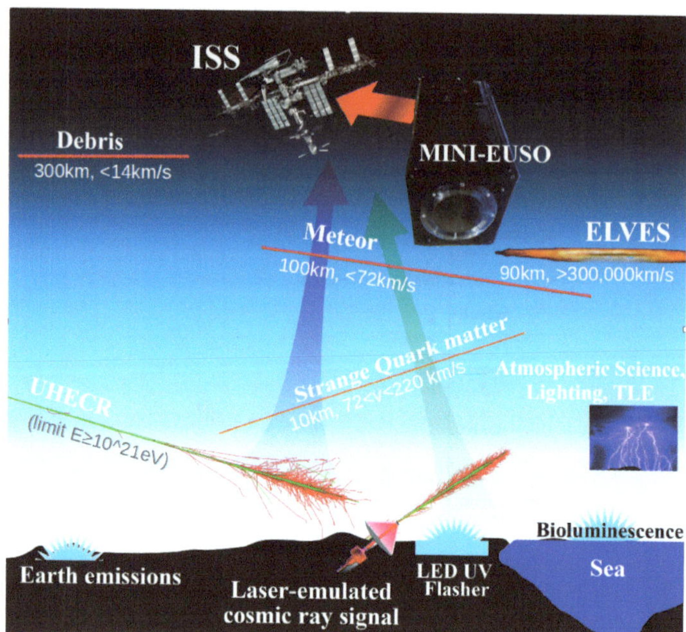

Figure 4. A summary of the Mini-EUSO scientific objectives. The detector is able to observe a large variety of different phenomenon with different intensities and durations, from the very fast atmospheric events, such as TLEs and in particular ELVES, to the slow terrestrial emissions, both natural and anthropogenic. Image adapted from [1] (© reproduced with permission from AAS).

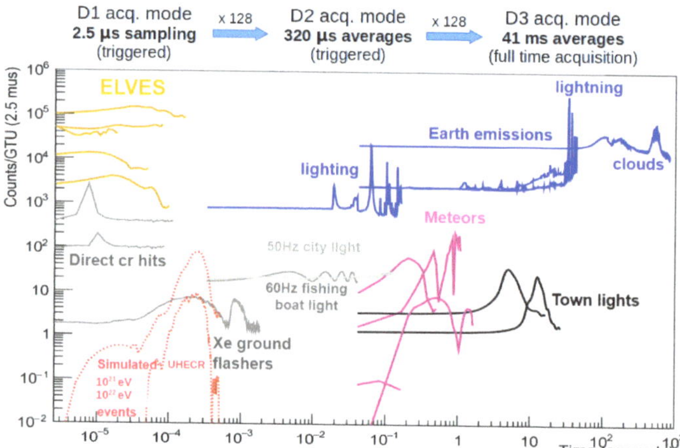

Figure 5. Time profile of different phenomena detected by Mini-EUSO. All the curves are referred to real observed data, except for the simulated UHECR events at 10^{21}–10^{22} eV. Curves are arbitrary scaled along the y axes for illustration purposes. Image adapted from [1] (© reproduced with permission from AAS).

The temporal and spatial profiles of the different signals detected by Mini-EUSO allow us to classify such signals. In the D1 mode, fast events are identified, such as direct cosmic ray hits, ELVES, the flashing of Xenon ground flashers. In the D2 acquisition mode, instead, it is possible to distinguish the modulation of the artificial lights within small town and villages. Lastly, using data acquired in the D3 time scale, meteors can be studied, interstellar meteors can be searched for, and Earth's night-time UV emissions, both natural and anthropogenic, can be mapped.

The main event kinds, ordered from fastest to slowest, are discussed in the following subsections.

3.1. Direct Hits

Direct cosmic ray hits occur when cosmic rays interact with the photocathode or the BG3 filter of the focal surface directly, through either direct ionisation or emission of Cherenkov light. Typically, these kinds of event last a few GTUs and release a high signal in one or a few pixels. These signals display a distinctive pattern characterized by a rapid rise followed by an exponential decrease resulting from the de-excitation of the physically hit components. Figure 6, left panel, shows a direct cosmic ray hit with its typical exponential decrease.

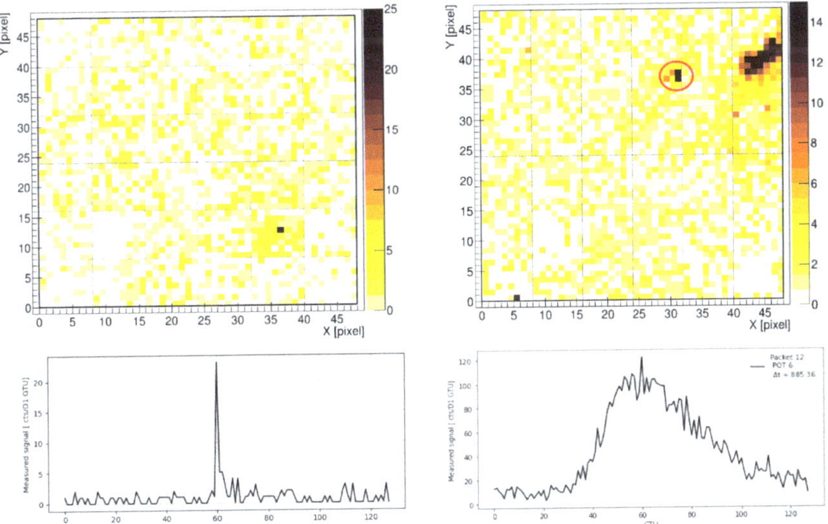

Figure 6. (Left): A direct cosmic ray detected by Mini-EUSO: A low-energy ($\simeq$GeV) cosmic ray hits the detector's focal surface perpendicularly, causing a bright signal in a single pixel (the number of photoelectron counts/GTU is indicated on the Z-axis). The corresponding plot at the bottom shows the light-curve of the hit pixel. **(Right)**: An EAS-like event (or, more properly, a short light transient, SLT) detected by Mini-EUSO. This particular event was detected just off the coast of Sri Lanka that appears as a luminous area in the upper right corner of the focal surface. The SLT shows up as a small group of pixels (in the red circle) and presents a bi-Gaussian light curve characterised by a faster rise and slower decay. The shown light curve is the sum of all 6 pixels above the threshold (POT in the legend) in the packet.

3.2. UHECRs

The high threshold energy and the short exposure of Mini-EUSO resulted in no detection of UHECRs so far. However, the detection of SLTs indirectly confirms the capability of the JEM-EUSO technology to potentially identify UHECRs from space, since they exhibit similarities from the point of view of light profile, intensity, duration and pixel pattern on the focal plane, although all of these features do not coincide simultaneously in a single event. Of more importance is the fact that Mini-EUSO has proved that these events cannot be misinterpreted as true EAS-induced signals, eliminating concerns for future observations. For further details refer to [23].

A SLT event is any flashing signal lasting more than 200 μs not originated by a ground flasher (see Section 3.3). In Figure 6, right panel, an example of the signal generated on the focal plane by a SLT together with its light curve is shown. The light-curve present a bi-Gaussian shape, with a faster rise and a slower decay, with a relatively long signal. It appears reasonable to assume, although no study has yet confirmed it, that the origin of these fast flashing lights is related to thunderstorm activity in the atmosphere. This event was compared to several simulated EAS events with different energy and zenith angle, some of which are presented in Figure 7. Of the simulated EAS events, none of them match either the image topology or the duration of the SLT light profile (Figure 6, right panel). Specifically, the light footprint on the focal surface from the SLT event (Figure 6, top-right panel) is consistent with a nearly vertical event (Figure 7, top-middle panel, in the red circle), but the SLT time duration ($\sim$80 GTUs) (Figure 6, bottom-right panel) far exceeds the time required by a vertical EAS for developing in the atmosphere and reaching the Earth's surface ($\sim$30 GTUs) (Figure 7, bottom-middle panel), while it is more similar to the time required by an inclined EAS (Figure 7, bottom-right panel).

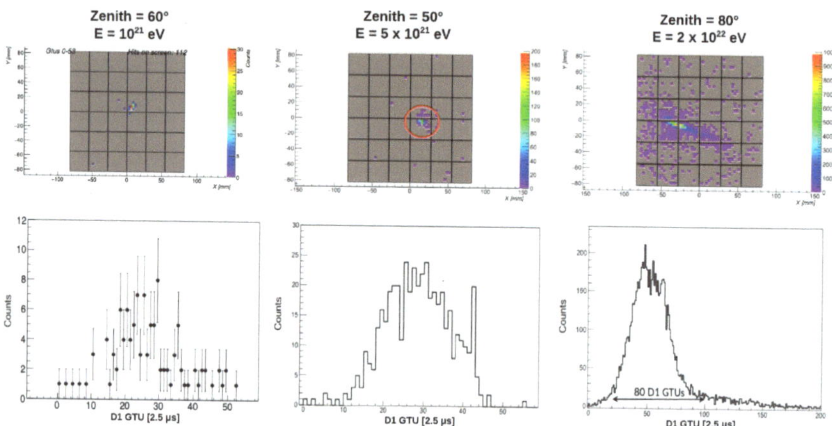

Figure 7. ESAF (EUSO Simulation and Analysis Framework) [24] simulation of proton-generated EASs of different energies and zenithal angles. (**Left**): Due to the high energy threshold, an EAS event of 10^{21} eV is at the limit of Mini-EUSO's triggering capability. (**Middle**): At about 5×10^{21} eV, the signal of shower with a zenith angle of a $50°$ is distinctly visible and has a duration of $\sim$30 GTUs ($\sim$75 µs). The signal stops when the EAS reaches the ground. (**Right**): With a zenith angle of $80°$ and energy of 2×10^{22} eV, the light curve is not truncated and the signal lasts for $\sim$80 GTUs ($\sim$200 µs).

3.3. Ground Flashers

Ground flashers, typically equipped with Xenon lights, are used as warning signals to aircraft, alerting them to the presence of structures like buildings or towers. These flashers vary in terms of their brightness and duration of blinking, typically lasting for a few hundred µs (see Figure 5). Mini-EUSO often detects these flashers multiple times as they traverse its field of view [25]; this characteristic makes them easily distinguishable even if they have similar signals to SLTs.

3.4. ELVES

ELVES (emission of light and very low-frequency perturbations due to electromagnetic pulse sources), which are part of the transient luminous events (TLEs) family, are observed as rapidly expanding luminous rings in the ionosphere. The typical ELVES lifetime is about 0.5 ms; this means that several 2.5 µs frames (on average 200) are associated with each event. In addition, the high spatial resolution (5 km at the ionosphere altitude) of the instrument and the fact that the telescope observes the ring expansion from above allow for a detailed analysis of the ELVES morphology (e.g., position, maximum radius, expansion velocity). In the available dataset, 37 ELVES have been identified, including single-ringed ELVES and multiple ELVES [26].

3.5. Artificial Light Modulation

A modulation of the artificial lights can be identified in Mini-EUSO data in the D2 time scale with a frequency of 50 or 60 Hz. This kind of modulation is more evident in smaller towns and villages where all the lights are linked to a single transformer. In contrast, larger cities, with different sections connected to various transformers and different phase arrangements, exhibit less pronounced modulation. Figure 5 shows the light modulation patterns observed in India, Canada, and from some fishing boats in the Indian Ocean.

3.6. Lightning

Lightning are transient atmospheric events lasting $\simeq$1 s which can partially or completely illuminate the focal surface, sometimes leading to the activation of the high voltage safety system [27]. When the satellite passes over regions characterized by high lightning activity, these events can extend over several hundred seconds. The time profiles of various lightning strikes observed in both D2 and D3 mode are shown in Figure 5.

3.7. Meteors

Meteors are observed by Mini-EUSO by looking for linear tracks moving in the field of view in the D3 time scale. The signals produced by meteors vary in intensity and duration according to their mass, speed and incidence angle. While looking for meteors, we also search for interstellar meteors and nuclearites.

To the best of our knowledge, Mini-EUSO is the first space-based mission to allow for a systematic study of meteors, mainly including the measurement of meteor light-curves and the determination of the meteor flux over a wide range of magnitudes. Furthermore, in some cases at least, it allows for the calculation of the original heliocentric orbits of meteoroids.

Mini-EUSO is capable of measuring meteor events with magnitude up to +5, with a significant statistic ($\simeq$2.4 meteors/min). In the present dataset, Mini-EUSO has successfully classified approximately 24,000 events as meteor events. More details about the analysis and complete results can be found in [28], together with the results from interstellar meteor search.

3.8. Night-Time UV Earth Emissions

Seen from space, the Earth's night-time UV emissions move through the field of view, and thus on the focal plane, with an apparent velocity equal to the ISS orbital speed (approximately 7.7 km/s). Consequently, a specific point on the Earth's surface remains visible from Mini-EUSO for about 50 s (equivalent to 1000 frames in D3 mode) making it possible to create ground maps with excellent spatial resolution and low statistical fluctuations. The temporal profile of a single pixel shows a gradual increase in luminosity when a village or town enters in its field of view (see Figure 5), the duration of this increase depends on the size of the source. Figure 8 shows the map of UV emissions over Italy and part of Europe reconstructed by Mini-EUSO.

For a comprehensive review of Mini-EUSO's capabilities in the reconstruction of UV maps refer to [27], while the complete dataset (.png, .dat and .kmz files) of the UV maps published in the previous article is available for downloading in [29].

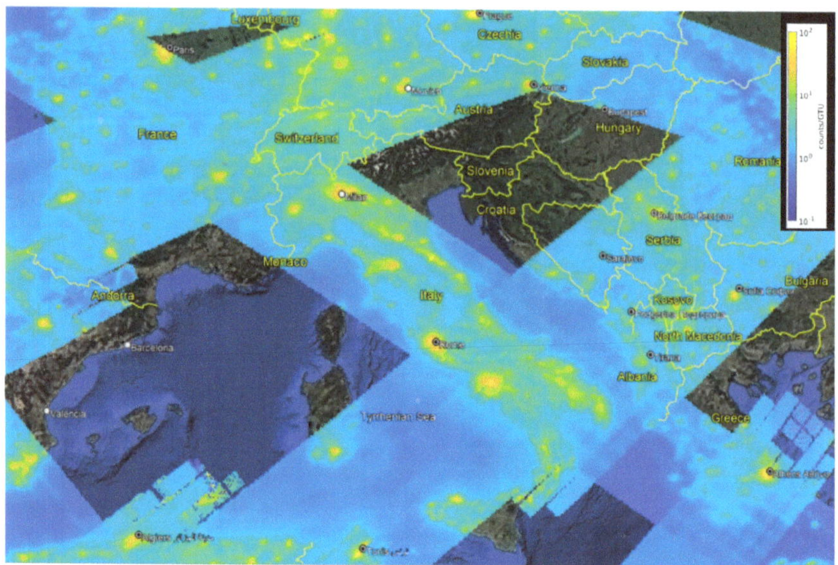

Figure 8. The map of night−time UV emission over Italy and part of Europe. Coastlines are well reconstructed and the main cities are easily identifiable. UV intensity is usually $\simeq$1 count/pixel/GTU on oceans and lands without urbanisation.

4. Conclusions

In this work, the pre-flight activities and the in-flight operations of the Mini-EUSO instrument are described. The telescope has been actively collecting data on board the ISS for a duration of four years so far, and it continues to behave nominally. Its performance during this time has proven to be immensely valuable in assessing the detection capabilities of future and larger space detectors, such as K-EUSO or POEMMA. Mini-EUSO measured the Earth's UV background with unprecedented accuracy, contributing significantly to estimating the real duty cycle of a future mission. Some preliminary results were also reported about the observation of ELVES, and meteors. These results prove the multidisciplinary nature of a UHECR space-based observatory.

Funding: This work was supported by the Italian Space Agency through the agreements n. 2016-1-U.0 and n. 2020-26-Hh.0, by the French space agency CNES, and by the National Science Centre in Poland through the grants 2017/27/B/ST9/02162 and 2020/37/B/ST9/01821.

Data Availability Statement: The data about Earth night-time UV emissions, related to the article [29], are openly available in the Mendeley repository at http://doi.org/10.17632/57fmn7rh4n.4 (accessed on 20 March 2023).

Acknowledgments: This research has been supported by the Interdisciplinary Scientific and Educational School of Moscow University "Fundamental and Applied Space Research" and by Russian State Space Corporation Roscosmos. The article has been prepared based on research materials collected in the space experiment "UV atmosphere". The author thanks the Altea-Lidal Collaboration for providing the orbital data of the ISS.

Conflicts of Interest: The authors declare no conflicts of interest.

Abbreviations

The following abbreviations are used in this manuscript:

EAS	Extensive Air Shower
ELVES	Emission of Light and Very Low-Frequency perturbations due to Electromagnetic Pulse Sources
EM	Engineering Model
EMI	Electromagnetic Interference
EMC	Electromagnetic Compatibility
ESAF	EUSO Simulation and Analysis Framework
FM	Flight Model
GTU	Gate Time Unit
ISS	International Space Station
JEM-EUSO	Joint Exploratory Missions for an Extreme Universe Space Observatory
MAPMT	Multi-Anode PhotoMultiplier Tube
Mini-EUSO	Multiwavelength Imaging New Instrument for the Extreme Universe Space Observatory
PDM	Photon Detector Module
PMMA	Poly(Methyl Methacrylate)
SiPM	Silicon PhotomMultiplier
SLT	Short Light Transient
SSD	Solid State Disk
TLE	Transient Luminous Events
UHECR	Ultra-High-Energy Cosmic Ray
UV	Ultraviolet

References

1. Bacholle, S.; Barrillon, P.; Battisti, M.; Belov, A.; Bertaina, M.; Bisconti, F.; Blaksley, C.; Blin-Bondil, S.; Cafagna, F.; Cambie, G.; et al. Mini-EUSO Mission to Study Earth UV Emissions on board the ISS. *Astrophys. J. Suppl.* **2021**, *253*, 36. [CrossRef]
2. Parizot, E.; Casolino, M. Overview of the JEM-EUSO program for the study of ultra-high-energy cosmic-rays from space. In Proceedings of the 38th International Cosmic Ray Conference (ICRC2023), Nagoya, Japan, 26 July–3 August 2023; Volume 444, p. 208 [CrossRef]
3. Linsley, J. Evidence for a Primary Cosmic-Ray Particle with Energy 10^{20} eV. *Phys. Rev. Lett.* **1963**, *10*, 146–148. [CrossRef]
4. Aab, A.; Abreu, P.; Aglietta, M.; Albuquerque, I.F.M.; Albury, J.M.; Allekotte, I.; Almela, A.; Castillo, J.A.; Alvarez-Muñiz, J.; Anastasi, G.A.; et al. Cosmic-Ray Anisotropies in Right Ascension Measured by the Pierre Auger Observatory. *Astrophys. J.* **2020**, *891*, 142. [CrossRef]
5. Abbasi, R.U.; Abe, M.; Abu-Zayyad, T.; Allen, M.; Azuma, R.; Barcikowski, E.; Belz, J.W.; Bergman, D.R.; Blake, S.A.; Cady, R.; et al. Evidence of Intermediate-scale Energy Spectrum Anisotropy of Cosmic Rays E $\geq$ 1019.2 eV with the Telescope Array Surface Detector. *Astrophys. J.* **2018**, *862*, 91. [CrossRef]
6. Abdellaoui, G.; Abe, S.; Adams, J.H.; Ahriche, A.; Allard, D.; Allen, L.; Alonso, G.; Anchordoqui, L.; Anzalone, A.; Arai, Y.; et al. EUSO-TA - First results from a ground-based EUSO telescope. *Astropart. Phys.* **2018**, *102*, 98–111. [CrossRef]
7. Adams, J.H.; Ahmad, S.; Allard, D.; Anzalone, A.; Bacholle, S.; Barrillon, P.; Bayer, J.; Bertaina, M.; Bisconti, F.; Blaksley, C.; et al. A Review of the EUSO-Balloon Pathfinder for the JEM-EUSO Program. *Space Sci. Rev.* **2022**, *218*, 3. [CrossRef] [PubMed]
8. Abdellaoui, G.; Abe, S.; Adams, J.H.; Ahriche, A.; Allard, D.; Allen, L.; Alonso, G.; Anchordoqui, L.; Anzalone, A.; Arai, Y.; et al. Ultra-violet imaging of the night-time earth by EUSO-Balloon towards space-based ultra-high energy cosmic ray observations. *Astropart. Phys.* **2019**, *111*, 54–71. [CrossRef]
9. Abdellaoui, G.; Abe, S.; Adams, J.; Allard, D.; Alonso, G.; Anchordoqui, L.; Anzalone, A.; Arnone, E.; Asano, K.; Attallah, R.; et al. EUSO-SPB1 mission and science. *Astropart. Phys.* **2024**, *154*, 102891. [CrossRef]
10. Eser, J.; Olinto, A.V.; Wiencke, L. Overview and First Results of EUSO-SPB2. In Proceedings of the 38th International Cosmic Ray Conference (ICRC2023), Nagoya, Japan, 26 July–3 August 2023; Volume 444, p. 397 [CrossRef]
11. Scotti, V.; Osteria, G. The EUSO-SPB2 mission. *Nucl. Instrum. Methods Phys. Res. Sect. A Accel. Spectrometers Detect. Assoc. Equip.* **2020**, *958*, 162164. [CrossRef]
12. Klimov, P.A.; Panasyuk, M.I.; Khrenov, B.A.; Garipov, G.K.; Kalmykov, N.N.; Petrov, V.L.; Sharakin, S.A.; Shirokov, A.V.; Yashin, I.V.; Zotov, M.Y.; et al. The TUS Detector of Extreme Energy Cosmic Rays on Board the Lomonosov Satellite. *Space Sci. Rev.* **2017**, *212*, 1687–1703. [CrossRef]
13. Barghini, D.; Bertaina, M.; Cellino, A.; Fenu, F.; Ferrarese, S.; Golzio, A.; Ruiz-Hernandez, O.I.; Klimov, P.; Montanaro, A.; Salsi, A.; et al. UV telescope TUS on board Lomonosov satellite: Selected results of the mission. *Adv. Space Res.* **2022**, *70*, 2734–2749. [CrossRef]

14. Klimov, P.; Battisti, M.; Belov, A.; Bertaina, M.; Bianciotto, M.; Blin-Bondil, S.; Casolino, M.; Ebisuzaki, T.; Fenu, F.; Fuglesang, C.; et al. Status of the K-EUSO Orbital Detector of Ultra-High Energy Cosmic Rays. *Universe* **2022**, *8*, 88. [CrossRef]

15. Olinto, A.V. POEMMA (Probe Of Extreme Multi-Messenger Astrophysics) Roadmap Update. In Proceedings of the 38th International Cosmic Ray Conference (ICRC2023), Nagoya, Japan, 26 July–3 August 2023; Volume 444, p. 1159. [CrossRef]

16. Blin, S.; Barrillon, P.; de La Taille, C.; Dulucq, F.; Gorodetzky, P.; Prevot, G. SPACIROC3: 100 MHz photon counting ASIC for EUSO-SPB. *Nucl. Instrum. Meth.* **2018**, *A912*, 363–367. [CrossRef]

17. Belov, A.; Bertaina, M.; Capel, F.; Faust, F.; Fenu, F.; Klimov, P.; Mignone, M.; Miyamoto, H. The integration and testing of the Mini-EUSO multi-level trigger system. *Adv. Space Res.* **2018**, *62*, 2966–2976. [CrossRef]

18. Turriziani, S.; Ekelund, J.; Tsuno, K.; Casolino, M.; Ebisuzaki, T. Secondary cameras onboard the Mini-EUSO experiment: Control software and calibration. *Adv. Space Res.* **2019**, *64*, 1188–1198. [CrossRef]

19. Casolino, M.; Cambie', G.; Marcelli, L.; Reali, E. SiPM development for space-borne and ground detectors: From Lazio-Sirad and Mini-EUSO to Lanfos. *Nucl. Instruments Methods Phys. Res. Sect. A Accel. Spectrometers Detect. Assoc. Equip.* **2021**, *986*, 164649. [CrossRef]

20. Marcelli, L.; Barghini, D.; Battisti, M.; Blaksley, C.; Blin, S.; Belov, A.; Bertaina, M.; Bianciotto, M.; Bisconti, F.; Bolmgren, K.; et al. Integration, qualification, and launch of the Mini-EUSO telescope on board the ISS. *Rend. Lincei. Sci. Fis. Nat.* **2023**, *34*, 23–35. [CrossRef]

21. Novotny, V. Energy spectrum of cosmic rays measured using the Pierre Auger Observatory. In Proceedings of the PoS(ICRC2021), Berlin, Germany, 12–23 July 2021; Volume 395, p. 324. [CrossRef]

22. Shin, H. The measurements of the cosmic ray energy spectrum and the depth of maximum shower development of Telescope Array Hybrid trigger events. In Proceedings of the PoS(ICRC2021), Berlin, Germany, 12–23 July 2021; Volume 395, p. 305. [CrossRef]

23. Marcelli, L. Mini Euso Experiment. In Proceedings of the 38th International Cosmic Ray Conference (ICRC2023), Nagoya, Japan, 26 July–3 August 2023; Volume 444, p. 1. [CrossRef]

24. Abe, S.; Adams, J.H., Jr.; Allard, D.; Alldredge, P.; Anchordoqui, L.; Anzalone, A.; Arnone, E.; Baret, B.; Barghini, D.; Battisti, M.; et al. Developments and results in the context of the JEM-EUSO program obtained with the ESAF Simulation and Analysis Framework. *Eur. Phys. J. C (Part. Fields)* **2023**, *83*, 1028. [CrossRef]

25. Adams, J.; Christl, M.; Csorna, S.; Sarazin, F.; Wiencke, L. Calibration for extensive air showers observed during the JEM-EUSO mission. *Adv. Space Res.* **2014**, *53*, 1506–1514. [CrossRef]

26. Romoli, G. Study of multiple ring ELVES with the Mini-EUSO telescope on-board the International Space Station. In Proceedings of the 38th International Cosmic Ray Conference (ICRC2023), Nagoya, Japan, 26 July–3 August 2023; Volume 444, p. 223. [CrossRef]

27. Casolino, M.; Barghini, D.; Battisti, M.; Blaksley, C.; Belov, A.; Bertaina, M.; Bianciotto, M.; Bisconti, F.; Blin, S.; Bolmgren, K.; et al. Observation of night-time emissions of the Earth in the near UV range from the International Space Station with the Mini-EUSO detector. *Remote. Sens. Environ.* **2023**, *284*, 113336. [CrossRef]

28. Barghini, D. Meteorites Recovery and Orbital Elements of Near-Earth Objects from the Observation of Bright Meteors. Ph.D. Thesis, University of Turin, Torino, Italy, 2023.

29. Marcelli, L.; Bolmgren, K.; Barghini, D.; Battisti, M.; Blaksley, C.; Blin, S.; Belov, A.; Bertaina, M.; Bianciotto, M.; Bisconti, F.; et al. Dataset of night-time emissions of the Earth in the near UV range (290–430 nm), with 6.3 km resolution in the latitude range $-51.6 < L < +51.6$ degrees, acquired on board the International Space Station with the Mini-EUSO detector. *Data Brief* **2023**, *48*, 109105. [CrossRef] [PubMed]

Review

Use of Silicon Photomultipliers in the Detectors of the JEM-EUSO Program

Francesca Bisconti on behalf of the JEM-EUSO Collaboration

Istituto Nazionale di Fisica Nucleare—Sezione di Roma Tor Vergata, Via della Ricerca Scientifica 1, 00133 Rome, Italy; francesca.bisconti@roma2.infn.it

Abstract: The JEM-EUSO program aims to study ultra-high energy cosmic rays from space. To achieve this goal, it has realized a series of experiments installed on the ground (EUSO-TA), various on stratospheric balloons (with the most recent one EUSO-SPB2), and inside the International Space Station (Mini-EUSO), in light of future missions such as K-EUSO and POEMMA. At nighttime, these instruments aim to monitor the Earth's atmosphere measuring fluorescence and Cherenkov light produced by extensive air showers generated both by very high-energy cosmic rays from outside the atmosphere and by neutrino decays. As the two light components differ in duration (order of microseconds for fluorescence light and a few nanoseconds for Cherenkov light) they each require specialized sensors and acquisition electronics. So far, the sensors used for the fluorescence camera are the Multi-Anode Photomultiplier Tubes (MAPMTs), while for the Cherenkov one, new systems based on Silicon PhotoMultipliers (SiPMs) have been developed. In this contribution, a brief review of the experiments is followed by a discussion of the tests performed on the optical sensors. Particular attention is paid to the development, test, and calibration conducted on SiPMs, also in view to optimize the geometry, mass, and weight in light of the installation of mass-critical applications such as balloon- and space-borne instrumentation.

Keywords: JEM-EUSO; SiPM; cosmic rays; extensive air-showers; fluorescence detector; Cherenkov detector; space instruments

Citation: Bisconti, F., on behalf of JEM-EUSO Collaboration. Use of Silicon Photomultipliers in the Detectors of the JEM-EUSO Program. *Instruments* **2023**, *7*, 55. https://doi.org/10.3390/instruments7040055

Academic Editor: Antonio Ereditato

Received: 12 October 2023
Revised: 24 November 2023
Accepted: 24 November 2023
Published: 14 December 2023

1. Introduction

In the last years, different types of semiconductor radiation detectors have been developed for fundamental science experiments. Silicon Photomultipliers (SiPMs) represent a novel category of photodetectors [1–3], with single photon detection capability and detection efficiencies reaching up to red ∼60% (at peak wavelength of ∼450 nm) under normal operating conditions, and achieving single-photon time resolutions in the range of redtens to hundreds of picoseconds, reddepending on the channel size [4]. Thanks to significant advancements in SiPM technology in recent years, these detectors can be closely compared to traditional PhotoMultiplier Tubes (PMTs), which have long been dominant in the realm of photon detection. SiPMs offer numerous advantages, such as significantly lower operating voltage, a lightweight and durable structure, and immunity to magnetic fields [5]. They exhibit robustness against excessive incident light; instead of damaging the collecting anode as in a PMT, a SiPM saturates, drawing constant current without harm. Nevertheless, SiPM characteristics are strongly influenced by temperature [6].

The Joint Exploratory Missions for an Extreme Universe Space Observatory (JEM-EUSO) Program [7] consists of a series of telescopes and experiments operated on the ground (EUSO-TA [8]), on stratospheric balloons (EUSO-Balloon [9], EUSO-SPB1 [10], and EUSO-SPB2 [11]), and in the space inside the International Space Station (ISS-Mini-EUSO [12]). They all contribute in growing the knowledge and extending the capability to develop the technology to detect Ultra-High Energy Cosmic Rays (UHECRs) from space (aboard orbiting satellites) by observing the fluorescence and Cherenkov light emitted during their path through the Earth's atmosphere and the Cherenkov emission from Earth skimming neutrinos. Such a future space-based experiment is the Probe of Extreme Multi-Messenger Astrophysics (POEMMA) Observatory [13], which will allow for the detection of Earth-skimming neutrinos and UHECRs via stereo measurements by two co-orbital telescopes with tilting capability in order to extend the observation area.

The use of Multi-Anode Photo-Multiplier Tubes (MAPMTs) in the indirect detection of UHECR via the observation of fluorescence and Cherenkov light is well established, as the focal surfaces of the JEM-EUSO fluorescence telescopes are made of MAPMTs. In Figure 1 the so-called Photon-Detection Module (PDM) with the focal surface made of MAPMTs (model R11265-M64 [14], 2.62×2.62 cm^2, with 2.88×2.88 mm^2 pixels) is visible on the left ($\sim$17 $\times$ 17 cm^2 in total), while on the right two MAPMTs of the same model are represented together with an example of SiPM array (model S12642-0808PA-50 [15], 2.58×2.58 cm^2, with 3×3 mm^2 pixels), with similar size and same number of channels.

The PDM is the basic module of the focal surface of the detectors of the JEM-EUSO program. EUSO-TA, EUSO-Balloon, and EUSO-SPB1 and Mini-EUSO host one PDM, while EUSO-SPB3 hosted three PDMs. POEMMA is foreseen to host several tens of PDMs. One PDM is composed of 3×3 Elementary Cells (ECs), each one made of 2×2 MAPMTs. Each MAPMT is covered with a BG3 filter [16] to limit its sensitivity to the UV region (290–430 nm) and reduce the background of photons outside this optimal frequency band.

Figure 1. A PDM with the focal surface made of 36 MAPMTs (**left**), image taken from [17]; 2 MAPMTs and a SiPM array of similar size and number of channels (**right**), composition of images taken from [18] (MAPMTs) and adapted from Hamamatsu product information [15] (SiPM array).

The time resolution of most of the detectors of the JEM-EUSO program is 2.5 μs, which was defined for the observations from space and kept also for the ground-based EUSO-TA and the first balloon-based EUSO-Balloon and EUSO-SPB1, in order to test the original design of a future space-based detector. It was reduced to 1 μs for the most recent balloon-based experiment EUSO-SPB2, to operate a mission with a more appropriate time resolution for observations from the stratosphere. These time lengths are called within the JEM-EUSO Collaboration Gate Time Units (GTUs).

Terrestrial detectors with SiPM have been developed in several experiments, such as the Cherenkov Telescope Array (CTA) [19], the First Auger Multi-pixel photon counter camera for the Observation of Ultra-high-energy air Showers (FAMOUS) [20], IceTop-Gen2 Scintillator Upgrade [21], etc.

The evaluation of SiPMs for non-terrestrial UHECR telescopes was performed and is still under study within the JEM-EUSO Collaboration. Several telescopes host additional cameras based on SiPMs to study the performance. In this contribution, these detectors will be described, paying particular attention to the SiPM cameras.

2. Detection of UHECRs and Skimming Neutrinos via the Fluorescence Technique

Cosmic rays are extraterrestrial particles, of which about 90% are hydrogen nuclei (protons), 9% are helium nuclei, and the remaining 1% is composed of heavier nuclei and electrons. Their energy range is from about 10^9 to 10^{20} eV or more. Over this energy range, the cosmic ray flux varies by many orders of magnitude, between 1 particle m^{-2} s^{-1} at low energies ($E \sim 10^9$ eV), 1 particle m^{-2} yr^{-1} at intermediate energies ($E \sim 10^{15}$ eV), 1 particle km^{-2} yr^{-1} at high energies ($E \sim 10^{19}$ eV), and 1 particle km^{-2} per century at extreme energies ($E \sim 10^{20}$ eV).

Up to 10^{15} eV cosmic rays can be observed directly with detectors mounted on balloons or space-based experiments. At higher energies, the flux of particles is so low that huge detectors would be necessary to measure them directly. Therefore, indirect detection methods have been developed, observing the secondary cosmic rays, i.e., particles produced by the interaction of primary cosmic rays with the atmospheric molecules and originating the so-called Extensive Air Showers (EASs). Above 5×10^{19} eV, a strong suppression in the flux is observed, consistent with the Greisen–Zatsepin–Kuzmin (GZK) cutoff [22,23] that considers the interaction of protons with photons from the Cosmic Microwave Background (CMB), and with the photo-disintegration of nuclei. The origin of the suppression is, however, still debated and subject of studies. Moreover, protons in this energy range could directly point back to their sources, since their deflection due to magnetic fields would be small and would give indications about their origin.

Other astrophysical observable of interest are neutrinos. The Earth can be used as a large volume to let neutrinos interact and produce particles that, in turn, would decay and generate upward-going EASs with an upward-going Cherenkov cone that could be observed from above, with balloon- and space-based telescopes.

The main goals of the JEM-EUSO program are the detection of trans-GZK cosmic rays with high statistics, the study of their arrival directions and the anisotropy, the identification and study of the sources, and the detection of the Earth-skimming neutrinos. Building a large-scale detector observing the Earth's atmosphere from an orbiting satellite, provides the advantage of observing large areas of the atmosphere at one time, increasing the exposure of the detector for the observation of UHECRs. Moreover, orbiting around the Earth would provide a uniform coverage of the sky.

The detectors of the JEM-EUSO program use an indirect method to observe UHE-CRs and skimming neutrinos by measuring the fluorescence and Cherenkov light along the development of EASs in the atmosphere. This kind of telescope can only be operated at night-time, with good weather conditions, and in places with low artificial light background. A duty cycle of about 10–15% is expected for ground-based fluorescence detectors, intended as the fraction of time in which UHECRs can be observed, which is limited mainly by sunlight but also by the presence of other steady backgrounds like night-glow and moonlight. The fluorescence light is emitted by the relaxation of nitrogen molecules that have been excited by the interaction with charged particles in the EASs, mainly electrons and positrons. The fluorescence spectrum has lines in the UV band at discrete wavelengths in the range from 290 nm to 430 nm, with the most intense emission at around 337.1 nm [24]. The overall fluorescence emission along the shower development is isotropic. The Cherenkov light is produced when charged particles move through a medium with a velocity that is higher than the velocity of light in that medium. The particle ionizes the medium through its path, leaving a locally excited path behind. The following relaxation causes the emission of Cherenkov light within a cone that has an opening angle usually within $\sim 1.4°$ and varying with the refractive index of the medium and the speed of the particle. The emission spectrum of Cherenkov radiation is continuous and the photon yield is higher in the UV than in the visible band.

3. SiPMs in the JEM-EUSO Experiments

3.1. Brief Description of the SiPMs and Comparison with MAPMTs

A SiPM is composed of a matrix of identical microcells, each one consisting of a so-called Geiger-mode Avalanche Photodiode (G-APD) or Single-Photon Avalanche Diodes (SPAD) and a quenching resistor connected in series [1–3]. The microcells are connected in parallel to a bias voltage. SiPMs are also known as Multi-Pixel Photon Counters (MPPCs), naming a microcell as pixel. The pixel of a SiPM should not be confused with the pixel of a MAPMT: in a MAPMT every output channel gets the signal from one pixel; in a SiPM there can be several hundreds or even thousands of pixels per output channel. Every pixel produces the same signal when it is hit by a photon and the sum of the pixel responses gives the channel output. The typical dimension of a SiPM sensor is between $1 \times 1\,\text{mm}^2$ and $6 \times 6\,\text{mm}^2$ and the number of microcells per device ranges from several hundreds to several tens of thousands. Microcells vary between $10 \times 10\,\mu\text{m}^2$ and $100 \times 100\,\mu\text{m}^2$ in size (as examples: Onsemi MICROFC-10010-SMT with channel of $1 \times 1\,\text{mm}^2$ and microcell of $10 \times 10\,\mu\text{m}^2$, and MICROFC-60035-SMT with channel of $6 \times 6\,\text{mm}^2$ and microcell of $35 \times 35\,\mu\text{m}^2$ [25]; Hamamatsu S13360-1325CS with channel of $1.3 \times 1.3\,\text{mm}^2$ and microcell of $25 \times 25\,\mu\text{m}^2$, and S13360-6075CS with channel of $6 \times 6\,\text{mm}^2$ and microcell of $75 \times 75\,\mu\text{m}^2$ [26]).

SiPMs present an opportunity for constructing modular detector surfaces due to their compact and lightweight design. They operate with low voltage and exhibit insensitivity to magnetic fields [5]. Noteworthy properties are a rapid response time (in the order of redtens to hundreds of picoseconds [4]), the ability for single photon detection, high photon detection efficiency, and resilience to light-induced damage. However, with their semi-conductive nature and clustered G-APDs, SiPMs suffer from significant noise sources: reddark-count rate (DCR), afterpulses, and crosstalk red [1–3]. G-APD performance depends on temperature, affecting breakdown voltage and quench resistor due to semiconductor band gap sensitivity. redDark-count rate, resulting from thermal effects, can be reduced by decreasing the temperature and the operating voltage: the latter lowers the potential for a thermally excited electron to initiate an avalanche, but would also decrease gain and PDE. Afterpulses, secondary peaks caused by trapped electrons in silicon defects, occur after the main signal, typically at 1 photoelectron level. Afterpulse probability increases with overvoltage and pixel size. redCrosstalk occurs when a photon is detected in one microcell and the avalanche pulse in this microcell can trigger (with a certain probability) avalanches in the neighboring microcells, creating two or three times the signal of a microcell in a SiPM, even though the original photon was only one. As crosstalk events coincide with the original photon-induced signal, they are indistinguishable, introducing the possibility of fake signals in multi-photoelectron signals.

To build large sensitive areas, SiPMs can be arranged in arrays. Furthermore, some models are available with the Through Silicon Via (TSV) technology. In single SiPMs the cathode is wired on the sides of the SiPM active surface and arrays of single SiPMs would have a large dead space between them. In the SiPM arrays TSV the cathode is etched through the silicon wafer in the middle of the SiPM, building an electrical interconnection from the surface to the back of the SiPM device. This allows to reduce the gaps between adjacent SiPM channels.

At the time of the preparation of the first relatively large-scale camera with SiPMs, two SiPM arrays were studied: the former model S12642-0808PA-50 with a top protective layer of epoxy resin [15] and the newer S13361-3050AS-08 with a top protective layer of silicone [27], both from Hamamatsu. The different top protective layer gives a different sensitivity of the SiPMs to the UV light with wavelengths lower than 320 nm, which causes a loss of photons in the case of epoxy resin. In Figure 2 the characteristic fluorescence spectrum of nitrogen molecules is shown in the wavelength range 295–430 nm.

The Photon Detection Efficiency (PDE) for a MAPMT prototype from 2009 (Hamamatsu R11265-00-M64 [14]), a SiPM array with epoxy resin, and a SiPM array with silicone are overlapped to the fluorescence spectrum and limited to its wavelength range.

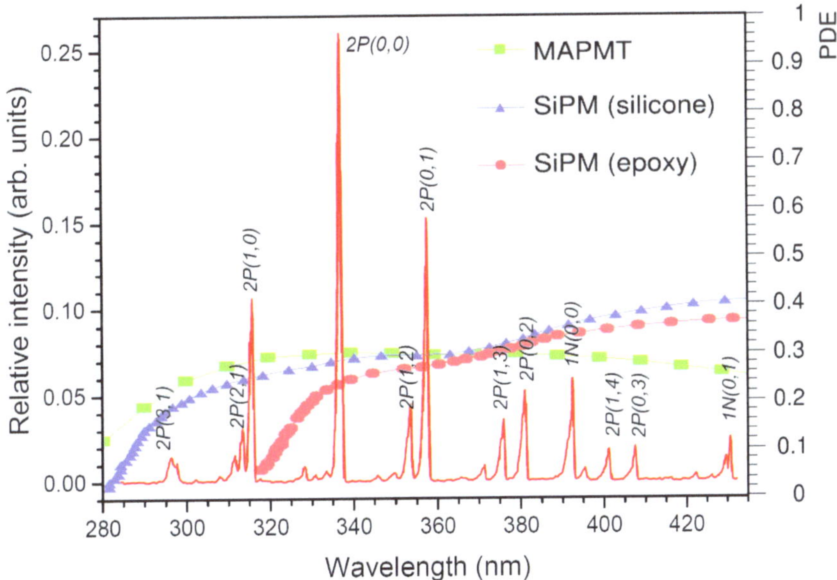

Figure 2. Fluorescence spectrum of nitrogen relaxation in the UV band from 280 nm to 435 nm at 800 hPa (about 2 km) measured by the AIRFLY Collaboration, taken from Ref. [24]. The area is scaled to unity. This shows that 25% of the spectrum intensity is due to the main line at 337.1 nm. PDEs of MAPMTs (calculated as the quantum efficiency present on the product datasheet and the collection efficiency of 80%, see the text) and SiPM arrays taken from Hamamatsu product information [14,15,27]. Image taken from Ref. [28].

The PDE for MAPMTs is defined as the quantum efficiency multiplied by the collection efficiency. The quantum efficiency of a MAPMT channel is the number of photoelectrons emitted by the photocathode divided by the number of incident photons; the collection efficiency is the probability that photoelectrons will land on the effective area of the first dynode of a MAPMT channel, assumed to be 80%. The PDE for SiPMs is the quantum efficiency multiplied by the fill factor and the avalanche triggering probability. In this case, the quantum efficiency is the number of electron-hole pairs generated by photons divided by the number of incident photons on the photosensitive area and the fill factor takes into account the dead area between one cell and the next one. In the plot it is visible that the SiPM with epoxy resin is not sensitive to the first three lines of the fluorescence spectrum, and MAPMTs and SiPM with silicone layer have a similar trend at the lower wavelengths up to about 380 nm, beyond which the SiPM array models become more efficient than MAPMTs. As anticipated, there are several differences in using SiPMs instead of MAPMTs, which include advantages and disadvantages and are indicated in Table 1.

Table 1. Differences between SiPMs and MAPMTs [29,30].

Characteristics	SiPMs	MAPMTs
Operation Voltage	$\sim$60 V (min. $\sim$30 V [30])	$\sim$1000 V
Gain	10^5–10^7	10^5–10^7
PDE	20–60%	20–40%
Spectrum	300–900 nm (peak 450 nm)	300–650 nm (peak 340 nm)
DCR	$\sim$$10^5$ cps/mm^2	$\sim$$10^2$ cps/mm^2
Behaviour in magnetic fields	good	bad
Temperature insensitivity	no	yes
Robustness and compactness	yes	no

One main difference is the operation voltage, which is lower for SiPMs than MAPMTs. They are not sensitive to magnetic fields and resistant to bright conditions: they saturate without damage, while the anodes of PMTs get damaged. Moreover, they are compact and light. However, the thermal noise is an issue for SiPMs that must be mitigated by operating a cooling system to maintain the silicon junction at a lower temperature. The dark-count rate of SiPMs at room temperature is about three orders of magnitude higher than that of traditional PMTs. The production of SiPMs is automated, while MAPMTs are assembled by hand. This lets us foresee a reduction of the cost of SiPMs with time and area, making the construction of large telescopes financially feasible in addition to providing scientific advantages.

3.2. EUSO-SPB1 (SiECA)

The Extreme Universe Space Observatory on a Super Pressure Balloon (EUSO-SPB1—formerly simply called EUSO-SPB) [10] was the first experiment of the JEM-EUSO program that operated with a SiPM camera onboard. It was installed on a Super Pressure Balloon (SPB) developed by NASA for the 2017 campaign. It was launched on 24 April 2017 at 23:51 UTC from Wanaka, New Zealand, as a mission of opportunity on a NASA Super Pressure Balloon test flight planned to circle the southern hemisphere supported by a fast stratospheric air circulation that develops twice a year at about 33 km (7 mbar) above the southern ocean. This circulation flows easterly in the southern fall and westerly in the southern spring. SPBs are designed to float at a constant displacement volume and consequently, at a constant altitude for months, complete an orbit every few weeks and terminate on land.

The scientific goals were to make the first observations of UHECRs by looking down on the atmosphere with a UV fluorescence detector from the near space altitude of 33 km; and measure background UV light at night over ocean and clouds. Unfortunately, after 12 days 4 h, the flight was terminated prematurely in the Pacific Ocean, about 300 km SE of Easter Island, due to a leak in the balloon. Despite the setbacks, the EUSO-SPB1 instrument operated successfully while aloft and returned about 60 GB of data.

The EUSO-SPB1 optics was made of two 1 m^2 Poly(methyl methacrylate) (PMMA) plastic Fresnel lenses, which focus the light on a main focal surface made of a PDM, providing a field of view of 11.1° × 11.1°. The design of the detector is visible in Figure 3. The figure includes pictures fo the telescope during the launch phase.

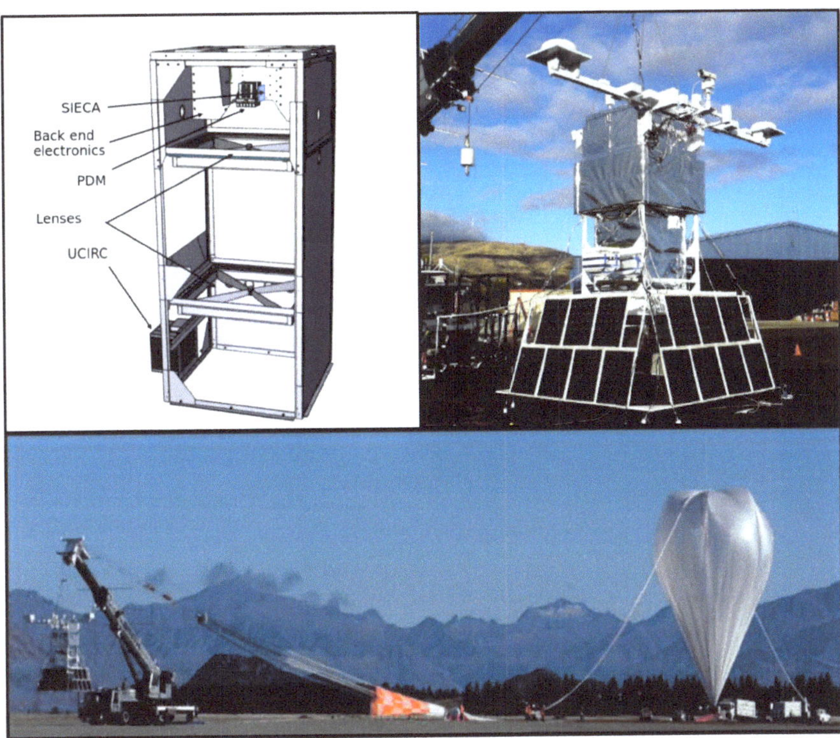

Figure 3. Images for the EUSO-SPB1 experiment. Sketch of the gondola with the PDM and an indication of the SiECA position (**top-left**). Pictures of EUSO-SPB1 before the launch (**top-right,bottom**). Images at the top taken from Ref. [31] and image at the bottom taken from Ref. [10].

An infrared (IR) camera system was developed and deployed to capture IR images of the area below EUSO-SPB1. The objective was to identify clouds and approximate the heights of cloud tops. The analysis of clouds is crucial for assessing the exposure of EUSO-SPB1, as high-altitude clouds diminish the available aperture of fluorescence telescopes operating at high altitudes [32]. The University of Chicago Infrared Camera (UCIRC) [33] was equipped with two identical IR cameras oriented toward the same region, featuring a field of view of $24° \times 30°$. Each camera was equipped with an IR filter. One filter transmitted IR light in the range 11.5–12.9 mm, while the other transmitted IR light in the range 9.6–11.6 mm. These specific ranges were chosen because they closely align with the typical black body peak for clouds. The methodology for measuring cloud color temperature, from which cloud top height can be derived, is elaborated in Ref. [34].

As an R&D test, the EUSO-SPB1 focal surface also included the Silicon Elementary Cell Add-on (SiECA) camera [35], with a 256 channel SiPM array that was mounted next to the PDM made of MAPMTs, on the same focal plane, and flown in a stand-alone sampling mode. This camera is described in more detail in the next section.

SiECA

The SiECA camera was built to test SiPMs as possible replacements for the MAPMTs used up to that time in the focal surface of telescopes of the JEM-EUSO program. The SiECA camera ($\sim 5 \times 5 \, cm^2$, with $\sim 3 \times 3 \, mm^2$ pixels) is visible in Figure 4 on the left side. On the right side of the same figure, it is installed next to the PDM ($\sim 17 \times 17 \, cm^2$, with $\sim 2.88 \times 2.88 \, mm^2$ pixels).

Figure 4. The SiECA camera (**left**); The SiECA camera assembled next to the EUSO-SPB1 PDM (with the noise-influenced EC of the PDM highlighted in red) (**right**). Images taken from Ref. [35].

Being placed next to the PDM, it would have allowed the observation of UHECR events with both the detectors. An example of a UHECR event simulated with the ESAF software [36] is visible in Figure 5 on the left, showing a proton event of energy 1.1×10^{19} eV and zenith angle 22.31° detected by both SiECA and the PDM. The signal was integrated over 41 GTUs, in order to have the track visible in a single frame. The panel on the right side of the same figure refers to a full camera test with a non-uniform light source, where the signal was integrated in 1 GTU. In both images, gaps between the MAPMTs of the PDM and between the SiPM arrays of SiECA are neglected.

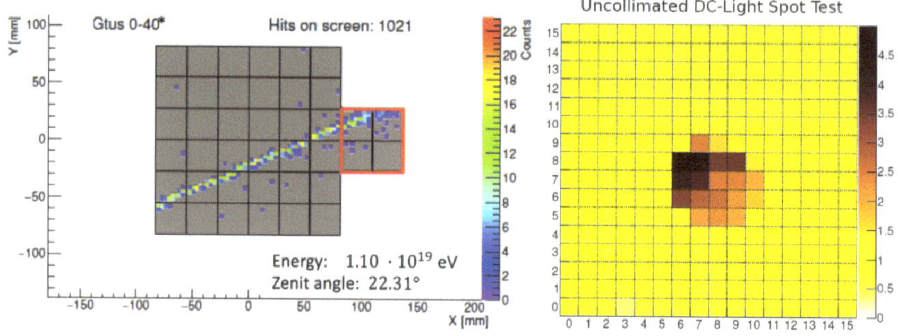

Figure 5. Simulation and measurement in the laboratory of the SiECA response. Simulation of a proton event of energy 1.1×10^{19} eV and zenith angle 22.31° detected by both SiECA and the PDM (**left**). The signal has been integrated over 41 GTUs and no background is added to the plot. Image taken from Ref. [37]. Full camera test with non-uniform light source. The response is the average photons detected per GTU (**right**). Image taken from Ref. [35]. Gaps between the MAPMTs of the PDM and between the SiPM arrays of SiECA are neglected in both images.

To make a full assessment of the currently available technology, hardware components were selected from available devices already on the market. Four 64-channel S13361-3050AS-08 SiPM arrays from Hamamatsu were arranged in a square of similar spacing to the EC of the JEM-EUSO PDMs. Details about the calibration of these sensors are available in Ref. [38]. The camera was biased by eight bias voltage generators (Hamamatsu C11204-02), and read out by eight Application Specific Integrated Circuit (ASIC) boards (Weeroc Citiroc 1A). The acquisition was controlled by a Field Programmable Gate Array (FPGA—Spartan6). Power was delivered by a commercial DC-DC converter to step down the available battery supply (26–32 V) to the required 5 V. Use of 3D printed mounting brace (off-white plastic between SiECA and PDM) provided thermal and electrical isolation.

Preliminary evaluation of the measurements made by SiECA during flight show that all channels of the camera are responding and sensitive to low-intensity light. Complications during the flight due to a leakage from the SPB led to an unexpected trajectory, descending during night and rising during day. Moreover, due to electrical interference, seemingly from the SiECA camera, causing instability in the EC of the PDM highlighted in red in Figure 4 on the right, SiECA was often switched off during the flight. Figure 6 represents the trajectory of EUSO-SPB1, and green circles indicate the periods in which SiECA operated.

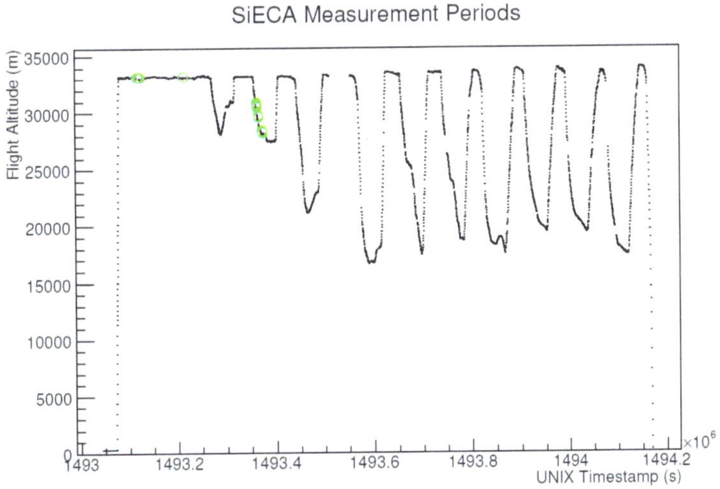

Figure 6. EUSO-SPB1 altitude with SiECA operation periods indicated in green circles. Descents indicate night cold cycles, then rising with the heat from the Sun. Image taken from Ref. [35].

During the short flight, SiECA collected nearly 400 events, mainly due to electronic noise, but none of them were related to UHECR events. However, the development of the SiECA camera and subsequent test flight on the EUSO-SPB1 mission has provided extensive information about the operation of SiPM at the low temperatures and pressures of the upper atmosphere. The SiECA camera was reproduced for further tests on ground [35].

3.3. EUSO-SPB2 (Cherenkov Telescope)

The Extreme Universe Space Observatory on a Super Pressure Balloon 2 (EUSO-SPB2) [11] is the successor experiment of EUSO-SPB1 and a pathfinder for the space-based mission POEMMA [13], a proposed dual satellite mission for the detection of UHECRs with energy above 1 EeV and very-high energy neutrinos, with energy above 1 PeV.

POEMMA will detect UHECRs via the fluorescence light emitted by EASs in the atmosphere. It will observe neutrinos by measuring the Cherenkov light emitted by EASs produced by the interaction or decay of charged leptons in the atmosphere after a neutrino propagates through the Earth and interacts near the surface. To perform both kinds of observations, each POEMMA telescope features a hybrid focal surface. As a pathfinder for POEMMA, EUSO-SPB2 was also designed to detect UHECRs and neutrinos, but with two different telescopes: a Fluorescence Telescope (FT) and a Cherenkov Telescope (CT), see Figure 7. The main goal for the FT was to observe UHECRs from above via the fluorescence technique for the first time, pointing directly downward. The CT was intended to measure cosmic rays with the direct Cherenkov technique for the first time, when pointed above the limb of the Earth, and to measure optical background for neutrino searches and search for astrophysical neutrinos when pointed below the limb.

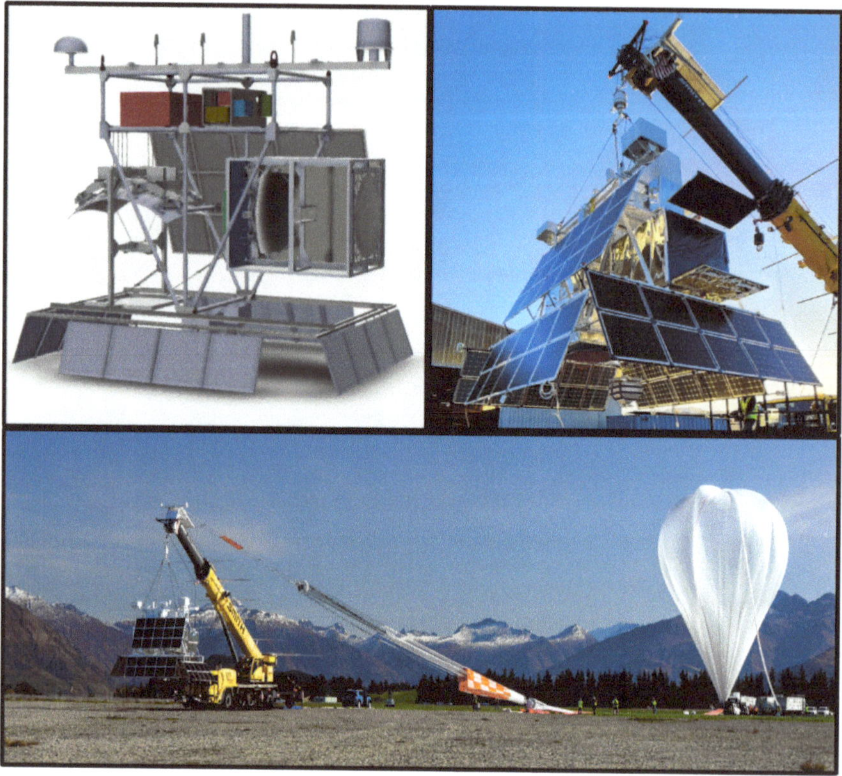

Figure 7. Images for the EUSO-SPB2 experiment. Sketch of the gondola with the fluorescence (**left in the sketch**) and the Cherenkov (**right in the sketch**) detectors (**top-left**). Pictures of EUSO-SPB2 before the launch (**top-right,bottom**). Images taken from Refs. [39–41].

The FT [42] was a 1 m diameter modified Schmidt telescope, focusing the light on three PDMs with MAPMTs (Hamamatsu R11265-M64-203 [14]) read out by ASIC boards (SPACIROC3) [43] that allowed a double pulse resolution of 6 ns and an integration time of 1 μs. The trigger [44] operated individually on each PDM. The field of view was $36° \times 12°$ with a fixed nadir pointing direction. A central data processor [45] connected the three PDMs and enabled a synchronized readout if one PDM was triggered. More information about the performance and the calibration of the FT is available in Refs. [39,42].

The CT [41] was a 1 m diameter modified Schmidt telescope featuring a bifocal alignment of four mirror segments and an elongated camera constructed with SiPM arrays designed for observing the Earth's limb. A more in-depth discussion of this telescope is provided in the next section, and additional information can be referenced in Refs. [41,46].

Additionally, the EUSO-SPB2 mission terminated prematurely, concluding after merely 1 day, 12 h, and 53 min, in the Pacific Ocean. This premature termination was attributed to a balloon leak. Nevertheless, the collected data indicates that all instruments were successfully activated at the floating altitude and operated effectively throughout the duration of the flight.

Cherenkov Telescope of EUSO-SPB2

The CT [41] optics is based on a Schmidt catadioptric system, with a 1 m^2 light collection area segmented into four identical mirrors, and with a focal length of 860 mm. A corrector lens at the entrance of the telescope controls aberrations. The effective aperture area is reduced to 0.785 m^2 taking into account the shadowing from the camera, transmission losses from the corrector plate, and reflection/scattering losses at the mirror. A particularity of the telescope is its bi-focal optics, achieved by rotating the optical axis of the lower and upper row of mirrors 0.4° relative to each other [47]. The bi-focal optics projects the image twice on the camera with a horizontal offset of 12 mm. This allows the discrimination between signals from noise or cosmic rays that interact directly with the instrument and that leave only one image, and the signals coming from outside the detector that, reflected by the bi-focal optics, produce a double image.

The camera focal plane is curved with a radius of 850 mm, see Figure 8. It hosts 8×4 SiPM arrays (Hamamatsu S14521-6050AN-04 [48]) in the horizontal and vertical, 4×4 channels (6.4×6.4 mm^2 each), for a total of 512 channels. With respect to other SiPM array models with smaller pixels (e.g., the one used for SiECA) the size of a single pixel in the CT worsens the spatial resolution. However, having fewer pixels in the telescope reduced the power consumption required by the read-out electronics. The used SiPM array has a broad wavelength sensitivity in the range 200–1000 nm, with a peak PDE of 50% at 450 nm. The wide spectral response reaches into the IR, which is ideal for our purpose, because, due to absorption and scattering effects, only the red components of the Cherenkov light arrive at the telescope from far-away showers, such as the ones expected from Earth-skimming neutrinos. At the operating voltage, direct optical crosstalk is only 1.5% and the temperature dependence of the gain is only $\sim$0.5%/°C.

The overall field of view is $12.8° \times 6.4°$ in the horizontal and vertical and can be pointed during the flight from horizontal to 10° below the Earth's limb.

The raw signals from each 4×4 SiPM array are routed into a Sensor Interface and Amplifier Board (SIAB), with an integration time of 10 ns. The 32 SIABs receive power and communication through the backplane. On a SIAB, two Multipurpose Integrated Circuit (MUSIC) chips shape and amplify the SiPM signals. The MUSIC chip is an 8-channel, low-power ASIC designed explicitly for SiPM applications in Cherenkov telescopes [49], which monitors the current output of each SiPM channel which is in turn digitized with a 24-bit Analog to Digital Converter (ADC). The temperature of the SiPMs is controlled with a thermistor mounted to the back of each SiPM array and will be used to offline correct temperature-dependent gain drifts of the SiPMs.

Figure 8. Half assembled Cherenkov camera of EUSO-SPB2 (Lego figures for scale). Image taken from Ref. [41].

The readout of the SiPMs is initiated whenever the bi-focal trigger condition is met [41]. As 90% of the light from a point source at infinity is contained in a 3 mm diameter circle on the focal surface, a typical EAS induced by an Earth-skimming tau-neutrino would usually produce a Cherenkov signal that illuminates only one pixel in the camera. With the bi-focal optics, the Cherenkov signal will be imaged into two pixels separated by one pixel. In Figure 9, the comparison between a simulated EAS (top) and an accidental-triggered event is shown.

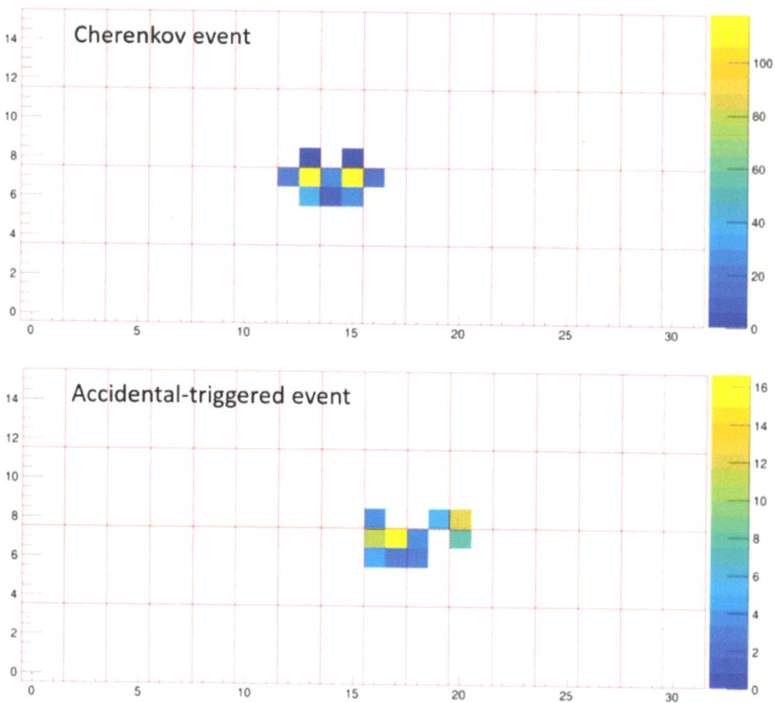

Figure 9. Camera event display of a simulated air-shower event (**top**) and accidental-triggered event (**bottom**). Values on the x and y axes are for pixels, and on the color scale are for counts. Image taken from Ref. [41].

The signals from all the 512 SiPM channels are saved in packets of 512 data frames (5.12 μs long, as the integration time of one frame is 10 ns) centered around the trigger, allowing to target very fast and bright signals such as the Cherenkov emission from EASs. The total power consumption of the Cherenkov telescope is estimated to be about 180 W during operation.

Information regarding the commissioning, calibration, and performance of the CT can be found in Ref. [46]. Throughout the flight, the CT conducted observations over 2 nights, recording approximately 30,000 candidate events, excluding events that did not meet the bifocal condition. The analysis of these events is currently ongoing. Multiple trigger scans were conducted to characterize the background trigger rate when observing both above and below the horizon. The CT was the first to attempt the observation of Cherenkov light from a suborbital altitude, and information about background light and expected trigger rates, including accidentals and actual events, was limited. As of the present writing, data obtained from these scans are still being analyzed to determine the actual rates for both accidentals and candidate events.

3.4. Mini-EUSO (Ancillary Sensors and Day/Night Transition)

The Multiwavelength Imaging New Instrument for the Extreme Universe Space Observatory (Mini-EUSO) [12,50] is the first telescope of the JEM-EUSO Program observing the Earth from space, onboard the ISS, through a nadir-facing UV-transparent window in the Russian Zvezda module. Mini-EUSO is capable of observing EASs generated by UHECRs with an energy above 10^{21} eV and detecting artificial showers generated with lasers from the ground. Other main scientific objectives of the mission are the search for nuclearites and strange quark matter, the study of atmospheric phenomena such as transient luminous events, meteors, and meteoroids, the observation of sea bioluminescence, and of artificial satellites and space debris. It is also mapping the nighttime UV emissions from the Earth, to study the background for the detection of the phenomena described before. The instrument was launched on 22 August 2019, from the Baikonur Cosmodrome and operates a few nights of data taking per month since October 2019. In Figure 10 Mini-EUSO is visible in the hands of a cosmonaut that operated it (Figure 10, left) and once connected to the UV-transparent window of the Zvezda module (Figure 10, right).

Figure 10. Pictures of Mini-EUSO onboard the ISS. Mini-EUSO in the hand of a cosmonaut (**left**) and connected to the UV-transparent window of the Zvezda module (**right**). Images taken from Ref. [12] (© AAS. Reproduced with permission).

It is based on an optical system employing two Fresnel lenses of 25 cm diameter and a main focal surface composed of a PDM with MAPMTs. The overall field of view is $44° \times 44°$.

The detector saves triggered transient phenomena with a sampling rate of 2.5 µs and 320 ms, as well as continuous acquisition at 40.96 ms scale. The 2.5 µs resolution is the highest resolution of Mini-EUSO, also called D1-GTU. Every D1-GTU data are read by the 36 ASICs (1 per MAPMT) and sent to the PDM FPGA board for acquisition and processing. Data are stored by the FPGA in a circular buffer of 128 GTUs. When the trigger conditions are met, 128 data frames centered at the trigger are stored. The trigger algorithm, described in Ref. [51], looks for fluctuations above the average value (dynamically updated) in each pixel. The excess signal must persist for more than 8 D1-GTU (20 µs) in any given pixel. The lens size limits the energy threshold to particles above 10^{21} eV, which so far have not been observed. The algorithm worked correctly, triggering on-ground Xenon flashers [52] and ELVES [53]. The 320 µs time resolution (D2-GTU) corresponds to the sum of 128 D1-GTUs and is calculated by the PDM acquisition board. If the trigger conditions are met, data are stored in a similar manner to D1-GTU. The 40.96 ms time resolution (D3-GTU) corresponds to the sum of $128 \times 128 = 16,384$ D1-GTUs and is calculated by the PDM acquisition board. Data frames are saved continuously without a trigger system, to perform a continuous monitoring of the UV emission of the Earth. It is used for the observation of meteors [54], the search for Strange Quark Matter [55], and for mapping of the night-time terrestrial UV emissions [50].

At the corners of the aperture plane, Mini-EUSO houses two cameras, one in the near-infrared (NIR; 1500–1600 nm) and one in the visible (VIS; 400–780 nm) band, to provide additional information in different frequency ranges. Data are acquired independently of the PDM [56] in 4 s exposure frames.

At the focal surface, next to the PDM, there are a few ancillary cameras, some of them with SiPMs, and are better discussed in the next section.

Ancillary Sensors and Day/Night Transition

As Mini-EUSO is designed to operate at nighttime, the CPU handles cycling between day and night based on the measurements performed by the UV sensors located in the same focal plane of the PDM. It hosts a single-pixel SiPM (Hamamatsu C13365) and two UV photodiodes (Analog Devices AD8304ARUZ with logarithmic response in the range 190–1000 nm, Lapis Semiconductor ML8511 with linear response in the wavelength range 280–400 nm), as visible at the bottom of Figure 11. They are used for information on the day/night transition. The ML8511 UV sensor is normally used for this purpose, although, all three sensors can be used. The photodiode with a linear response is less sensitive than the others, while the SiPM is very sensitive and gets saturated right away. Having multiple ancillary sensors allows us to measure in different illumination conditions. Moreover, they also serve for redundancy, in case one of them (or the electronics) breaks down.

Figure 11. The Mini-EUSO focal surface. The main camera is the PDM with 36 MAPMTs. On top of the PDM there is a 64-channel SiPM array, at the bottom of the PDM there are two UV-light sensors and a single-pixel SiPM. Image adapted from Ref. [12] (© AAS. Reproduced with permission).

Figure 12 shows the light measured by the ML8511 UV sensor as a function of time during a session of data taking. It is possible to see the transition between day and night every ∼45 min (about half the period of an ISS orbit around the Earth). To avoid fluctuations at the day/night terminator line, 2 thresholds are used to determine the transition from day to night: 60 ADC (Analog to Digital Conversion) counts (blue line); and 100 ADC counts (orange line) from night to day.

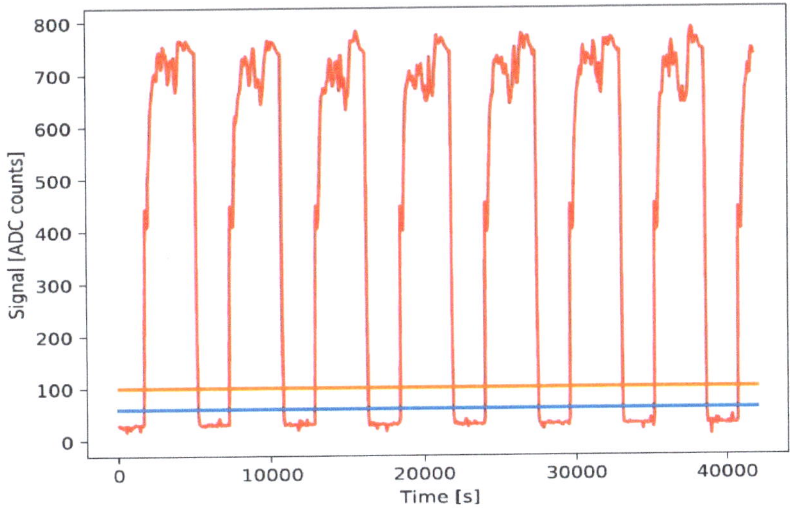

Figure 12. Measurements of the ML8511 UV sensor as a function of time. Mini-EUSO operates at nighttime when the sensor measures a value below 60 ADC counts. To avoid fluctuations at the day–night terminator line, 2 thresholds are used to determine the transition from day to night (60 ADC counts, blue line) and vice-versa (100 ADC counts, orange line). Image taken from [12] (© AAS. Reproduced with permission).

Mini-EUSO also includes a 64-channel SiPM array (Hamamatsu C14047-3050EA08) read independently of the PDM with a multiplexer. The goal of this camera is to test the SiPM array in space.

4. Conclusions

Several experiments of the JEM-EUSO program hosted and host SiPMs: EUSO-SPB1 with the fluorescence camera SiECA; EUSO-SPB2 with the focal surface of the Cherenkov Telescope; Mini-EUSO with the SiPM array and the single-pixel SiPM to evaluate day/night transition. The experience acquired over time was essential for the following experiments, in terms of interfacing them with the other apparatus made with MAPMTs, and in terms of development of the electronics necessary for their operation.

5. Future Directions

At the time of writing, it is foreseen to build in the near future the successor of EUSO-SPB2, called "POEMMA-Balloon with Radio" (or PBR). It will be a telescope with a double focal surface for the observation of the fluorescence and Cherenkov emissions, tiltable from 0° to 90°. It will host auxiliary detectors for the detection of strange quark matter and cameras sensitive to different wavelength ranges: IR, radio, gamma-ray, and X-ray.

For this purpose, there are several activities going on concerning SiPMs, such as the characterization of a few models of SiPMs and the development of the ASIC boards for the Cherenkov cameras. For the development of the gamma-ray detector for the detection, for example, of atmospheric events such as the terrestrial gamma flashers, some designs are under study to discriminate between gamma-rays and charged particles, using SiPMs as sensors.

All the effort is made to obtain stable and reliable detectors in view of future space-based experiments, like POEMMA.

Funding: This research was funded by Basic Science Interdisciplinary Research Projects of RIKEN and JSPS KAKENHI Grant (22340063, 23340081, and 24244042), by the ASI-INAF agreement n.2017-14-H.O, by the Italian Ministry of Foreign Affairs and International Cooperation, by the Italian Space Agency through the ASI INFN agreements Mini-EUSO n. 2016-1-U.0, EUSO-SPB1 n. 2017-8-H.0, OBP n. 2020-26-Hh.0, and EUSO-SPB2 n. 2021-8-HH.0, by NASA award 11-APRA-0058, 16-APROBES16-0023, 17-APRA17-0066, NNX17AJ82G, NNX13AH54G, 80NSSC18K0246, 80NSSC18K0473, 80NSSC19K0626, 80NSSC19K0627, 80NSSC18K0464 and 80NSSC22K1488 in the USA, by the French space agency CNES, by the Deutsches Zentrum für Luft- und Raumfahrt, the Helmholtz Alliance for Astroparticle Physics funded by the Initiative and Networking Fund of the Helmholtz Association (Germany), by Deutsche Forschungsgemeinschaft (DFG, German Research Foundation) under Germany Excellence Strategy—EXC-2094-390783311, by Slovak Academy of Sciences MVTS JEM—EUSO, by National Science Centre in Poland grants 2017/27/B/ST9/02162, 2020/37/B/ST9/01821, and 2022/45/B/ST2/02889 the Polish National Agency for Academic Exchange within Polish Returns Programme no. PPN/PPO/2020/1/00024/U/00001, by Mexican funding agencies PAPIIT-UNAM, CONACyT and the Mexican Space Agency (AEM), as well as VEGA grant agency project 2/0132/17, by grant S2018/NMT-4291 (TEC2SPACE-CM) "Desarrollo y explotación de nuevas tecnologías para instrumentación espacial en la Comunidad de Madrid", and by State Space Corporation ROSCOSMOS and the Interdisciplinary Scientific and Educational School of Moscow University "Fundamental and Applied Space Research".

Data Availability Statement: No new data were created or analyzed in this study. The data presented in this study are available in this article.

Conflicts of Interest: The author declares no conflict of interest. The funders had no role in the design of the study; in the collection, analyses, or interpretation of data; in the writing of the manuscript; or in the decision to publish the results.

Abbreviations

The following abbreviations are used in this manuscript:

ADC	Analog to Digital Converter
CMB	Cosmic Microwave Background
CT	Cherenkov Telescope (of EUSO-SPB2)
CTA	Cherenkov Telescope Array
redDCR	redDark Count Rate
EAS	Extensive Air Shower
EC	Elementary Cell
EUSO-SPB1	Extreme Universe Space Observatory on a Super Pressure Balloon (formerly EUSO-SPB)
EUSO-SPB2	Extreme Universe Space Observatory on a Super Pressure Balloon 2
FAMOUS	First Auger Multi-pixel photon counter camera for the Observation of Ultra-high-energy air Showers
FPGA	Field Programmable Gate Array
FT	Fluorescence Telescope (of EUSO-SPB2)
G-APD	Geiger-mode Avalanche Photodiode
GTU	Gate Time Unit
GZK	Greisen–Zatsepin–Kuzmin
IR	Infrared
ISS	International Space Station
JEM-EUSO	Joint Exploratory Missions for an Extreme Universe Space Observatory
MAPMT	Multi-Anode Photo-Multiplier Tubes
Mini-EUSO	Multiwavelength Imaging New Instrument for the Extreme Universe Space Observatory
MPPC	Multi-Pixel Photon Counters
MUSIC	Multipurpose Integrated Circuit
NIR	Near-Infrared
PDE	Photon Detection Efficiency
PDM	Photon-Detection Module
PMMA	Poly(methyl methacrylate)
POEMMA	Probe of Extreme Multi-Messenger Astrophysics

SIAB	Sensor Interface and Amplifier Board
SiECA	Silicon Elementary Cell Add-on
SiPM	Silicon Photomultipliers
SPB	Super Pressure Balloon
TSV	Through Silicon Via
UCIRC	Chicago Infra Red Camera
UHECR	Ultra-High Energy Cosmic Rays
UV	Ultraviolet
VIS	Visible

References

1. Renker, D.; Lorenz, E. Advances in solid state photon detectors. *J. Instrum.* **2009**, *4*, P04004. [CrossRef]
2. Acerbi, F.; Gundacker, S. Understanding and simulating SiPMs. *Nucl. Instrum. Methods Phys. Res. A* **2019**, *926*, 16–35. [CrossRef]
3. Gundacker, S.; Heering, A. The silicon photomultiplier: Fundamentals and applications of a modern solid-state photon detector. *Phys. Med. Biol.* **2020**, *65*, 17TR01. [CrossRef] [PubMed]
4. Nemallapudi, M.V.; Gundacker, S.; Lecoq, P.; Auffray, E. Single photon time resolution of state of the art SiPMs. *J. Instrum.* **2016**, *11*, P10016. [CrossRef]
5. Eckert, P.; Schultz-Coulon, H.C.; Shen, W.; Stamen, R.; Tadday, A. Characterisation Studies of Silicon Photomultipliers. *Nucl. Instrum. Meth.* **2010**, *A620*, 217–226. [CrossRef]
6. Ferri, A.; Acerbi, F.; Gola, A.; Paternoster, G.; Piemonte, C.; Zorzi, N. A comprehensive study of temperature stability of Silicon PhotoMultiplier. *J. Instrum.* **2014**, *9*, P06018. [CrossRef]
7. Parizot, E.; Casolino, M.; Picozza, P.; Ebisuzaki, T.; Bertaina, M.E.; Fuglesang, C.; Haungs, A.; Kajino, F.; Klimov, P.; Olinto, A.; et al. The JEM-EUSO Program for UHECR Studies from Space. *EPJ Web Conf.* **2023**, *283*, 06007. [CrossRef]
8. Abdellaoui, G.; Abe, S.; Adams, J.H., Jr.; Ahriche, A.; Allard, D.; Allen, L.; Alonso, G.; Anchordoqui, L.; Anzalone, A.; Arai, Y.; et al. EUSO-TA–First results from a ground-based EUSO telescope. *Astropart. Phys.* **2018**, *102*, 98–111. [CrossRef]
9. Adams, J.H., Jr.; Ahmad, S.; Allard, D.; Anzalone, A.; Bacholle, S.; Barrillon, P.; Bayer, J.; Bertaina, M.; Bisconti, F.; Blaksley, C.; et al. A Review of the EUSO-Balloon Pathfinder for the JEM-EUSO Program. *Space Sci. Rev.* **2022**, *218*, 3. [CrossRef]
10. Abdellaoui, G.; Abe, S.; Adams, J.H., Jr.; Allard, D.; Alonso, G.; Anchordoqui, L.; Anzalone, A.; Arnone, E.; Asano, K.; Attallah, R.; et al. EUSO-SPB1 Mission and Science. *Astropart. Phys.* **2023**, *154*, 102891. [CrossRef]
11. Eser, J.; Olinto, A.V.; Wiencke, L., for the JEM-EUSO Collaboration. Overview and First Results of EUSO-SPB2. *arXiv* **2023**, arXiv:2308.15693.
12. Bacholle, S.; Barrillon, P.; Battisti, M.; Belov, A.; Bertaina, M.; Bisconti, F.; Blaksley, C.; Blin-Bondil, S.; Cafagna, F.; Cambiè, G.; et al. Mini-EUSO mission to study Earth UV emissions on board the ISS. *Astrophys. J. Suppl. Ser.* **2021**, *253*, 2. [CrossRef]
13. Olinto, A.V.; Krizmanic, J.; Adams, J.H.; Aloisio, R.; Anchordoqui, L.A.; Anzalone, A.; Bagheri, M.; Barghini, D.; Battisti, M.; Bergman, D.R. et al. (POEMMA Collaboration). The POEMMA (Probe of Extreme Multi-Messenger Astrophysics) observatory. *J. Cosmol. Astropart. Phys.* **2021**, *2021*, 007.
14. Hamamatsu Corporation. *Datasheet 64-Channel MAPMT R11265-00-M64*; Hamamatsu Corporation: Hamamatsu City, Japan, 2009.
15. Hamamatsu Corporation. *Datasheet 64-Channel SiPM Array S12642-0808PA-50*; Hamamatsu Corporation: Hamamatsu City, Japan, 2014.
16. Schott Corporation. *Datasheet of the BG3 Filter*; Schott Corporation: Mainz, Germany, 2020.
17. Bisconti, F. for the JEM-EUSO Collaboration. EUSO-TA fluorescence detector. XXV ECRS 2016 Proceedings–eConf C16-09-04.3. *arXiv* **2016**, arXiv:1701.07091.
18. Karus, M. Development of a Calibration Stand for Photosensors for Extremely High-Energy Cosmic Ray Research. Ph.D. Thesis, Karlsruhe Institute of Technology, Karlsruhe, Germany, 2019.
19. Ambrosi, G.; Bissaldi, E.; Giglietto, N.; Giordano, F.; Ionica, M.; Paoletti, R.; Rando, R.; Simone, D.; Vagelli, V. for the CTA Consortium. Silicon Photomultipliers and front-end electronics performance for Cherenkov Telescope Array camera development. *Nucl. Instrum. Meth. A* **2017**, *845*, 8–11. [CrossRef]
20. Niggemann, T.; Assis, P.; Brogueira, P.; Bueno, A.; Eichler, H.M.; Ferreira, M.; Hebbeker, T.; Lauscher, M.; Mendes, L.; Middendorf, L.; et al. Status of the Silicon Photomultiplier Telescope FAMOUS for the Fluorescence Detection of UHECRs. In Proceedings of the 33rd International Cosmic Ray Conference (ICRC2013), Rio de Janeiro, Brazil, 2–9 July 2013.
21. Tosi, D. for the IceCube-Gen2 Collaboration. Surface detectors for IceCube-Gen2. *J. Instrum.* **2021**, *16*, P08057. [CrossRef]
22. Greisen, K. End to the cosmic-ray spectrum? *Phys. Rev. Lett.* **1966**, *16*, 748. [CrossRef]
23. Zatsepin, G.T.; Kuzmin, V.A. Upper limit of the spectrum of cosmic rays. *Sov. Phys. JETP* **1966**, *4*, 78–80.
24. Ave, M.; Bohacova, M.; Buonomo, B.; Busca, N.; Cazon, L.; Chemerisov, S.D.; Conde, M.E.; Crowell, R.A.; Di Carlo, P.; Di Giulio, C. et al. (AIRFLY Collaboration). Spectrally resolved pressure dependence measurements of air fluorescence emission with AIRFLY. *Nucl. Instrum. Methods Phys. Res. Section A* **2008**, *597*, 41–45. [CrossRef]
25. Onsemi Corporation. *Datasheet of C Series SiPMs*; Onsemi Corporation: Scottsdale, AZ, USA, 2014.
26. Hamamatsu Corporation. *Datasheet of S13360 Series SiPMs*; Hamamatsu Corporation: Hamamatsu City, Japan, 2021.

27. Hamamatsu Corporation. *Datasheet 64-Channel SiPM Array S13361-3050AS-08*; Hamamatsu Corporation: Hamamatsu City, Japan, 2015.

28. Bisconti, F. Performance of the Cosmic Ray Fluorescence Detector EUSO-TA. Ph.D. Thesis, Karlsruhe Institute of Technology, Karlsruhe, Germany, 2017.

29. Hamamatsu Corporation; New Jersey Institute of Technology. *Low Light Detection: PMT v. SiPM*; Webinar by S. Piatek; Hamamatsu Corporation: Hamamatsu City, Japan, 2016.

30. Onsemi Corporation. *Introduction to the Silicon Photomultiplier (SiPM)*; Application Note; Onsemi Corporation: Scottsdale, AZ, USA, 2011.

31. Bacholle, S., for the JEM-EUSO Collaboration. The EUSO-SPB Instrument. *Proc. Sci.* **2017**, *301*, 384.

32. Adams, J.H., Jr.; Ahmad, S.; Albert, J.N.; Allard, D.; Anchordoqui, L.; Andreev, V.; Anzalone, A.; Arai, Y.; Asano, K.; Ave Pernas, M.; et al. The JEM-EUSO observation in cloudy conditions. *Exp. Astron.* **2015**, *40*, 135. [CrossRef]

33. Allen, L.; Rezazadeh, M.; Meyer, S.; Olinto, A.V., for the JEM-EUSO Collaboration. UCIRC: Infrared Cloud Monitor for EUSO-SPB1. *Proc. Sci.* **2017**, *301*, 436.

34. Tabone, I.; Bertaina, M.; Carli, D.; Ferrarese, S.; Cremonini, R.; Cassardo, C., for the JEM-EUSO collaboration. The WRF model contribution to the Cloud Top Height retrieval in EUSO-Balloon experiment. *Proc. Sci.* **2015**, *236*, 642.

35. Painter, W. Development of a SiPM Camera for Detection and Measurement of Fluorescence Emission from Extensive Air-Showers Generated by Ultra High Energy Cosmic Rays. Ph.D. Thesis, Karlsruhe Institute of Technology, Karlsruhe, Germany, 2019.

36. Abe, S.; Adams, J.H.; Allard, D.; Alldredge, P.; Anchordoqui, L.; Anzalone, A.; Arnone, E.; Baret, B.; Barghini, D.; Battisti, M.; et al. Developments and results in the context of the JEM-EUSO program obtained with the ESAF Simulation and Analysis Framework. *Eur. Phys. J. C* **2023**, *83*, 1028. [CrossRef]

37. Haungs, A.; Painter, W.; Bertaina, M.E.; Bortone, A.; Menshikov, A.; Renschler, M., for the JEM-EUSO Collaboration. SiECA: Silicon Photomultiplier Prototype for Flight with EUSO-SPB. *Proc. Sci.* **2017**, *301*, 442.

38. Renschler, M.; Painter, W.; Bisconti, F.; Haungs, A.; Huber, T.; Karus, M.; Schieler, H.; Weindl, A. Characterization of Hamamatsu 64-channel TSV SiPMs. *Nucl. Instrum. Methods Phys. Res. Sect. A* **2018**, *888*, 257–267. [CrossRef]

39. Kungel, V.; Battisti, M.; Filippatos, G.; Heibges, T.; Kuznetsov, E.; Mese, M.; Meyer, S.S.; Parizot, P.; Scotti, V.; Sternberg, P.; et al., for the JEM-EUSO Collaboration. EUSO-SPB2 Fluorescence Telescope Calibration and Field Tests. *Proc. Sci.* **2023**, *444*, 468.

40. NASA Super Pressure Balloon Mission Terminated Due to Anomaly. Available online: https://blogs.nasa.gov/superpressureball oon/category/2023-campaign/euso/ (accessed on 1 September 2023).

41. Bagheri, M.; Bertone, P.; Fontane, I.; Gazda, E.; Judd, E.G.; Krizmanic, J.F.; Kuznetsov, E.N.; Miller, M.J.; Nachtman, J.; Onel, Y.; et al., for the JEM-EUSO Collaboration. Overview of Cherenkov Telescope on-board EUSO-SPB2 for the Detection of Very-High-Energy Neutrinos. *Proc. Sci.* **2021**, *395*, 1191.

42. Filippatos, G., for the JEM-EUSO Collaboration. EUSO-SPB2 Fluorescence Telescope in-flight performance and preliminary results. *Proc. Sci.* **2023**, *444*, 251.

43. Blin, S.; Barrillon, P.; de La Taille, C.; Dulucq, F.; Gorodetzky, P.; Prévôt, G., for the JEM-EUSO Collaboration. SPACIROC3: 100 MHz photon counting ASIC for EUSO-SPB. *Nucl. Instrum. Meth. A* **2018**, *912*, 363. [CrossRef]

44. Filippatos, G.; Battisti, M.; Belov, A.; Bertaina, M.; Bisconti, F.; Eser, J.; Mignone, M.; Sarazin, F.; Wiencke, L. Development of a cosmic ray oriented trigger for the fluorescence telescope on EUSO-SPB2. *Adv. Space Res.* **2022**, *70*, 2794. [CrossRef]

45. Osteria, G.; Scotti, V.; Perfetto, F., for the JEM-EUSO Collaboration. The Data Processor of the EUSO-SPB2 Telescopes. *arXiv* **2023**, arXiv:2310.03012.

46. Romero Matamala, O.F., for the JEM-EUSO Collaboration. Commissioning, Calibration, and Performance of the Cherenkov Telescope on EUSO-SPB2. *Proc. Sci.* **2023**, *444*, 1041.

47. Kungel, V.; Bachman, R.; Brewster, J.; Dawes, M.; Desiato, J.; Eser, J.; Finch, W.; Huelett, L.; Olinto, A.V.; Pace, J. et al., for JEM-EUSO Collaboration. EUSO-SPB2 Telescope Optics and Testing. *Proc. Sci.* **2021**, *395*, 412.

48. Hamamatsu Corporation. *Datasheet 16-Channel SiPM Array S14521-6050AN-04*; Hamamatsu Corporation: Hamamatsu City, Japan, 2018.

49. Gómez, S.; Gascón, D.; Fernández, G.; Sanuy, A.; Mauricio, J.; Graciani, R.; Sanchez D. MUSIC: An 8 channel readout ASIC for SiPM arrays. *Proc. SPIE* **2016**, *9899*. [CrossRef]

50. Casolino, M.; Barghini, D.; Battisti, M.; Blaksley, C.; Belov, A.; Bertaina, M.; Bianciotto, M.; Bisconti, F.; Blin, S.; Bolmgren, K.; et al. Observation of night-time emissions of the Earth in the near UV range from the International Space Station with Mini-EUSO detector. *Remote Sens. Environ.* **2023**, *284*, 113336. [CrossRef]

51. Battisti, M.; Barghini, D.; Belov, A.; Bertaina, M.; Bisconti, F.; Bolmgren, K.; Cambiè, G.; Capel, F.; Casolino, M.; Ebisuzaki, T. Onboard performance of the level 1 trigger of the mini-EUSO telescope. *Adv. Space Res.* **2022**, *70*, 2750. [CrossRef]

52. Battisti, M.; Bertaina, M.; Arnone, E.; Pretto, G.; Sammartino, G., for the JEM-EUSO Collaboration. Analysis of EAS-like events detected by the Mini-EUSO telescope. *Proc. Sci.* **2023**, *444*, 353.

53. Romoli, G., for the JEM-EUSO Collaboration. Study of multiple ring ELVES with the Mini-EUSO telescope on-board the International Space Station. *Proc. Sci.* **2023**, *444*, 223.

54. Barghini, D.; Battisti, M.; Belov, A.; Bertaina, M.E.; Bertone, S.; Bisconti, F.; Capel, F.; Casolino, M.; Cellino, A.; Ebisuzaki, T.; et al. Analysis of meteors observed in the UV by the Mini-EUSO telescope onboard the International Space Station. *Eur. Sci. Congr.* **2021**, *15*, EPSC2021-243.

55. Casolino, M., for the JEM-EUSO Collaboration. Search for Strange Quark Matter from the International Space Station with the Mini-EUSO experiment. *Proc. Sci.* **2023**, *444*, 1410.
56. Turriziani, S.; Kelund, J.; Tsuno, K.; Casolino, M.; Ebisuzaki, T. Secondary cameras onboard the Mini-EUSO experiment: Control Software and Calibration. *Adv. Space Res.* **2019**, *64*, 1188–1198. [CrossRef]

 *instruments*

Article

A Configurable 64-Channel ASIC for Cherenkov Radiation Detection from Space

Andrea Di Salvo [1,*], Sara Garbolino [1], Marco Mignone [1], Stefan Cristi Zugravel [1,2], Angelo Rivetti [1], Mario Edoardo Bertaina [1,3] and Pietro Antonio Palmieri [1,3]

[1] Section of Torino, Istituto Nazionale di Fisica Nucleare, INFN, 10125 Turin, Italy
[2] Department of Electronics and Telecommunications, Politecnico di Torino, 10129 Turin, Italy
[3] Physics Department, University of Torino (UniTO), 10124 Turin, Italy
* Correspondence: andrea.disalvo@to.infn.it

Abstract: This work presents the development of a 64-channel application-specific integrated circuit (ASIC), implemented to detect the optical Cherenkov light from sub-orbital and orbital altitudes. These kinds of signals are generated by ultra-high energy cosmic rays (UHECRs) and cosmic neutrinos (CNs). The purpose of this front-end electronics is to provide a readout unit for a matrix of silicon photo-multipliers (SiPMs) to identify extensive air showers (EASs). Each event can be stored into a configurable array of 256 cells where the on-board digitization can take place with a programmable 12-bits Wilkinson analog-to-digital converter (ADC). The sampling, the conversion process, and the main digital logic of the ASIC run at 200 MHz, while the readout is managed by dedicated serializers operating at 400 MHz in double data rate (DDR). The chip is designed in a commercial 65 nm CMOS technology, ensuring a high configurability by selecting the partition of the channels, the resolution in the interval 8–12 bits, and the source of its trigger. The production and testing of the ASIC is planned for the forthcoming months.

Keywords: ASIC; CMOS; cosmic rays; Cherenkov light; SiPM

Citation: Di Salvo, A.; Garbolino, S.; Mignone, M.; Zugravel, S.C.; Rivetti, A.; Bertaina, M.E.; Palmieri, P.A. A Configurable 64-Channel ASIC for Cherenkov Radiation Detection from Space. *Instruments* **2023**, *7*, 50. https://doi.org/10.3390/instruments7040050

Academic Editors: Matteo Duranti and Valerio Vagelli

Received: 20 October 2023
Revised: 21 November 2023
Accepted: 26 November 2023
Published: 7 December 2023

1. Introduction

Ultra-high energy cosmic rays (UHECRs) and neutrinos (UHENUs) passing through the atmosphere generate extensive air showers (EASs). The relativistic particles produced in the cascade emit Cherenkov light collimated with the direction of the EAS propagation which can be used to track the direction and the energy of the parent UHECR/UHENU. A telescope based on an optical system that focuses the light on a focal plane made of SiPMs can image such light and derive the EAS parameters. The signal induced in the sensor from sub-orbital height such as ~30–40 km or low Earth orbit (LEO), such as a ~500 km height, has a time extension of tens of nanoseconds. This result shows a dependance from the angle between the EAS direction and the axis marked out by the telescope according to simulations. This very short time extension of the signal demands a sampling rate of at least 100 MHz to achieve the required time resolution. In addition, the store of a waveform associated to the event allows for the discrimination of the EAS-related event from those originated by direct cosmic ray hits. Due to these requirements, the implementation of an ASIC is mandatory and its design is inspired by present and future projects in the field of UHECR and UHE neutrino astronomy such as Extreme Universe Space Observatory-Super Pressure Balloon 2 (EUSO-SPB2) [1], an on-board stratospheric balloon platform, or Terzina [2] and POEMMA [3] space-based missions.

The EUSO-SPB2 mission used a NASA Super Pressure Balloon for a test flight and it flew on 13 May 2023 from Wanaka (New Zealand). Unfortunately, the balloon developed a hole in the envelope and was terminated over the Pacific Ocean after only about 37 h of flight. The payload is composed of two telescopes, one devoted to fluorescence light measurements from UHECR EAS with energy above 1 EeV by pointing a multi-anode photo-multiplier tube (MAPMT) camera to nadir. The other telescope is reserved for the Cherenkov emission of CR EAS, with an energy target above 1 PeV, by collecting the Cherenkov light with a focal surface made of SiPMs. The latter instrument is based on a modified Schmidt telescope of 1m diameter, where four mirrors focus the light in two points on the camera area. This bifocal alignment is adopted to reduce the background noise discriminating the light that comes from outside the telescope tube (namely, two spots) and a direct cosmic ray (one spot). The SiPMs are provided by Hamamatsu and the pixel camera is formed by 512 units. The integration time is 10 ns and the system can readout 512 frames centered around the trigger point with a field-of-view (FoV) of 6.4° in zenith and 12.8° in azimuth. The entire telescope can be rotated from a horizontal level to 10° below the terrestrial limb. The JEM-EUSO collaboration is currently planning a new NASA SPB mission named POEMMA-Balloon with Radio (PBR) with a targeted launch in 2026. The payload will be a single telescope hosting both a fluorescence and a Cherenkov camera as conceived for the POEMMA mission.

NUSES [4] is an orbital mission whose target is the study of the Sun–Earth environment and the analysis of cosmic radiation. The satellite is composed of two payloads called Ziré [5] and Terzina [2] to detect cosmic rays with energies below 250 MeV and UHECRs beyond 100 PeV, respectively. NUSES is designed to work for three years orbiting at a $\sim$550 km altitude. Terzina is equipped with a Cherenkov telescope to detect the light from EAS generated by UHECRs in the atmosphere, pointing at the terrestrial limb. The mission will also be used to monitor the ground emissions for a characterization of the light intensity. Considering the available volume, a dual mirror architecture has been selected to maximize the focal length, which is $\sim$925 mm. The sensor area is made up of 10 SiPMs of 8×8 pixels arranged in two rows of five tiles each. The FoV is 7.2° (zenith) and 2.5° (azimuth), leading to a cross-section of 140×360 km^2.

The Probe Of Extreme Multi-Messenger Astrophysics (POEMMA) is composed of two identical satellites flying in formation at an altitude of 525 km with the ability to observe overlapping regions during moonless nights at angles ranging from nadir to just above the limb of the Earth, but also with independent pointing strategies to exploit at the maximum of the scientific program of the mission. Each telescope is composed of a wide (45°) FoV Schmidt optical system with an optical collecting area of over 6 m^2. The focal surface (FS) of POEMMA is composed of a hybrid of two types of cameras: about 90% of the FS is dedicated to the POEMMA Fluorescence Camera (PFC), while the POEMMA Cherenkov camera occupies the crescent moon shaped edge of the FS, which images the limb of the Earth. The PFC is composed of 55 JEM–EUSO PDMs based on MAPMTs for a total of $\sim$130,000 channels. The gate time unit (GTU) for the PFC is 1 μs. The much faster POEMMA Cherenkov camera is composed of silicon photo-multipliers.

PBR is still a conceptual study but it will be largely inspired by EUSO-SPB2. EUSO-SPB2, Terzina, and PBR represent three different pathfinder missions currently under development (Terzina and PBR) or just terminated (EUSO-SPB2) by the scientific community planning the future POEMMA mission.

In the context of UHECR and UHE neutrino astronomy, an ASIC was implemented to realize an entire acquisition chain, from the signal acquisition to its on-board conversion and readout. In the following section, the main features of this chip are described.

2. Materials and Methods

In this section, an overview of the conceived camera structure and the ASIC architecture, whose concepts are previously explored in [6], is provided. It could represent a viable solution for the forthcoming PBR and Terzina payloads, appropriately re-adapted for the specific needs of each experiment.

2.1. Camera Architecture

Figure 1a depicts a simplified representation of the camera hosting the SiPMs tiles and the board developed for the ASICs. This design matches the exact needs of Terzina but it could represent at the same time one module of the PBR camera, which is expected to be formed by a larger number of SiPM tiles (3–6 times larger, the exact number still being under definition). These boards are connected to each other through a high-speed high-density socket (blue place holders) while the connections to the FPGA board are ensured with a bank of shielded twisted pairs used for the low-voltage differential signaling (LVDS) signals (orange ones). The latter are 12 differential pairs for each ASIC and they are used both to send configuration and commands to the chip and to receive data. Figure 1b illustrates the layout of the entire camera made of the two boards. The red square is a benchmark to highlight the orientation of the tiles and the connectors.

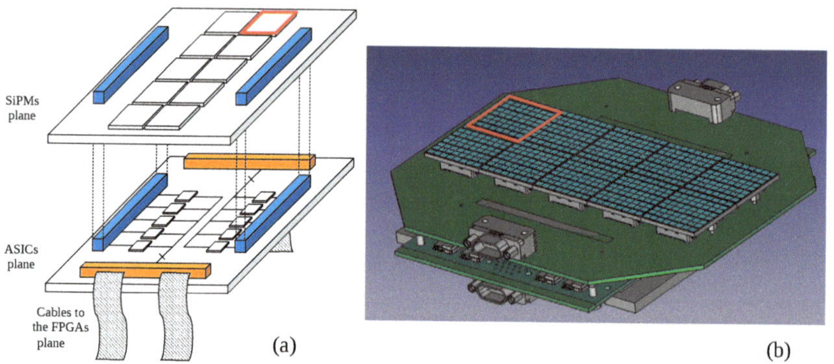

Figure 1. (**a**) Block diagram of a tower-like structure formed by the SiPMs plane and the ASICs board, (**b**) layout of the design where the red square is a SiPM tile used as reference.

2.2. ASIC Architecture

Figure 2 shows a block diagram representation through the ASIC hierarchy. In the upper part of the image, the general partition of the ASIC is illustrated. The main digital logic is implemented in the End-Of-Column (EOC) block and in the same area, two serializers, an SPI module, and the configuration registers are located. This unit manages the configuration of the chip as well as the stages of the finite state machines (FSM) to realize the sampling, the digitization, and the readout of the data. These tasks are distributed along the 64 channels whose circuitry is schematized in the bottom part of the picture. The SiPMs induce a current signal, which is amplified by a dedicated stage. The output of the amplifier is then split between a pair of comparators and 256 cells. The first branch is used to locally generate a trigger and this information is merged into the EOC. If an event is detected, the ASIC builds a hitmap to be sent out to the FPGA and it raises a flag. In the simplest case, if the event is accepted and confirmed by the FPGA, the digital conversion can take place, as well as the readout. These steps are achieved by distributing the second branch among the array of cells. The channel can be configured to use smaller partitions of cells based on a minimum group of 32 units, which is called a section. The user can choose between 32, 64, or 256 cells operating the derandomization of the Poissonian distributed events. Indeed, in this way, the data processing involving the sampling, the conversion, and the readout can be carried out in parallel, strongly reducing the waste of time. In the blue box of Figure 2, a schematic of the analog cell is represented. Each cell is equipped with a Wilkinson ADC, which is composed of a capacitor (C) and a comparator. The capacitor stores the analog information of the signal during the sampling stage, closing S_0 and S_2 switches, and a pointer is used to control S_0 cell by cell. If the digitization is enabled, both S_0 and S_2 are opened while the bottom plate of C is connected to the output of a ramp generator with S_1. During the data conversion, both the ramp generator and a Gray counter work with the same phase. In other terms, the output of the ramp generator is increased by the least significant bit (LSB) while the Gray counter is incremented by one at each clock cycle. When the condition $V_{IN} \geq V_{BL}$ is achieved, where V_{BL} is a voltage threshold, the output of the comparator enables the local storage of the current Gray counter value. Another key feature of the system is the programmability of the resolution in the range between 8 bits and 12 bits. This characteristic also allows for the use of the ASIC in applications where a high granularity is not a severe requirement. Moreover, the dead time due to the digitization can be considerably reduced. However, the maximum time needed for digitizing is $2^{N}T_{clk}$, where N is 12 bits and T_{clk} is the clock period used by the digital logic, which is equal to 5 ns because of the working frequency of 200 MHz, namely $\sim$20.5 μs are required to complete the process with the nominal resolution. After the digital conversion, a data packet is available for the transmission and the dedicated serializer receives the data to build an event. The stream is composed of 8 bits for the header used for the alignment with the FPGA, 6 bits reserved for the information of the packet, 16 bits providing a timestamp, and 9 bits of address followed by the event. Because of the configurability of the parameters, the data length depends on the partition and the resolution chosen. For instance, if the user selects the 32-cells partition and a resolution of 12 bits, the total length of the data stream will be 440 bits per channel, including the headers. At ASIC level, 64 channels contribute to the event, thus this is described by 28,160 bits. Since the serializers work in double data rate at 400 MHz, in terms of time this takes slightly more than 35 μs to transmit the entire data packet. The ASIC is designed in a commercial 65 nm CMOS technology.

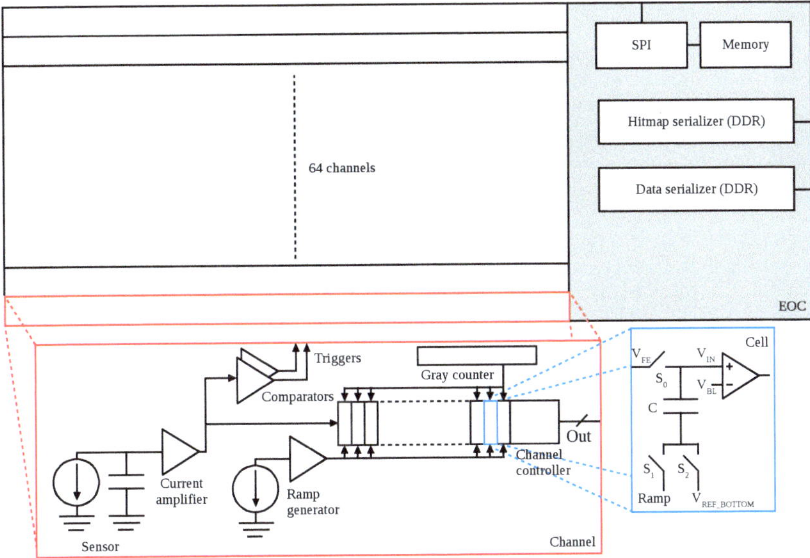

Figure 2. Block diagram representation of the ASIC, a channel (red box), and a cell (blue box).

Hitmap Generator

In order to reduce both the digitization and the readout times, a hitmap generator was implemented. In Figure 2, the two comparators were previously pointed out as the source of the internal triggers defining two distinct thresholds. As the point spread function of the optics has a size comparable with the pixel size, the double threshold will act in selecting both occurrences in which the spot size is localized in one pixel or distributed on more pixels. In the first case, a signal passing the high threshold in one pixel will be enough for triggering. In the other case, a lower threshold will be used for triggering, but will be conditioned to the presence of two or more nearby pixels above a low threshold. These signals are collected at the EOC level to discriminate between the cases reported in Figure 3.

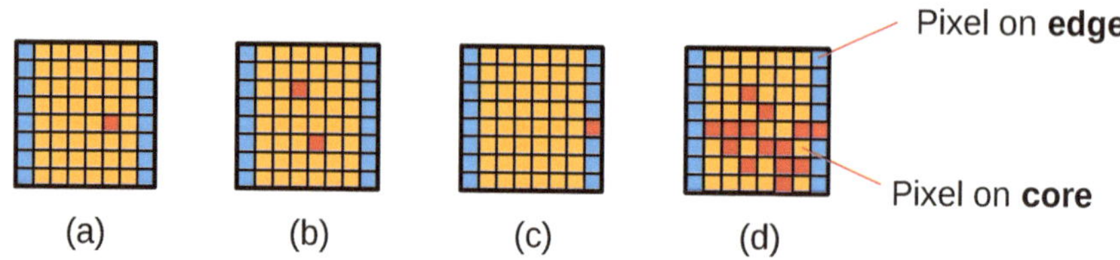

Figure 3. Possible hitmaps configurations: (**a**) single pixel, (**b**) coincidence, (**c**) edge-pixel and (**d**) light pollution where the orange area indicates the core-pixels, the blue regions illustrate the edge-pixels and the red boxes show the hit pixels.

(a)   The hit is concentrated into a single pixel, candidating a direct cosmic ray or a possible neutrino-event if no bi-focality is implemented in the optical system;

(b)   The event is split into two corresponding pixels in the case of a system which adopts the bi-focality to discriminate a signal from a UHECR or UHE neutrino with respect to a direct cosmic ray hit in the detector;

(c)   An event occurs at the edge of the SiPM tile, suggesting an adjacent event in the nearby tile;

(d)   A light pollution due to a city can flare a large area of the tile.

These combinations are taken into account with a dedicated FSM designed in the EOC. Each pixel has a configuration bit to set it as a core-pixel or edge-pixel. The core implements a combinatorial circuitry based on fast-OR chains to detect two or more coincidences. The edge area carried out a similar digital detector with a programmable feature. The user can select three modes for the generation of the hitmap.

- TIME WINDOW: a time window is defined using a programmable 6-bits register with a step of 5 ns. At the end of this period, a hitmap is generated, nevertheless a physical event occurred or not;

- HIGH THRESHOLD + COINCIDENCE + EDGE DRIVEN: the high threshold is continuously checked. If it is verified, a read hitmap request is sent to the FPGA, otherwise the coincidence condition is tested within another programmable window. If this case also fails, the pixels at the edges are monitored and a read request is generated as well;

- FPGA REQUEST: the last mode is reserved for the active interaction with the FPGA, which can require the local generation of a hitmap. This feature results in being useful when the user wants an entire snapshot of the entire focal plane.

If the event is detected, a read request warns the FPGA that a new hitmap is available. The FPGA can reject the request, thus the FSM inside EOC comes back to a monitoring state, otherwise the signal can be accepted and the ASIC prepares the hitmap data packet for the serialization. The stream is made of 98 bits, segmented in a header part and a data part. The time required for the transmission of each hitmap is equal to ~122 ns. If the FPGA is not responding, after an acknowledged period, the ASIC cleans its own registers and continues the sampling stage.

3. Results

The implementation of the circuitry described in the previous section was carried out using electronic design automation (EDA)/computer-aided design (CAD) tools for both the analog implementation and the digital one.

Figure 4 shows the physical implementation of a section. In the picture, the single cell unit is also magnified to highlight the analog and digital areas. The capacitor C, where the analog information coming from the SiPMs is locally stored during the sampling, is easily recognizable. On the digital side, the routing connects the 12-bits length latches dedicated to registering the digitized value to the rest of the logic. The size of a cell is 21 μm $\times$ 56 μm.

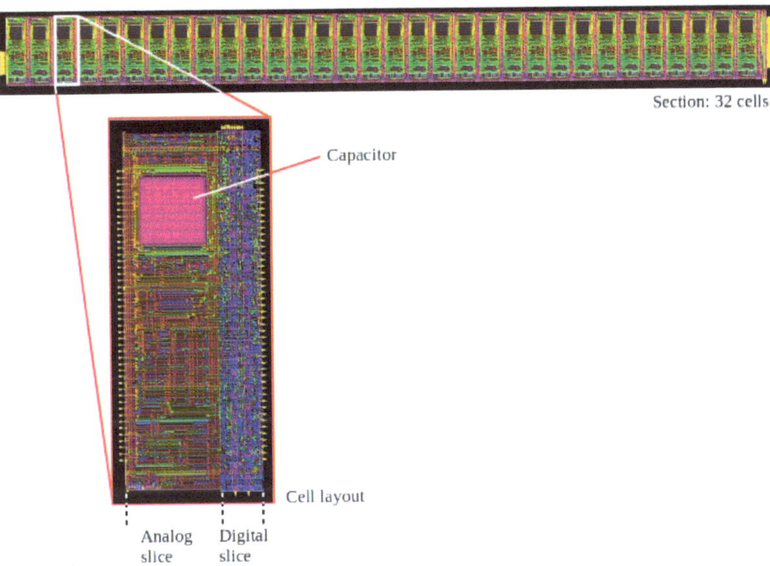

Figure 4. Layout of a section and schematic view of a cell.

Table 1 reports a very preliminary estimation of the digital power. These results are obtained using the toggle rate of the behavioral simulation. The table collects the main digital blocks implemented at the channel level of hierarchy. Since, in this evaluation, the routing contribution is not considered, the final power consumption may be incremented.

Table 1. Power estimation for the digital blocks integrated into the channel. The leftmost two columns refer to the power and area of a single block, N indicates the number of units, and the rightmost two columns report the total power and area, respectively.

Block	Power (μW)	Area (μm^2)	N	Power$_T$ (μW)	Area$_T$ (μm^2)
Cell	2.5	352	256	640.0	90,225
Section	2.5	96	8	20.0	772
Section controller	31.4	840	4	125.4	3360
Gray counter	42.4	376	1	42.4	376
Gray decoder	2.9	357	1	2.9	357
Channel controller	430.0	8083	1	430.0	8083
Total				1260.7	103,173

4. Discussion

This section is dedicated to some considerations of the features implemented in the ASIC.

The partition of the cell array into smaller segments was adopted to derandomize the signals detected by the SiPMs. The latter are Poissonian-distributed, thus the probability of receiving n events is.

$$P_n = \mu^n \frac{e^{-\mu}}{n!},\tag{1}$$

where μ is the average number of events in the time window. Let us suppose we now have an event rate of 100 kHz and a dead-time of 8 μs. This means that the probability of losing an event is:

$$P_{loss} = 1 - e^{-0.8} \simeq 0.55. \tag{2}$$

By changing the acquisition strategy, N segments can be used and the probability is evaluated as:

$$P_{loss} = 1 - \sum_{n=0}^{N} \mu^n \frac{e^{-\mu}}{n!}. \tag{3}$$

If N = 4 segments are considered, P_{loss} would be equal to about 0.5 %. This is a significant reduction compared to the 1-segment case. Moreover, another benefit derived from using sections concerns the reduced time transmission. Indeed, for short-on-time events, a smaller partition becomes convenient. This advantage is further increased by the configurability of the resolution. This feature has a direct impact on both the time conversion itself and on the data transmission. Since the conversion time depends on $2^N T_{clk}$, where N is the number of bits and T_{clk} is the frequency of the clock—which is equal to 5 ns—the minimum time required for the conversion is 1.28 µs (8 bits) and the maximum time required is ∼20.5 µs (12 bits). These results also affect the data transmission because of the length of the words appended in the data stream. Nevertheless, several combinations are possible as well; let us consider the best and worst cases in terms of time required to complete the data sending. For the first one, a partition of 32 cells and a resolution of 8 bits are assumed. With this configuration, the transmission time in DDR takes ∼23.8 µs considering all the data packets for 64 channels and including the headers budget. Conversely, selecting the 256-cells mode and the maximum possible resolution of 12 bits, the same process requires ∼249 µs. Finally, an estimation of the total time required by the conversion and the data readout in the two cases is possible. In the first configuration, the system takes slightly more than 25 µs to complete these tasks, while in the second case, ∼270 µs are needed. Naturally, the cross combinations between segmentation and resolution define a trade-off that must be evaluated considering the application requirements.

5. Conclusions

In the context of UHECR and UHE neutrino astronomy, a 64-channel ASIC for an SiPMs readout carried out in a commercial 65 nm CMOS technology has been described. The chip is designed to be interfaced with a 8 × 8 pixel matrix, providing a hitmap to an FPGA for preliminary pattern recognition. A channel can be partitioned into sections of 32, 64, or 256 cells to derandomize the signal at 200 MHz. The main benefit of this architecture is a considerable saving of time during the processing of the data. Another key feature is represented by the configurability of the resolution in the interval 8–12 bits. The readout is realized with two dedicated serializers, one reserved to the hitmap and the other to the data. The serializers work at 400 MHz in the double data rate and the data stream sent out to the FPGA has a variable length according to the parameters selected during the configuration of the ASIC.

Author Contributions: Conceptualization, A.D.S., M.E.B. and A.R.; methodology, A.D.S., S.G. and A.R.; software, A.D.S., S.G., M.M., S.C.Z. and P.A.P.; validation, A.R.; investigation, A.D.S., S.G., M.M., S.C.Z. and P.A.P.; resources, M.E.B. and A.R.; data curation, A.D.S., M.M. and P.A.P.; writing—original draft preparation, A.D.S.; writing—review and editing, A.D.S. and M.E.B.; visualization, A.D.S. and M.M.; supervision, A.D.S. and M.E.B.; project administration, M.E.B. and A.R.; funding acquisition, M.E.B. All authors have read and agreed to the published version of the manuscript.

Funding: The front-end electronics described in this paper are part of the development performed within the ASI-INFN agreement for EUSO-SPB2 n.2021-8-HH.0 and its amendments.

Conflicts of Interest: The authors declare no conflict of interest.

References

1. Eser, J.; Olinto, A.V.; Wiencke, L. Overview and First Results of EUSO-SPB2. In Proceedings of the 38th International Cosmic Ray Conference (ICRC2023), Nagoya, Japan, 26 July–3 August 2023; Volume 444.
2. Burmistrov, L.; NUSES Collaboration. Terzina on board NUSES: a pathfinder for EAS Cherenkov Light Detection from space. *arXiv* **2023**, arXiv:2304.11992.
3. Olinto, A.V.; Krizmanic, J.; Adams, J.H.; Aloisio, R.; Anchordoqui, L.A.; Anzalone, A.; Bagheri, M.; Barghini, D.; Battisti, M.; Bergman, D.R.; et al. The POEMMA (Probe of Extreme Multi-Messenger Astrophysics) observatory. *J. Cosmol. Astropart. Phys.* **2021**, *2021*, 007 [CrossRef]
4. De Mitri, I.; Di Santo, M. The NUSES space mission. *J. Phys. Conf. Ser.* **2023**, *2429*, 012007. [CrossRef]
5. De Mitri, I.; Di Santo, M. The Ziré instrument on board the NUSES space mission. In Proceedings of the 38th International Cosmic Ray Conference (ICRC2023), Nagoya, Japan, 26 July–3 August 2023; Volume 139.
6. Tedesco, S.; Di Salvo, A.; Rivetti, A.; Bertaina, M. A 64-channel waveform sampling ASIC for SiPM in space-born applications. *J. Instrum.* **2023**, *18*, C02022. [CrossRef]

instruments

Article

A Silicon-Photo-Multiplier-Based Camera for the Terzina Telescope on Board the Neutrinos and Seismic Electromagnetic Signals Space Mission

Leonid Burmistrov *,† on behalf of the NUSES Collaboration

Département de Physique Nuclèaire et Corpusculaire, Université de Genève, 1205 Genève, Switzerland
* Correspondence: leonid.burmistrov@unige.ch
† Current address: Ecole de Physique, Quai Ernest-Ansermet 24, 1205 Genève, Switzerland.

Abstract: NUSES is a pathfinder satellite project hosting two detectors: Ziré and Terzina. Ziré focuses on the study of protons and electrons below 250 MeV and MeV gamma rays. Terzina is dedicated to the detection of Cherenkov light produced by ultra-high-energy cosmic rays above 100 PeV and ultra-high-energy Earth-skimming neutrinos in the atmosphere, ensuring a large exposure. This work mainly concerns the description of the Cherenkov camera, composed of SiPMs, for the Terzina telescope. To increase the data-taking period, the NUSES orbit will be Sun-synchronous (with a height of about 550 km), thus allowing Terzina to always point toward the dark side of the Earth's limb. The Sun-synchronous orbit requires small distances to the poles, and as a consequence, we expect an elevated dose to be received by the SiPMs. Background rates due to the dose accumulated by the SiPM would become a dominant contribution during the last two years of the NUSES mission. In this paper, we illustrate the measured effect of irradiance on SiPM photosensors with a variable-intensity beam of 50 MeV protons up to a 30 Gy total integrated dose. We also show the results of an initial study conducted without considering the contribution of solar wind protons and with an initial geometry with Geant4. The considered geometry included an entrance lens as one of the options in the initial design of the telescope. We characterize the SiPM output signal shape with different μ-cell sizes. We describe the developed parametric SiPM simulation, which is a part of the full Terzina simulation chain.

Keywords: SiPM; UHECR; Cherenkov telescope

Citation: Burmistrov, L.; on behalf of the NUSES Collaboration. A Silicon-Photo-Multiplier-Based Camera for the Terzina Telescope on Board the Neutrinos and Seismic Electromagnetic Signals Space Mission. *Instruments* 2024, *8*, 13. https://doi.org/10.3390/instruments8010013

Academic Editor: Antonio Ereditato

Received: 26 October 2023
Revised: 18 January 2024
Accepted: 31 January 2024
Published: 20 February 2024

1. Introduction

The NUSES (Neutrinos and Seismic Electromagnetic Signals) space mission aims to explore new technological and scientific pathways in cosmic-ray and multi-messenger astrophysics [1–3]. The NUSES satellite will have a ballistic trajectory without orbital control. At the beginning of life (BoL), it will operate at an altitude of 535 km. The high inclination of the 97.8 deg (LTAN = 18:00) orbit will allow a Sun-synchronous location of the satellite along the day–night boundary. The NUSES satellite will host two scientific apparatuses, namely, Zirè [4,5] and Terzina [2,6], and will operate for at least three years.

Zirè consists of a scintillating fiber tracker, a stack of plastic scintillator counters, an array of LYSO crystals, an active VETO system, and a Low-Energy Module (LEM). It will perform spectral measurements of electrons, protons, and light nuclei below a few to hundreds of MeV. Zirè will also test innovative detection techniques for 0.1–10 MeV photons and monitor the Van Allen radiation belt.

Terzina is a telescope specifically designed for the detection of the Cherenkov light emitted by extensive air showers (EASs) induced by ultra-high-energy cosmic rays (UHE-CRs) and neutrinos in the Earth's atmosphere. We expect to detect proton-induced air showers with energies above 100 PeV for the first time from space. This constitutes a very relevant exploration for space-based instruments like POEMMA [7,8].

We briefly describe the telescope below, but this contribution is mainly devoted to the SiPM characterization, radiation tests, and simulations.

2. Terzina Telescope

Terzina is composed of an optical head unit; a focal plane assembly (FPA), including the photosensitive SiPM camera and the readout integrated circuits; a thermal control system; and an external harness and electronic unit, which is in a separate box, shielded from irradiation. The optical head unit is a Schmidt–Cassegrain near-UV–optical telescope.

The Terzina optical system achieves an effective area of about 0.1 m^2 and an equivalent focal length of 930 mm, with the diameter of the circle containing 80% of the photons being less than 1 mm^2. The results in this proceeding refer to the initial configuration. It consists of two hyperbolic mirrors (primary and secondary) and a corrector lens, integrated with the flat FPA in a hole at the center of the primary mirror. The corrective lens at the entrance of the telescope shields the internals from radiation; however, it is heavy, and we have considered moving it to the secondary mirror.

The Terzina photosensitive (detection) plane is composed of two rows of five SiPM arrays manufactured by FBK [9–11]. Each tile is made of 8 × 8 channels of 3 × 3 mm^2 pixels. The sensitive areas of a pixel are limited by the packaging to about 2.4 × 2.7 mm^2. We chose the NUV-HD-MT [10] (Near-Ultraviolet High-Density Metal Trench) technology provided by FBK for our application.

3. SiPM Signal Waveform Characterization

The design of the electronic readout chain and the SiPM simulation (Section 4) are dependent on the SiPM response and signal shape. As a result, we carried out SiPM characterization, which is covered in this part, along with a signal shape analysis.

Figure 1 (left panel) shows the bias and readout schematics of SiPMs. A SiPM operates at reverse bias voltage; however, it can be powered with positive and negative voltages, providing positive or negative output signals accordingly. This change of sign needs to be taken into account while designing and optimizing front-end electronics and the entire readout chain. In the case of single-photon operation or weak light fluxes, we are not able to see a signal with a conventional oscilloscope without an amplifier. Hence, we used a short-duration ($\sim$25 ps) light-pulse laser with a 370 nm wavelength to flash all the μ-cells at the same time. As a result, we obtained a $\sim$1 V signal with a shape roughly equal to a single μ-cell response. We measured a fast signal rising edge of $\sim$200 ps with a 2 GHz bandwidth oscilloscope.

A SiPM signal tail can be fitted with an exponent, as shown in Figure 1 (right panel), and its duration can be quantified as an exponent decay time (τ). This time is a function of the μ-cell capacitance, hence its size. We did not observe any significant change in the signal shape with the variation in the bias voltage.

We performed a set of measurements with three different configurations: with LED (450 nm) and an integration sphere, with a laser (370 nm) and an integration sphere, and with a laser only. Figure 2 (left panel) summarizes the obtained results. As expected, the shortest decay time was measured with the laser. As the integration sphere induces an additional time spread, it has been removed.

If we want to minimize the pulse duration only, we have to choose a SiPM with a small μ-cell size. However, SiPMs with small μ-cells have a lower PDE due to a smaller fill factor, and the signal is linearly dependent on the PDE. Hence, the final choice of the SiPM μ-cell of 30 µm is considered to be a balance between these parameters.

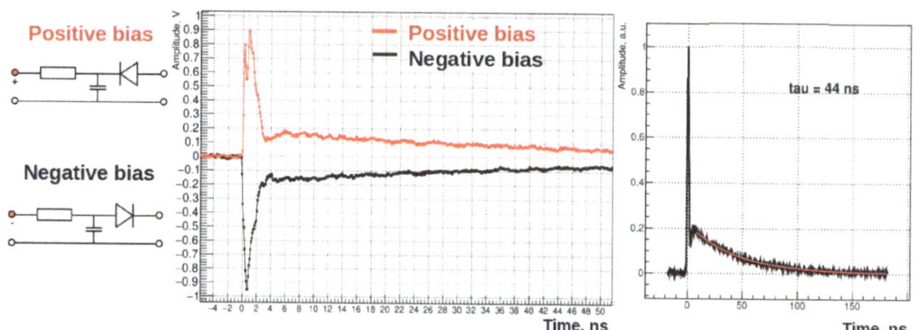

Figure 1. (**Left**): Schematics of bias (red) and readout of SiPM sensors. We used 100 nF capacitor and 8 kΩ resistor. (**Center**): SiPM response to a 370 nm laser with ∼25 ps pulse duration (FWHM) in saturation mode (all μ-cells produce an avalanche). The waveform was recorded with a 2 GHz oscilloscope. (**Right**): Fit with an exponent of the SiPM signal tail. We measured 44 ns decay time for 25 µm cell size.

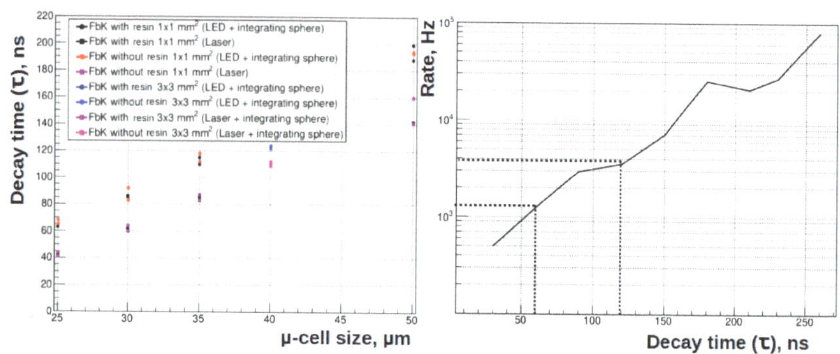

Figure 2. (**Left**): Measurements with an LED and laser, with/without an integration sphere, of the SiPM signal decay time as a function of the μ-cell size. (**Right**): Rate at the fixed threshold (7 p.e.'s) as a function of the SiPM signal decay time obtained with the parametric simulation of the SiPM response.

4. Parametric Simulation of the SiPM Response

The Terzina full simulation chain is a sophisticated instrument for assessing the experiment's physics performance. Considering that the SiPM response affects the performance, we have developed lightweight simulation software that can be easily integrated into our framework.

In this section, we briefly discuss the SiPM response simulation, while an exhaustive description of the SiPM physics can be found in reference [11]. Our parametric simulation [12] takes the following as an input:

- The pulse template of the SiPM response to a single p.e., where the amplitude is scaled linearly to increase the over-voltage;
- The probability of direct optical cross-talk (OCT) and after-pulse (AP) as a function of the SiPM over-voltage;
- After-pulse decay time;
- The root mean square error (RMSE) of the SiPM gain variation and the RMSE of the electronic noise.

The simulation is realized as a recursion of the physics processes: every time we generate the primary p.e., there is a probability of generating a secondary p.e. via OCT or AP (see Figure 3, left-top and left-bottom panels, respectively). The after-pulse process simulation takes into account a μ-cell recovery time, which strongly depends on the μ-cell area (and so too its capacitance mainly). The model does not take into account the over-voltage variations. The AP and OCT probabilities are functions of their mother avalanche amplitudes. We assume a linear dependency between the AP/OCT probability and the amplitude of the avalanche. (The initial probability of the AP/OCT is given for fully recovered μ-cell.)

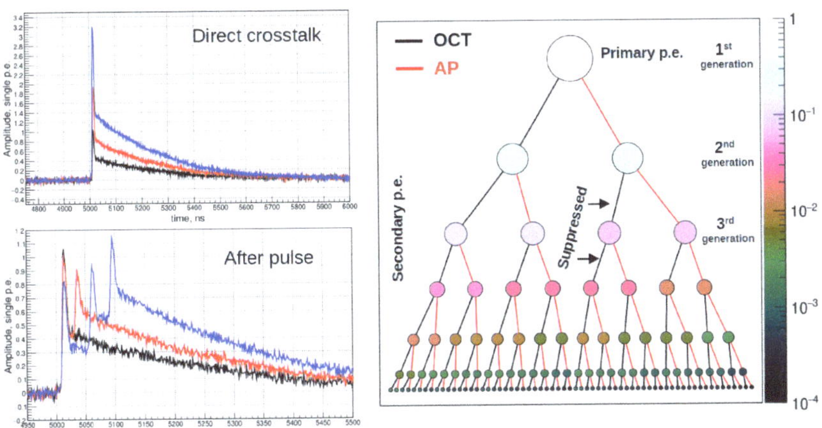

Figure 3. (**Left top**): Examples of simulated waveforms with the OCT processes only (the AP probability is set to 0). (**Left bottom**): Example of the waveform with the AP process only (the OCT probability is set to 0). (**Right**): The process history avalanche tree; the black line corresponds to OCT processes, while the red line corresponds to AP processes. For this diagram only, we set the equal probabilities of AP and OCT to 30% and the decay time constant of the μ-cell recovery time to 50 ns. Starting from the third generation, one can see the slight suppression from the right (AP) with respect to the left side (OCT). The AP/OCT probability with respect to the first generation is represented by the Z-axis.

The process history avalanche tree of a single p.e. generation is shown in the right panel of Figure 3. The AP and OCT probabilities depend on the history depth and the origin of the second p.e. generation. This is explained by the AP and OCT probability adjustment, which is a function of its mother avalanche amplitude: when a μ-cell is not fully recharged, it cannot generate a full avalanche, and so the probability of AP/OCT drops. To illustrate the net effect of the μ-cell recovery, we use the avalanche tree diagram. We set equal probabilities of AP and OCT to 30% and the decay time constant of the μ-cell recovery time to 50 ns. Starting from the third generation, one can see the slight suppression from the right (AF—red line) with respect to the left side (OCT—black line).

This simulation is a part of the full simulation chain of the Terzina telescope. It is used to estimate the trigger rates for the pure expected noise as a function of the electronic threshold. Our background is mostly sourced from the SiPM dark count rate (DCR) and the night-glow background (NGB).

Using this parametric simulation, we investigated two variables and their effects on the trigger rate for an identical electronic threshold:

- Signal decay time: see Figure 2 (right panel). We confirmed our expectation: in the case of the AC/DC coupling readout, by reducing the decay time of the SiPM signal, one can significantly reduce the fake rate while keeping the same sensitivity to the signal.
- Different bandwidths of the electronics. We found a significant rate variation with the preamplifier bandwidth. However, the front-end preamplifier is not completely defined; therefore, we do not claim the expected rates.

This study shows that changing from the 40 (FBK's most used μ-cell size) to the 30 µm cell size reduces the background rate by a factor of 4 (Figure 2, left).

5. Dose Estimation for Terzina Telescope

The trapped electrons and protons in the Van Allen Belt are responsible for only a part of the radiation damage to the SiPMs and electronics. In this work, we do not consider solar wind protons, which can have a large impact on the total received dose. The effect will also depend on the exact time of the flight, currently foreseen to be 2026, at a time close to the maximum of the solar cycle activity.

We used SPENVIS [13] in two different ways to simulate the expected background signal on the camera. First, SPENVIS itself can estimate an accumulated dose by assuming an oversimplified geometry. We chose SPENVIS geometry with two spheres, one internal and one external. The external one is an absorber or can be considered a shield made of fused silica, and the internal one, made of silicon, is a sensitive volume, where we measure the dose. The dose in silicon with a variable layer of fused silica is shown in Figure 4. For unprotected surfaces in orbit, the radiation level is as high as $\sim 10^6$ rad $= 10^4$ Gy in 3 yr of exposure.

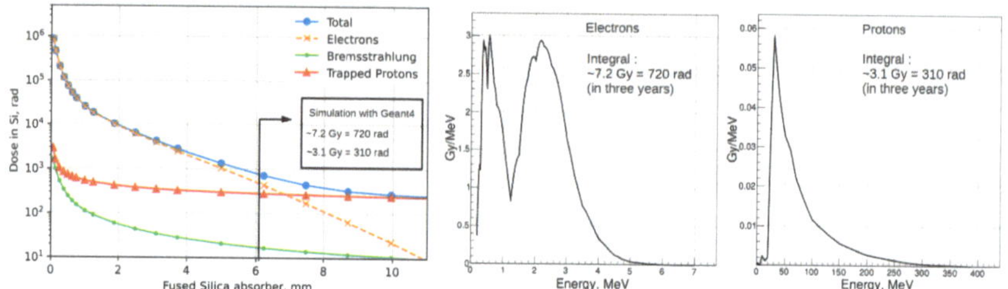

Figure 4. (**Left**): Dose due to trapped protons and electrons in silicon obtained with SPENVIS for 3 years in Terzina's orbit vs. thickness of fused silica shielding. (**Center,Right**): Accumulated dose in the aluminum volume located in the vicinity of the camera and its readout electronics as a function of particle energy (for electrons and protons, respectively). The total accumulated dose in 3 years in the aluminum plane is 7.2 Gy for electrons and 3.1 Gy for protons. The electrons after ~ 1 MeV produce more secondary gammas with high enough energy to deposit a dose in the SiPM camera.

Moreover, SPENVIS provides us with the fluxes of protons and electrons for orbits in space, trapped in the Van Allen Belt [14]. We injected the obtained SPENVIS electron and proton fluxes with an isotropic angular distribution as an input in our Geant4 [15,16] simulation of the telescope. In Figure 5, one can see the initial geometry of the Terzina telescope we consider. It consists of ~ 12 mm of the corrector lens made of fused silica (no crystalline quartz), primary and secondary mirrors made of aluminum with ~ 2 mm thick walls, and 2 mm of aluminum on the satellite walls. Figure 4 (right panel) shows the results of the simulation and its contribution as a function of particle energy. The estimated dose for three years of operation for only trapped protons and electrons is 7.2 Gy for electrons and 3.1 Gy for protons. (Depending on the particle type, the delivered total radiation dose causes different non-ionizing energy losses, hence causing different impacts on the SiPM DCR.) This estimation was made for the aluminum volume placed in the vicinity of the SiPM camera. The second peak in the distribution can be explained by the fact that electrons with energies higher than ~ 1 MeV produce more secondary gammas with sufficient energy to reach and create a dose in the SiPM camera (Figure 4). With the same simulation, we estimated the Cherenkov background photon rate generated in the corrector lens: 181 Hz/mm produced by electrons and 0.16 Hz/mm produced by protons.

The doses obtained with the Geant4-based simulation are in agreement with SPENVIS for ~ 6 mm of fused silica protection.

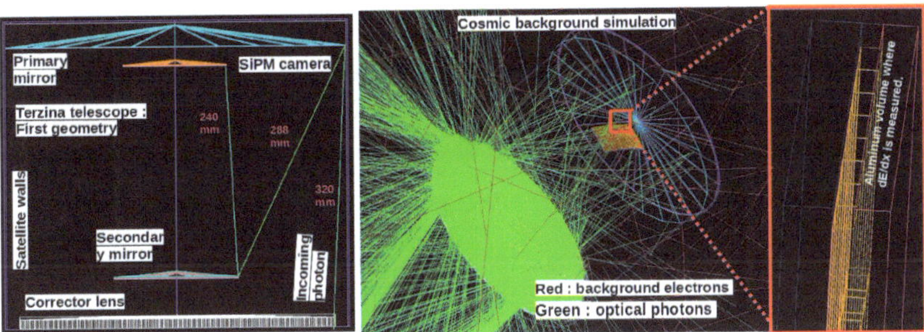

Figure 5. (**Left**): The initial geometry of the Terzina telescope. The input light (green trace) corrects its initial trajectory due to the corrector lens. Then, the primary spherical mirror reflects the photon toward the secondary mirror, which finally focalizes it on the SiPM camera. (**Center**): Example of background electrons with ~6 MeV energy, producing Cherenkov light in the corrector lens and inducing dE/dx losses in the SiPM camera. The image represents a 300 μs snapshot in space. The background electrons are in red, while the Cherenkov light is in green. One can notice that these background photons have a wide angular spread. (**Right**): Zoom on the SiPM camera showing the aluminum volume, exactly where we count the dose deposition. One can see the camera with separate pixels made of silica.

6. Irradiation of the SiPM with Protons

We performed the first proton irradiation test at IFJ PAN in Krakow [17] with a 50 MeV proton beam (Figure 6). The proton beam spot had a circular shape with a 35 mm diameter and homogeneity better than 5% with respect to the mean fluence. We tested SiPMs with different μ-cell sizes (25, 30, 35, 40, 50 μm) and channel sizes (1×1 mm^2 and 3×3 mm^2) with and without an entrance window (resin protective layer).

After every new step of irradiation, we measured the current–voltage characteristics (IV curves) to monitor increases in the DCR (Figure 6). In the reverse bias mode, the absolute voltage range was between 30 V and 50 V. In total, we performed eight irradiation sessions. After each session, the total doses received by the test samples were 1, 2, 3, 5, 7, 10, 20, and 30 Gy.

As expected, we observe an increase in the DCR with the accumulated dose. In Figure 6 (top panel), the black curve corresponds to the IV measurements taken before irradiation, and one can see a 2–3 $\times$ 10^{-8} A current at 42 V (~10 V over-voltage). At a 1 Gy accumulated dose, the current for the same over-voltage has increased by two orders of magnitude (5–6 $\times$ 10^{-6} A) with respect to the non-irradiated case. For a 30 Gy accumulated dose, the current at 42 V bias reaches 2–3 $\times$ 10^{-4} A. This large increase in the DCR, even for relatively small doses, degrades the sensitivity and increases the power consumption. This preliminary study shows the importance of understanding radiation damage and the precise estimation of the dose in an orbit. The increase in the DCR due to the dose is by far dominant with respect to other backgrounds, and it is the main limiting factor for the sensitivity of the Terzina telescope. Examples of single p.e. signals before and after irradiation are shown in Figure 6 (bottom panel). The red curve corresponds to a single p.e. signal before irradiation, and the black one is the waveform taken in dark conditions after receiving a 30 Gy dose in total. The amplifier used for this measurement is AC-coupled. One can see that the signal amplitudes are compatible with single, double, and even triple p.e.'s.

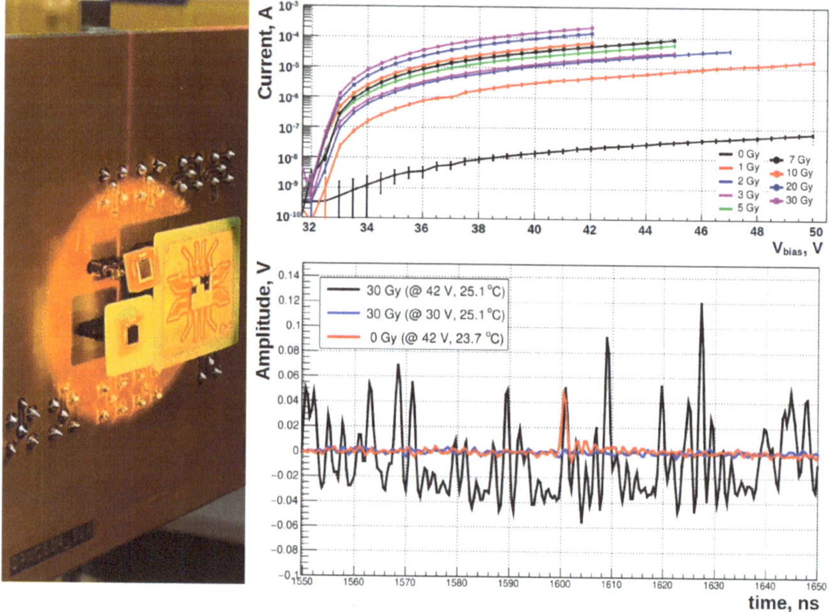

Figure 6. (**Left**): Photo of the SiPM samples installed in the IFJ PAN proton beam facility. The light spot indicates the proton beam location. (**Top**): The measured SiPM current as a function of the bias voltage for different accumulated doses (1×1 mm^2, 25 µm cell, without resin). (**Bottom**): The waveform of the SiPM signal recorded in dark conditions and corresponding to single p.e. signals before (red) and after irradiation (black). The curve in blue shows the output SiPM signal before the breakdown voltage. It demonstrates the stability of the signal baseline.

7. Background Created in a Window of a Photo Sensor

The Terzina telescope has a relatively small primary mirror; therefore, our expected signal is at the level of 7–10 p.e.'s [6]. To minimize light pollution, we require no scintillation material in the telescope. However, charged background particles (mainly electrons and positrons) can create Cherenkov light in the corrector lens or mirror substrate in the telescope. Even if a source of optical photons is located relatively far from the light sensors (see Figure 5), it can still create additional undesired noise in the SiPM camera. However, the background photons will be spread around the sensitive area (SiPM camera), making it easier to separate them from the signal. In other words, this light will be de-focalized, unlike signal photons focalized in two–three pixels.

Usually, photosensors contain a transparent window, which is a radiator of Cherenkov light. Unavoidably, PMTs suffer from this source of background. To evaluate the possible effect of this dangerous background, we used PMTs.

We carried out the test with 5.6 GeV electrons provided by the DESY [18] accelerator facility. The experimental setup contained a PMT and plastic scintillator for the trigger, installed after the PMT. We perform two tests: first (configuration A), the PMT was directly exposed to a perpendicular electron beam composed of individual electrons at a 2 kHz rate, and second (configuration B), the PMT was rotated by 90 degrees with respect to the beam (see Figure 7).

Taking into account the single-p.e. amplitude ($\sim$15 mV), we measured 19 p.e.'s and 75 p.e.'s created by Cherenkov light for two configurations, respectively. (This is consistent with back-of-the-envelope calculations taking into account the PMT (HAMAMATSU-R7378A) quantum efficiency (bialkali photocathode material) and window refractive index (synthetic silica).) The average window thickness of the PMT and its diameter are 3 mm and 25 mm, which gives us 6 p.e.'s/mm and 3 p.e.'s/mm for the two configurations.

The second configuration has twice fewer photons because half of the Cherenkov photons from the cone escape the window.

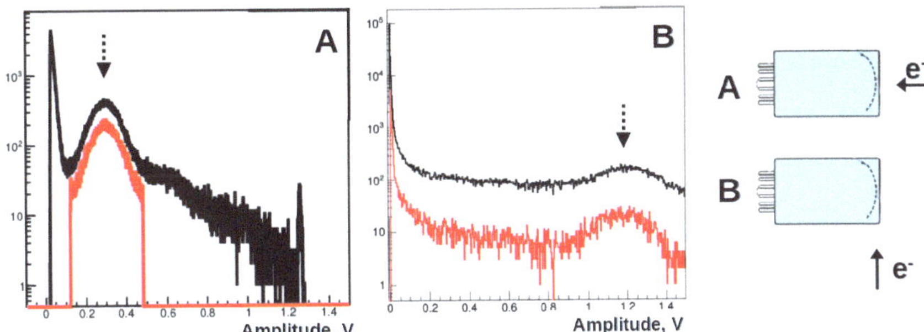

Figure 7. (**Left**): Amplitude of the signal (arrow) from the 5.6 GeV electrons impinging perpendicularly to the PMT surface. The black curve corresponds to all the measured events, while the red one requires a coincidence with a plastic scintillator, ensuring clean sample of electrons without secondaries. (**Right**): Amplitude of the signal from the 5.6 GeV electron impinging parallel to the PMT surface.

This kind of background is suppressed in the case of SiPM sensors since one can have a very thin window (resin protective layer), or it could even be removed completely.

8. Conclusions

The usage of SiPM sensors in space applications will grow in the future. They are light, with low power consumption, high PDE, and good time and spatial resolutions (see Appendix A). However, they are radiation- and temperature-sensitive devices. Therefore, the Terzina telescope design has to consider possible shields, thermostats, and SiPM annealing strategies. A description of the temperature dependence of the radiation damage annealing of a SiPM can be found in [19].

For 30 μm and 40 μm cell sizes, we measured the SiPM decay time at $\sim$60 ns and $\sim$110 ns, respectively (see Section 3). The background rate estimated using a parametric simulation of the SiPM response (see Section 4) decreases by a factor of four as a result of this notable reduction in signal duration. In Section 7, the background generated in the PMT/SiPM incoming window is explained. For ions, it is considerably more noticeable because the quantity of Cherenkov light increases squarely with the particle charge. Therefore, even a 0.1 mm thin window yields a sizable signal. We chose to employ naked sensors in order to cancel this dangerous background. Based on the study described in Sections 3, 4, and 7, we decided to use a bare sensor with a 30 μm cell size.

Even in low Earth orbit, the Earth's magnetic field loses some of its ability to shield objects from radiation. Therefore, the radiation damage will affect electronics, especially SiPMs. Geant4 [15] and SPENVIS [13] were used to assess the radiation dose levels (see Section 5) and Cherenkov light radiated by different optical elements. For only trapped protons and electrons, the dose for three years of operation is 3.1 Gy (protons) and 7.2 Gy (electrons). We calculated the Cherenkov background photon rate at 181 Hz/mm from electrons and 0.16 Hz/mm from protons using the same simulation.

In the case of crystalline materials (Si, for example), non-ionizing energy losses result in lattice dislocation and a rise in the SiPM DCR. Protons deliver significantly greater non-ionizing energy losses relative to electrons for the same total dose. Therefore, utilizing a 50 MeV proton beam supplied by the IFJ PAN radiation test facility in Krakow [17], we analyzed the rise in the SiPM DCR as a function of the total dose. A three-orders-of-magnitude increase was observed in the measured SiPM current at a 42 V bias voltage (about 10 V over-voltage) for a 3 Gy total irradiation dose (see Section 6).

Background rates due to the dose accumulated by the SiPM would become the dominant contribution during the last two years of the three-year NUSES mission (see Sections 5 and 6).

Author Contributions: The presented material is a result of the joint work of L.B. and the NUSES Collaboration. All authors have read and agreed to the published version of the manuscript.

Funding: NUSES is a joint project of the Gran Sasso Science Institute and Thales Alenia Space Italia, funded by the Italian Government (CIPE n. 20/2019), by the Italian Minister of Economic Development and the Abruzzo Region (MISE n. F/130087/00/X38), by the Italian Space Agency (ASI n. 15/2022), and by the Swiss National Foundation (SNF grant n. 178918). This project has received funding from the European Union's Horizon Europe Research and Innovation Programme under Grant Agreement No. 101057511 (EURO-LABS).

Data Availability Statement: No data was created for this research.

Acknowledgments: The NUSES Collaboration: R. Aloisio, C. Altomare, F. C. T. Barbato, R. Battiston, M. Bertaina, E. Bissaldi, D. Boncioli, L. Burmistrov, I. Cagnoli, M. Casolino, A. Cummings, N. D'Ambrosio, I. De Mitri, G. De Robertis, C. De Santis, A. Di Giovanni, A. Di Salvo, M. Di Santo, L. Di Venere, J. Eser, M. Fernandez Alonso, G. Fontanella, P. Fusco, S. Garbolino, F. Gargano, R. A. Giampaolo, M. Giliberti, A. Gola, F. Guarino, M. Heller, R. Iuppa, J. F. Krizmanic, S. Kusyk, A. Lega, F. Licciulli, F. Loparco, L. Lorusso, M. Mariotti, M. N. Mazziotta, S. Merzi, M. Mese, H. Miyamoto, T. Montaruli, E. Moretti, A. Nagai, R. Nicolaidis, F. Nozzoli, A. V. Olinto, D. Orlandi, G. Osteria, P. A. Palmieri, B. Panico, G. Panzarini, A. Parenti, L. Perrone, P. Picozza, R. Pillera, R. Rando, M. Rinaldi, A. Rivetti, V. Rizi, M. Ruzzarin, F. Salamida, E. Santero Mormile, V. Scherini, V. Scotti, D. Serini, I. Siddique, L. Silveri, A. Smirnov, R. Sparvoli, J. Swakoń, S. Tedesco, C. Trimarelli, D. Wróbel, L. Wu, P. Zuccon, S. C. Zugravel.

Conflicts of Interest: The authors declare no conflicts of interest.

Abbreviations

The following abbreviations are used in this manuscript:

AP	After-pulse
BoL	Beginning of life
CR	Cosmic rays
DCR	Dark count rate
EAS	Extensive air showers
EoL	End of life
FPA	Focal plane assembly
FWHM	Full-Width Half-Maximum
FWTM	Full-Width Tenth-Maximum
LEO	Low Earth orbit
LTAN	Local Time of Ascending Node
MA-PMT	Multi-Anode Photo-Multiplier Tube
NGB	Night-glow background
NUSES	Neutrinos and Seismic Electromagnetic Signals
NUV-HD-MT	Near-Ultraviolet High-Density Metal Trench SiPM
OCT	Optical cross-talk
PDE	Photon detection efficiency
p.e.	Photoelectron
PMT	Photo-Multiplier Tube
SiPM	Silicon Photo-Multiplier
RMSE	Root mean square error
UHECR	Ultra-High-Energy Cosmic Rays

Appendix A

Most of the astroparticle physics missions have adopted PMTs as photosensors, while recently, a few missions have started using SiPMs. A comparison between PMTs and SiPMs is given in Table A1 with selected parameters. One can notice that SiPMs are light and low-power-consuming devices with good time and spatial resolutions with respect to PMTs. SiPMs are ageless photodetectors since their *total integrated charge* has practically no limits. (The total integrated charge is defined here as the total sum of SiPM output charge over its life-time. It does not depend on the operational regime (i.e., saturation or single photon) and is not related to the SiPM dynamic range.) However, we need to emphasize that SiPMs are radiation- and temperature-sensitive devices. Additionally, to detect light signals on the level of a single photon, a pre-amplifier needs to be a part of the readout chain. At room temperature, the typical SiPM dark count rate (DCR) is about $\sim$100 kHz/mm^2, and the signals have a long falling edge defined by the quenching resistor and μ-cell capacitance.

Table A1. Photodetector comparative table.

Parameter	SiPM	PMT
Operation voltage	<100 V	$\sim$ 1000 V
Current	$\sim$1 µA	$\sim$100 µA
Power per cm^2	$\sim$1 mW	$\sim$100 mW
Weight per cm^2 of sensitive area	$\sim$10 g	$\sim$100 g
Total integrated charge	∞	$\sim$200 C
Single-p.e. time resolution	<100 ps	$\sim$1 ns
Spatial resolution	$\sim$mm	few mm [1]
Photon detection efficiency @ 400 nm	>50%	<50%
Temperature-sensitive	yes	no
Need of pre-amplifier	yes	optional
Radiation resistance	low	high
Signal FWTM [2]	$\sim$100 ns	$\sim$10 ns

[1] Multi-Anode PMTs or MA-PMTs can have small channels, 5×5 mm^2 or 3×3 mm^2. [2] SiPMs have a short rising edge and a very long falling edge.

The comparison between single-photoelectron (p.e.) (single p.e. (PMT) denotes a single photon that produces a primary photoelectron (p.e.) in a photocathode that reaches the first dynode of a PMT and triggers an avalanche; single p.e. (SiPM) is a single photon that triggers an avalanche) responses for a SiPM and a PMT and SiPM is shown in Figure A1. SiPMs can clearly separate single-p.e. events from the pedestal, while PMTs cannot. The PMT spectral shape is defined by the stochastic variation in the secondary electron emission, mainly from the first dynodes. The measurement for the PMT was performed with a pulsed LED (450 nm) signal. The majority of the triggers (LED pulses) contained no photons to measure a single photo-response, producing the pedestal in Figure A1(right). The simulation of the PMT, which describes the signal shape, can be found in reference [20]. The SiPM response shape is discussed in more detail in Section 4.

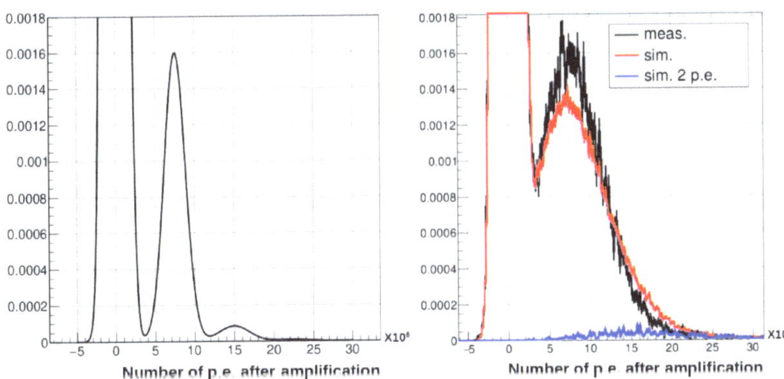

Figure A1. (Left): Single-p.e. response of a SiPM (amplitude spectrum). The first peak corresponds to a pedestal, the second is where only one p.e. is detected, and others are due to the primary p.e. triggering other micro-cells via either optical cross-talk or after-pulses. **(Right)**: For comparison, we show the spectrum of a PMT with its pedestal. The simulation is shown in red, while the black line corresponds to the measurements. The single-photoelectron peak is superimposed with rare contamination due to two primary p.e.'s, as shown by the simulation of the two-p.e. response in the blue line. The number of p.e.'s on the *x*-axis is after the amplification of the dynode stages (the peak corresponds to a gain of 7.5×10^6).

References

1. De Mitri, I. for the NUSES Collaboration. *J. Phys. Conf. Ser.* **2023**, *2429*, 012007. [CrossRef]
2. Nuses, R.; Aloisio, C.; Altomare, F.; Barbato, R.; Battiston, M.; Bertania, E.; Bissaldi, D.; Boncioli, L.; Burmistrov, I.; Cagnoli, M.; et al. The Terzina instrument on board the NUSES space mission. In Proceedings of the 38th International Cosmic Ray Conference (ICRC2023)-Cosmic-Ray Physics (Indirect, CRI), Nagoya, Japan, 26 July–3 August 2023; Volume 444; p. 391. [CrossRef]
3. Giovanni, A.D.; Santo, M.D.; on Behalf of the NUSES Collaboration. The NUSES space mission. In Proceedings of the 41st International Conference on High Energy physics (ICHEP2022)-Detectors for Future Facilities, R&D, Novel Techniques, Bologna, Italy, 6–13 July 2022; Volume 414, p. 354. Available online: https://pos.sissa.it/414/354 (accessed on 18 January 2024).
4. Mazziotta, M.N.; Pillera, R. The light tracker based on scintillating fibers with SiPM readout of the Zire instrument on board the NUSES space mission. In Proceedings of the 38th International Cosmic Ray Conference (ICRC2023)-Cosmic-Ray Physics (Direct, CRD), Nagoya, Japan, 26 July–3 August 2023; Volume 444, p. 083. [CrossRef]
5. Nuses, R.; Aloisio, A.; Altomare, B.; Barbato, B.; Battiston, B.; Bertania, B.; Bissaldi, B.; Boncioli, B.; Burmistrov, C.; Cagnoli, C.; et al. The Zire experiment on board the NUSES space mission. In Proceedings of the 38th International Cosmic Ray Conference (ICRC2023)-Cosmic-Ray Physics (Direct, CRD), Nagoya, Japan, 26 July–3 August 2023; Volume 444, p. 139. [CrossRef]
6. Burmistrov, L. for the NUSES Collaboration. Terzina on board NUSES: a pathfinder for EAS Cherenkov Light Detection from space. *arXiv* **2023**, arXiv:2304.11992. https://arxiv.org/abs/2304.11992.
7. Olinto, A.V.; Krizmanic, J.; Adams, J.H.; Aloisio, R.; Anchordoqui, L.A.; Anzalone, A.; Bagheri, M.; Barghini, D.; Battisti, M.; Bergman, D.R.; et al. The POEMMA (Probe of Extreme Multi-Messenger Astrophysics) observatory. *J. Cosmol. Astropart. Phys.* **2021**, *2021*, 7. [CrossRef]
8. Krizmanic, J. for the POEMMA Collaboration. POEMMA: Probe of extreme multi-messenger astrophysics. *Epj Web Conf.* **2019**, *210*, 06008. EDP Sciences. [CrossRef]
9. Available online: https://sd.fbk.eu/en/ (accessed on 18 January 2024).
10. Merzi, S.; Brunner, S.E.; Gola, A.; Inglese, A.; Mazzi, A.; Paternoster, G.; Penna, M.; Piemonte, C.; Ruzzarin, M. NUV-HD SiPMs with metal-filled trenches. *J. Instrum.* **2023**, *18*, P05040. [CrossRef]
11. Gola, A.; Acerbi, F.; Capasso, M.; Marcante, M.; Mazzi, A.; Paternoster, G.; Piemonte, C.; Regazzoni, V.; Zorzi, N. NUV-Sensitive Silicon Photomultiplier Technologies Developed at Fondazione Bruno Kessler. *Sensor* **2019**, *19*, 308. [CrossRef] [PubMed]
12. Available online: https://gitlab.com/nuses-satellite-full-simulation/terzina_wfsim (accessed on 18 January 2024).
13. Available online: https://www.spenvis.oma.be/ (accessed on 18 January 2024).
14. Available online: https://github.com/burmist-git/spenvis (accessed on 18 January 2024).
15. Agostinelli, S.; Allison, J.; Amako, K.A.; Apostolakis, J.; Araujo, H.; Arce, P.; Asai, M.; Axen, D.; Banerjee, S.; Barr, G.J.N.I.; et al. GEANT4: A simulation toolkit. *Nucl. Instrum. Meth.* **2003**, *A506*, 250–303. [CrossRef]
16. Available online: https://geant4.web.cern.ch/ (accessed on 18 January 2024).
17. Swakon, J.; Olko, P.; Adamczyk, D.; Cywicka-Jakiel, T.; Dabrowska, J.; Dulny, B.; Grzanka, L.; Horwacik, T.; Kajdrowicz, T.; Michalec, B.; et al. Facility for proton radiotherapy of eye cancer at IFJ PAN in Krakow. *Radiat. Meas.* **2010**, *45*, 1469–1471. [CrossRef]
18. Diener, R.; Dreyling-Eschweiler, J.; Ehrlichmann, H.; Gregor, I.M.; Kötz, U.; Krämer, U.; Meyners, N.; Potylitsina-Kube, N.; Schütz, A.; Schütze, P.; et al. The DESY II Test Beam Facility. *Nucl. Instruments Methods Phys. Res. Sect. Accel. Spectrometers Detect. Assoc. Equip.* **2019**, *922*, 265–286. [CrossRef]
19. De Angelis, N.; Kole, M.; Cadoux, F.; Hulsman, J.; Kowalski, T.; Kusyk, S.; Mianowski, S.; Rybka, D.; Stauffer, J.; Swakon, J.; et al. Temperature dependence of radiation damage annealing of Silicon Photomultipliers. *Nucl. Instruments Methods Phys. Res. Sect. Accel. Spectrometers, Detect. Assoc. Equip.* **2023**, *1048*, 167934. [CrossRef]
20. Available online: https://github.com/burmist-git/plume_PMT_sim_USBconv (accessed on 18 January 2024).

instruments

Article

Characterization of a Large Area Hybrid Pixel Detector of Timepix3 Technology for Space Applications

Martin Farkas [1,*], Benedikt Bergmann [2], Pavel Broulim [1], Petr Burian [1,2], Giovanni Ambrosi [3], Philipp Azzarello [4], Lukáš Pušman [1], Mateusz Sitarz [5], Petr Smolyanskiy [2], Daniil Sukhonos [4] and Xin Wu [4,†] on behalf of the PAN Collaboration

[1] Faculty of Electrical Engineering, University of West Bohemia, Univerzitní 26, 306 14 Pilsen, Czech Republic; broulimp@fel.zcu.cz (P.B.); petr.burian@cern.ch (P.B.); lukas.pusman@luvetech.cz (L.P.)
[2] Institute of Experimental and Applied Physics, CTU, Husova 240/5, 110 00 Prague, Czech Republic; benedikt.bergmann@utef.cvut.cz (B.B.)
[3] National Institue of Nuclear Physics, Perugia Department, Via A. Pascoli, 06123 Perugia, Italy; giovanni.ambrosi@pg.infn.it
[4] Département de Physique Nucléaire et Corpusculaire, Universite de Geneve, 24 Quai Ernest-Ansermet, CH-1205 Geneva, Switzerland; philipp.azzarello@unige.ch (P.A.); daniil.sukhonos@unige.ch (D.S.); xin.wu@cern.ch (X.W.)
[5] The Danish Centre for Particle Therapy, Aarhus University Hospital, Palle Juul-Jensens Boulevard 99, DK-8200 Aarhus, Denmark; sitarz@au.dk
* Correspondence: farkasm@fel.zcu.cz
† Collaborators of the Group PAN is provided in the Acknowledgments.

Abstract: We present the characterization of a highly segmented "large area" hybrid pixel detector (Timepix3, 512×512 pixels, pixel pitch 55 µm) for application in space experiments. We demonstrate that the nominal power consumption of 6 W can be reduced by changing the settings of the Timepix3 analog front-end and reducing the matrix clock frequency (from the nominal 40 MHz to 5 MHz) to 2 W (in the best case). We then present a comprehensive study of the impact of these changes on the particle tracking performance, the energy resolution and time stamping precision by utilizing data measured at the Super-Proton-Synchrotron (SPS) at CERN and at the Danish Center for Particle Therapy (DCPT). While the impact of the slower sampling frequency on energy measurement can be mitigated by prolongation of the falling edge of the analog signal, we find a reduction of the time resolution from 1.8 ns (in standard settings) to 5.6 ns (in analog low-power), which is further reduced utilizing a lower sampling clock (e.g., 5 MHz, in digital low-power operation) to 73.5 ns. We have studied the temperature dependence of the energy measurement for ambient temperatures between $-20°$ and $50\,°C$ separately for the different settings.

Keywords: PAN; Timepix3 low-power modes; Timepix3 Quad; Timepix3 temperature dependency; Timpepix3 space application

Citation: Farkas, M.; Bergmann, B.; Broulim, P.; Burian, P.; Ambrosi, G.; Azzarello, P.; Pušman, L.; Sitarz, M.; Smolyanskiy, P.; Sukhonos, D.; et al. Characterization of a Large Area Hybrid Pixel Detector of Timepix3 Technology for Space Applications. *Instruments* **2024**, *8*, 11. https://doi.org/10.3390/instruments8010011

Academic Editor: Alessandro Cianchi

Received: 26 October 2023
Revised: 20 January 2024
Accepted: 2 February 2024
Published: 14 February 2024

1. Introduction

The Timepix [1] detector made its debut in space inside the International Space Station (ISS) [2,3], where it was used as a compact radiation monitor [4]. The first Timepix detector exposed directly to the space environment was SATRAM (Space application of Timepix Monitor) [5,6] detector on board European Space Agency (ESA) satellite Proba-V [7]. Subsequently, Timepix detectors were used on various satellites, such as CubeSat VZLUSAT-1 [8] and RISESAT [9]. Based on SATRAMs previous 10 years of successful operation in low earth orbit, advanced and miniaturized space radiation monitors based on Timepix3 [10] and Timepix2 [11] technology have been developed for the European Space Agency. Large area Timepix3 detectors (512×512 pixels, 55 µm pitch) were proposed for the demonstrator of the penetrating particle analyzer [12] (mini.PAN), which is a compact magnetic spectrometer (MS) to precisely measure the cosmic ray flux, composition, spectral characteristics and directions. Mini.PAN employs position-sensitive (pixel and strip) detectors and (fast) scintillators to infer the particle type and velocity of GeV particles (and antiparticles) passing through the instrument's magnetic field by measuring their bending angles, charge deposition and the time of flight. It will measure the properties of cosmic rays in the 100 MeV/n to 20 GeV/n energy range in deep space with unprecedented accuracy, thus providing novel results to explore the mechanisms behind the origin, acceleration, and propagation of galactic cosmic rays and solar energetic particles.

While cutting-edge hybrid pixel detectors of the Timepix family provide high spatial and temporal resolution (currently down to ~2 ns), their implementation into space applications comes with challenges, such as their relatively large power consumption, creating heat, which has to be removed, or their high data throughput, which requires data compression and on-board processing capability. Moreover, careful electronics and chip carrier board design is required to withstand vibration and shock, and operate reliably in the radiation environments imposed by the missions. Another crucial environmental variable is temperature, which oscillates in a wide range of values (depending on the position of the spacecraft). The device needs to start up, keep working, and measure valid and correct data across a wide temperature range. Specific temperature range affecting the device depends on the mission (orbit, etc.) and the spacecraft design.

The aim of the present work is to develop a "large area" Timepix3 (area: ~8 cm^2 at 55 μm), investigate possibilities for a reduction of the power consumption, and study the impact of environmental parameters on the device performance.

2. Materials and Methods

2.1. Timepix3 Quad

Timepix3 was developed as the successor of Timepix in the Medipix3 collaboration. It features a pixel matrix of 256 × 256 pixels at a pixel pitch of 55 μm. Each of the pixels can measure simultaneously the deposited energy (ToT) and the time of arrival (ToA), the latter with precision of up to 1.6 ns. Additionally, Timepix3 implements the data-driven readout mode, where only the individual pixels triggered by ionizing radiation are read out (ToT and ToA measurements together with pixel position), while all other pixels are capable of measuring if triggered. In this mode, the minimal per-pixel dead time amounts to 19 clock cycles, which corresponds to 475 ns at the nominal 40 MHz matrix clock (see Section 2.2). Timepix3 ASICs are hybridized with the actual radiation-sensitive sensor using flip-chip bump bonding. In flip-chip bonding, the active side of the integrated circuit is directly attached or "flipped" onto a substrate or another die, rather than being mounted upright and wire-bonded.

Ionizing radiation interacting in the sensor creates free charge carriers, which drift through the fully depleted sensor until they are collected at the electrode of the opposite charge. During the drift motion they induce currents at the pixel electrodes. These pulses are shaped and amplified in a charge-sensitive amplifier (CSA) circuit. The voltage pulses at the output of CSA are then compared to a globally adjustable threshold level (THL). Once the voltage pulse crosses THL on its upwards slope, ToA is measured and the ToT measurement is started. The latter is stopped by the pulse crossing THL on its downwards slope. The sampling of ToT and "coarse" ToA is conducted with a matrix clock of 40 MHz distributed across the entire detector matrix. This clock is derived from the external reference clock, which is generated within the readout electronics (see Section 2.2). In order to improve the time stamping, a fast clock, created in local ring oscillators from the matrix clock, samples the time from the actual crossing of THL until the next rising edge of the coarse clock.

The pixel detector module, developed for the present studies, consists of 4 Timepix3 ASICs in 2 × 2 arrangement (quad), flip-chip bump bonded to a monolithic silicon sensor of 300 μm (see Figure 1). The sensitive area covers an area of ~7.9 cm^2 (with 512 × 512 pixels). The chipboard power supply includes a switching pre-regulator with a linear regulator to ensure precise voltage for the chips. The chips are glued to an aluminum heat sink to help manage and dissipate the heat generated during operation. In the current version, we use two data lines per chip (20 MHit/s/chip), so that the maximum data rate for the entire quad is 80 MHit/s.

Figure 1. Picture of the Timepix3 quad Chipboard.

2.2. Low-Power Modes of Timepix3 Quad

In particular, applications in space or in a vacuum, where power budgets and heat removal options are limited, low-power (LP) operation can be paramount. A previous work [13] outlines possibilities to reduce the power consumption of Timepix3 significantly and demonstrates that, in the best case, the single chip was operated at 250 mW. The power consumption of the Timepix3 detector can be divided into two primary components: the consumption of the analog part and the digital part.

The nominal power consumption of the analog part is 800 mW per chip. This power consumption was reduced to 55 mW per chip by tuning the digital-to-analog converter (DAC) settings. Changes to the main DAC registers responsible for the power consumption are shown in Table 1.

Power consumption on the digital part can be reduced by decreasing the clock frequency. Timepix3 has two main clock domains: system clock and matrix clock. The system clock is generated externally in the Katherine readout [14], while the matrix clock is generated inside the Timepix3 chip. The system clock is used by the Timepix3 peripherals like the Bus Controller and the Command Controller. These peripherals are only a small part of the Timepix3 chip, so the influence of the system clock on the detector's power consumption is minor. The matrix clock is used for the ToT/ToA measurements and has a major influence on power consumption.

In default mode, the matrix clock is derived from the internal Phase-Locked Loop (PLL) and this PLL needs an external reference clock (ClkInRefPLL) to be 40 MHz to function properly. This external reference clock is also provided by the Katherine readout. However, the PLL can be bypassed and the matrix clock can be fed directly by the reference clock. Bypassing the PLL and generating a matrix clock directly from the reference clock allows us to find a suitable clock frequency to meet our needs for performance and power consumption. The nominal power consumption of the digital part is 600 mW per chip. By lowering the matrix clock frequency (in this case to 5 MHz) we reduced the power consumption to 150 mW per chip. The matrix clock also influences the timing properties of the detector (ToA measurements). The fast clock (640 MHz) used for the fine ToA is generated internally from the matrix clock and is available only if we use the default value of the matrix clock (40 MHz). Without the fine ToA, the timing properties are limited by the sampling frequency of the clock externally fed to the ASIC. Moreover, since the clock frequency also determines the readout time, the per-pixel dead time increases.

For all presented measurements a custom-built improved Katherine readout was used with in-house built control software.

Table 1. The main changes in DACs responsible for the power consumption change from normal and analog low-power mode in Timepix3 Quad chips. For reproducibility, we give the entire set of used DACs in Appendix A Table A1.

DAC Register	Analog Low-Power	Standard Settings
Ibias_Preamp_ON	8 (1.294 V)	128 (1.157 V)
Ibias_DiscS1_ON	8 (1.294 V)	100 (1.059 V)
Ibias_DiscS2_ON	8 (1.294 V)	128 (0.333 V)
Ibias_PixelDAC	20 (1.066 V)	128 (0.942 V)

We consider dynamic switching between modes in order to best adapt the device performance to the available platform resources. For example, a user-specified average power budget could be achieved by varying between different modes and taking data at different temporal resolutions and readout speeds. We define the following 4 Timepix3 power modes (1 normal, 3 LP):

- **Normal mode**—full performance (standard DACs settings, default 40 MHz matrix clock). The nominal power consumption is $\approx$6 W;
- **Analog LP mode**—the DACs are set to LP mode, while the default 40 MHz matrix clock is used. The power consumption is $\approx$3 W;
- **LP20**—The DACs are set to LP mode (analog LP) and the matrix clock is reduced to 20 MHz. The power consumption is $\approx$2.2 W;
- **LP5**—Same as LP20, but at a matrix clock of 5 MHz. The power consumption is lower than $\approx$2 W.

To reduce the impact of the lower sampling frequency on the energy measurement performance (sampling error) at lower matrix clock frequencies, the *IKrum* DAC was set at 10 and 5 for LP20 and LP5, respectively, thus, elongating the analog pulses.

Measurement of the real Timepix3 quad power consumption in digital LP modes has revealed higher values compared to the extrapolation of four single Timepix3 chipboards (see Table 2). This can be explained by an overhead in the losses on the chipboard voltage regulators, which do not work as efficiently in low-power modes. It should be noted that future improved hardware chipboard design might overcome this issue.

Table 2. Comparison of the power consumption of the Timepix3 Quad in digital LP modes with the nominal values of four Timepix3 single chips.

LP Mode	Power Consumption (W)	
	Timepix3 Quad	4 × Timepix3
Normal	6	6
Analog LP	2.9	2.9
LP 20	2.2	1.7
LP 5	2	1.2

2.3. Experimental Setups

The functionality of the detector was tested with laboratory sources and in charged particle beams at CERN's super-proton-synchrotron (SPS) and the Danish Center for Particle Therapy (DCPT). At SPS, the 120 GeV hadron beam (90% pions), we focused on investigating the difference between normal and analog low-power modes. At DCPT, a more detailed study was conducted using proton beams of 80 MeV and 240 MeV. Here, the detector response was determined at all predefined power settings (normal, analog LP, LP20, and LP5). Since the data rate in the center of the clinical proton beam exceeded the capabilities of the chip in the data-driven mode, the measurements were taken at a lateral displacement of 12 cm. In the particle beams, measurements were taken at different impact angles in the range from 0 (perpendicular) to 90 degrees with respect to the sensor normal. For thermal tests, the detector was calibrated in a laboratory with a chip temperature of +50 °C (± 1 °C) (corresponds to an ambient temperature of $\sim$25 °C). The detector was then placed together with an ^{241}Am γ-source (peak energy 59.6 keV) of 300 kBq inside a climate chamber (Votsch VCV 7060 - 5, Vötsch Industrietechnik GmbH; Beethovenstrasse 34, Balingen-Frommern 72336, Germany). The detector was connected to the Katherine readout, positioned outside the climate chamber (the readout electronics remained unaffected by temperature changes), enabling us to measure only the detector's response to temperature variations. Measurements were taken for three power modes (normal, LP20, LP5) in an ambient temperature range of -20 °C–+50 °C. Before each measurement, we waited until thermal equilibrium was achieved. During all performed tests, the following environment variables were monitored: the ambient temperature, the temperature of the used Timepix3 detectors, and the temperature on the heat sink.

All measurements were taken using the data-driven mode of Timepix3. The sensor was fully depleted using a bias voltage of 60 V.

2.4. Data Analysis

The data are sent off the chip as a list of partially temporally unsorted pixel hits. These are then grouped into so-called "clusters" using temporal coincidence and spatial neighborhood conditions. For the present evaluations, a floating coincidence time window of 200 ns was used. Figure 2 shows a set of 1000 clusters measured at DCPT with 80 MeV protons.

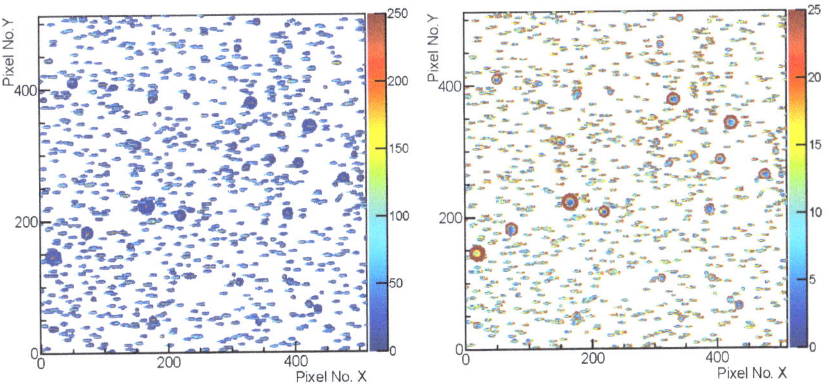

Figure 2. Detector response in the form of a 2D matrix of the energy deposition (keV) (**left**) and the relative time differences (ns) within a cluster (**right**). A set of 1000 typical clusters found in the 80 MeV proton beam at DCPT is depicted.

For each found cluster the following features are calculated:

- The **deposited energy** E_{dep} is defined as the sum of the energies measured in the pixels;
- The **cluster size** N_{cluster} is defined as the number of pixels within a cluster;
- The cluster **drift time difference** dt_{drift} is defined as the difference of the minimal and maximal timestamp measured within a cluster.

3. Results

3.1. Impact of Low-Power Settings on the Energy Measurement

The ToT-to-energy calibration was conducted with X-ray fluorescence lines of known energy and the 59.6 keV γ-line from a ^{241}Am source, as described in [14]. A stable noise-free operation at a threshold (THL) of 2.75 keV was found. For further measurements and anticipating the temperature effects, it has been conservatively set at THL = 4 keV.

The calibrated energy spectra of a ^{241}Am source are shown in Figures 3 and 4 for normal mode and low-power modes, respectively. Different colors give the spectra for different cluster sizes. While the best resolution is achievable utilizing single-pixel clusters, we determine the overall energy resolution by fitting the 59.6 keV peak with a Gaussian curve sitting on an error-function:

$$f(E) = A \exp\left[-\frac{(E - E_{\text{mean}})^2}{2\sigma^2}\right] + \frac{B}{2}\left[1 - \text{Erf}\left(\frac{E - E_{\text{mean}}}{\sqrt{2}\sigma}\right)\right], \tag{1}$$

where the Gaussian amplitude A, the error-function height B, the mean energy E_{mean} and the energy resolution σ were the fit parameters. Table 3 summarizes the energy resolutions found for the different power modes. Overall, consistent results were found at the different settings.

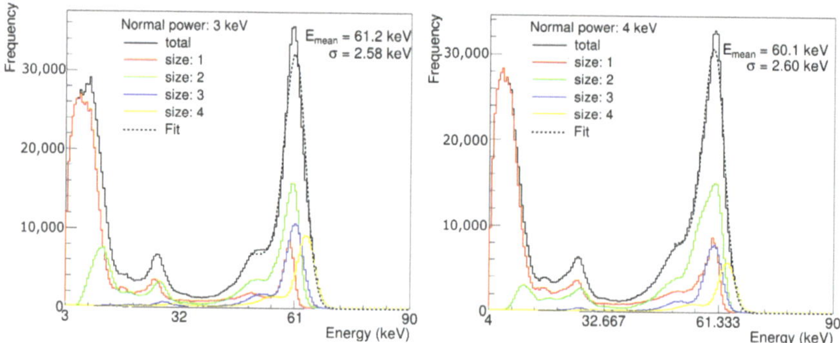

Figure 3. ^{241}Am photon spectrum measured at standard chip settings. The spectra of different cluster sizes are shown. The used detector THL was 3 keV (**left**) and 4 keV (**right**). The overall achieved energy resolution was determined by fitting a Gaussian sitting on an error-function to the 59.6 keV peak.

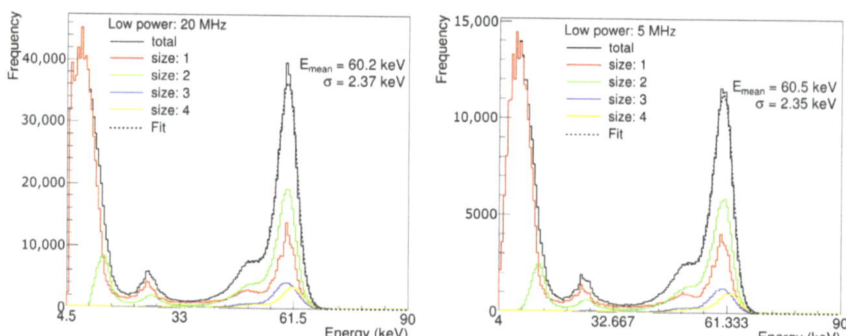

Figure 4. Same as Figure 3, but at low-power operation (LP20 and LP5). The used detector THL was 4 keV for both LP modes.

Table 3. Summary for energy calibration verification with ^{241}Am.

Settings	σ_{Energy} (keV)	Chip Temperature (°C)	Ikrum
Normal (THL 3 keV)	2.58	57	15
Normal (THL 4 keV)	2.60	57	10
LP (THL 4 keV, 20 MHz)	2.37	50	10
LP (THL 4 keV, 5 MHz)	2.35	50	2

The energy responses from the proton test are shown in Figure 5. The same behavior was found for the measurement with pions at SPS. In this measurement campaign, different energy deposition in the sensor was achieved by measurement at particle impact angles of 0, 50, and 70 degrees with respect to the sensor normal. For a thin absorber, the physics of the energy deposition are described by a Landau curve, which is, however, convolved with a Gaussian smearing describing the energy resolution of the detector (0 degree impact angle). At both higher impact angles (50 and 70 degrees) the sensor cannot be regarded as a thin layer anymore. Consequently, the tail towards higher energy is suppressed while the overall expected spectrum shape becomes Gaussian-like. From the presented measurements, we can conclude that the investigated LP modes do not have a significant effect on the overall energy measurement.

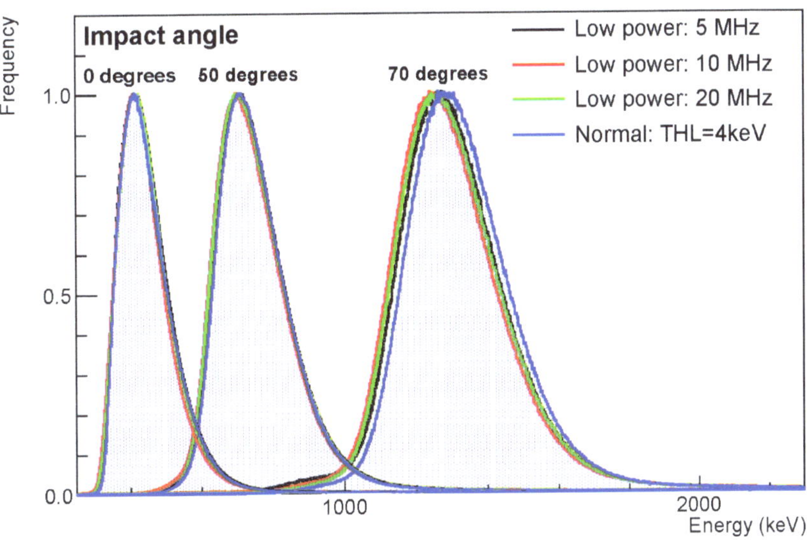

Figure 5. Energy deposition spectra of 80 MeV protons at the impact angles 0 degrees, 50 degrees and 70 degrees with respect to the sensor normal using different operational settings.

3.2. Impact of the Low-Power Settings on the Time Resolution

To determine the performance of the time measurement, we use data taken at a proton energy of 240 MeV and a 70-degree impact angle. We then measure the drift time across the entire sensor by calculating the difference of the maximal time differences within a cluster (see, e.g., [15] for a discussion about the drift time studies with Timepix3). The measured time distributions are shown in Figure 6 for the different modes investigated. The peaks were fitted with a Gaussian distribution to determine the widths σ_{fit}. Since the time is measured twice (maximum and minimum time), the real time resolution of the detector can be calculated as $\sigma_{\text{meas}} = \frac{\sigma_{\text{fit}}}{\sqrt{2}}$. The determined time resolution for the different modes is given in Table 4.

While at the standard clock (normal, analog low-power) the time resolution is given by the pulse shaping, in digital low power, the sampling frequency determines the achievable resolution. At 20 and 10 MHz, the measured values, i.e., 18 ns and 28 ns, are consistent with the resolutions calculated from the time binning as $\sigma_{\mathrm{pred}} = \frac{\Delta t}{\sqrt{12}}$ with the time stamping granularity $\Delta t = \frac{1}{f_{\mathrm{matrix\,clock}}}$, which are 14 ns and 29 ns, respectively. At 5 MHz, the measured resolution is slightly worse than the predicted one (74 ns versus 58 ns). This is caused by the frequency being at the edge of the supported range of the clock generator adding an additional clock instability.

Table 4. Time stamping precision for Timepix3 Quad in normal mode and LP modes.

Settings	Sampling Frequency (MHz)	σ_{meas} (ns)
Normal	640 (1.56 ns)	1.8
Analog LP	640 (1.56 ns)	5.9
LP 20	20 (50 ns)	17.7
LP 5	5 (200 ns)	73.5

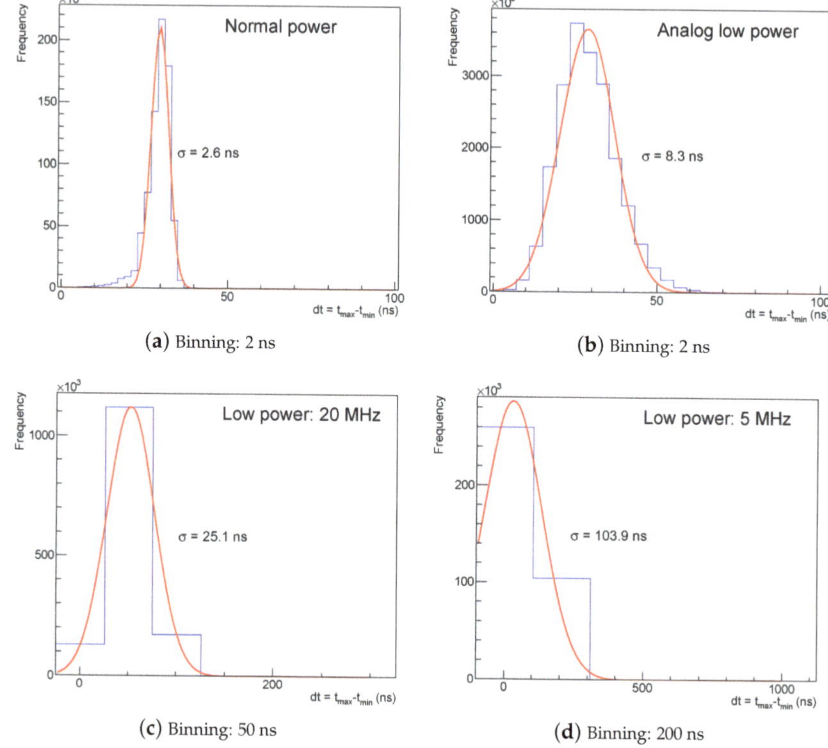

(**a**) Binning: 2 ns

(**b**) Binning: 2 ns

(**c**) Binning: 50 ns

(**d**) Binning: 200 ns

Figure 6. Histogram of measured drift time differences across the thickness of the sensor in normal mode (**a**), analog low-power mode (**b**), LP20 (**c**) and LP5 (**d**); The observed peaks were fitted with a Gaussian distribution to determine the achievable resolution of the time measurement. The time binning was chosen to resemble the sampling frequency of the selected clock frequency.

3.3. Impact of the Low-Power Settings on the Cluster Shape

For completeness, we present a qualitative comparison of clusters measured in normal (at different THL) and low-power settings (Figure 7a,b). A quantitative description is shown by the cluster size distributions for the 80 MeV and 240 MeV proton measurements at a 70-degree impact angle in Figure 7c,d, respectively. Overall, the cluster sizes decrease at analog low-power DAC settings. This effect is stronger for clusters with high stopping power (see the 80 MeV data sets) and is not visible at lower stopping power. This difference can be explained by looking at the behavior of the Krummenacher circuit in the pixel electronics. The capacitor in the amplification circuit is discharged at a constant rate $R_{\mathrm{discharge}} = \frac{dQ_{\mathrm{loss}}}{dt}$. The charge loss until the pulse reaches the peak position t_{peak} can thus be estimated as $Q_{\mathrm{loss}} = R_{\mathrm{discharge}} \times t_{\mathrm{peak}}$. Since the rise time in the low-power modes $t_{\mathrm{peak}}^{\mathrm{LP}}$ is slower than at normal settings $t_{\mathrm{peak}}^{\mathrm{normal}}$, charge losses are higher and the pulses in the periphery of the tracks that have energy just above the threshold are lost in LP settings. While this cluster shape difference in the different modes does not impact the overall performance of the detector, it should be carefully considered when using already trained neural networks used for particle identification [16].

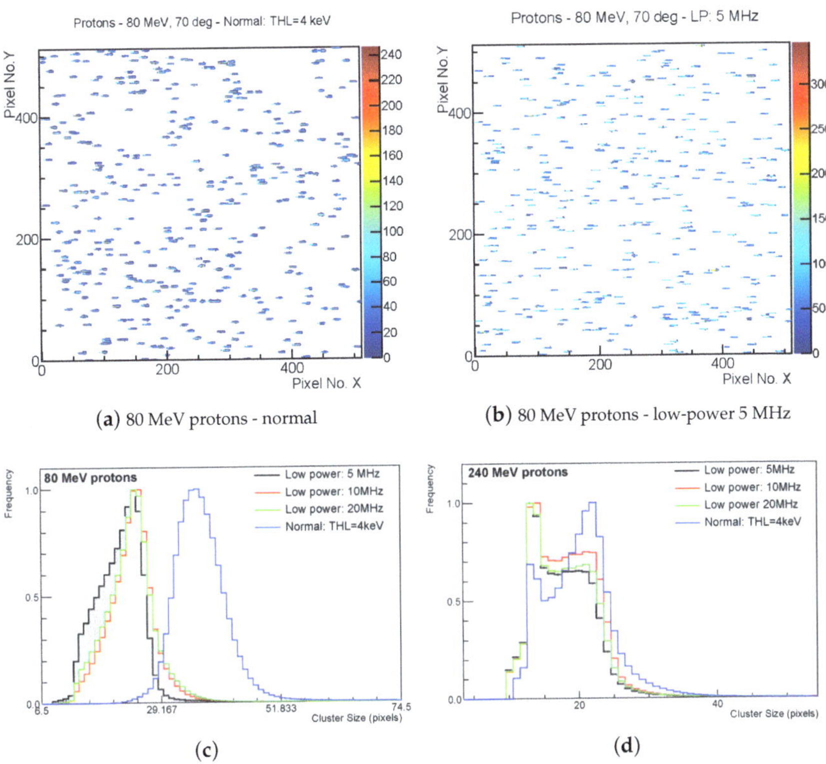

(**a**) 80 MeV protons - normal

(**b**) 80 MeV protons - low-power 5 MHz

(**c**)

(**d**)

Figure 7. (**a**,**b**) Qualitative comparison of the 2D projections of deposited energy after 80 MeV proton impact at 70 degrees in normal mode and LP5, respectively; (**c**,**d**) quantitative evaluation using the distributions of the observed cluster sizes for proton impact at 70 degrees at energy 80 and 240 MeV, respectively.

3.4. Effect of the Temperature on the Energy Measurement

Figure 8 shows the measured ^{241}Am γ-peak position at different chip temperatures. For these measurements, the ambient temperature range was $-20\,°C$–$50\,°C$. In all modes, the peak positions move towards lower energy at increasing temperatures. However, the peak shift at digital low-power settings shows a stronger dependency on chip temperature with a larger variation amongst the chips. Therefore, we recommend a determination of the calibration coefficients at different temperatures. The shift of measured energy based on temperature change was also demonstrated in papers [17,18], where a single Timepix3 chip in normal mode was tested.

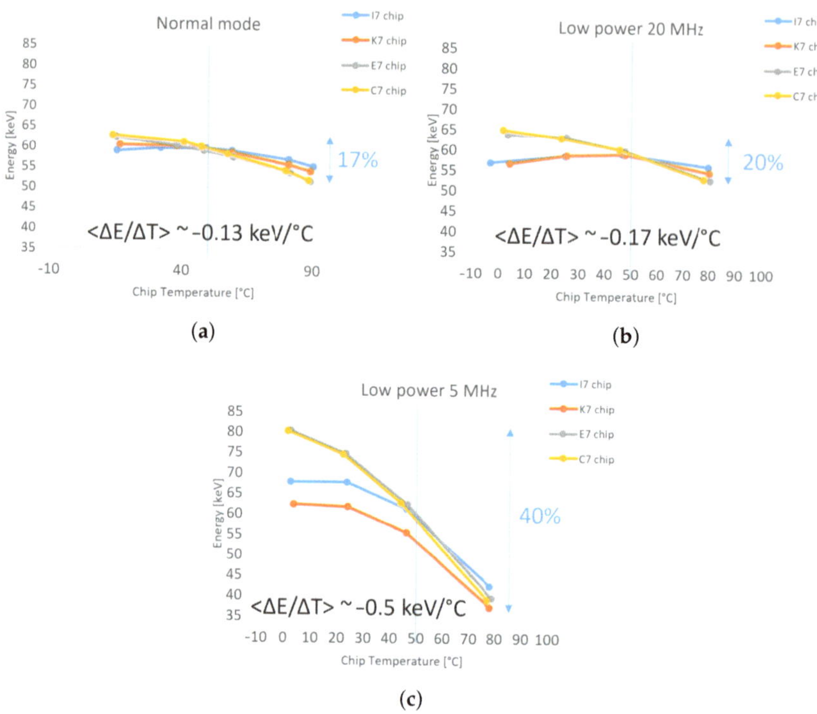

Figure 8. 59.6 keV γ-peak position as a function of the chip temperature at (**a**) normal mode, (**b**) LP20 and (**c**) LP5. The temperature dependence is shown separately for each Timepix3 chip within the quad assembly.

4. Conclusions

The manuscript presented the characterization of a large area Timepix3 detector with a 300 µm-thick silicon sensor, particularly for measurement in space. The requirements for cooling and limited energy resources were mitigated by implementing different power consumption modes. Since available resources can change during the lifetime of a mission, it is possible to dynamically change between these modes.

Thermal tests show that the measured energy is dependent on temperature and this dependency varies with different DACs settings and used clock frequency. The behavior between every single chip on the Timepix3 Quad varies by a small margin. This variation is stronger for low-power modes with a reduced matrix clock (very noticeable for LP5). In particular, the temperature dependence in the low-power modes is not negligible. Thus, in applications where the temperature could change more than ±5 °C from the calibration temperature, we recommend prior calibration at different temperatures and selecting the appropriate calibration coefficients.

A custom-built readout was used to test the functionality of the detector and its response to a 120 GeV/c hadron beam at the Super-Proton-Synchrotron (SPS) at CERN and to protons of 80 and 240 MeV at the Danish Center for Proton Therapy (DCPT). The power consumption of 6 W with standard settings was reduced to 3 W by changing the Timepix3 DACs. While these changes did not affect the energy measurement resolution, the time stamping precision was reduced from 1.8 ns to 5.9 ns due to the slower shaping of the rising edge of the analog pulse. Further reduction of the power consumption was achieved by reducing the matrix clock. At a matrix clock of 5 MHz, we achieved a power consumption of 2 W. While the sampling frequency effect on the energy measurement was successfully mitigated by adjustment of the pulse length (through the chips' IKrum DAC), the time stamping precision is determined by the used matrix clock in the deep digital low-power modes (LP20, LP5).

Author Contributions: Project administration and Funding acquisition G.A., P.A. and X.W.; Investigation, Resources P.B. (Pavel Broulim) and L.P.; Investigation, software P.B. (Petr Burian); Formal analysis, Writing—review & editing, Project administration and Funding acquisition B.B.; Formal analysis, Resources P.S.; Investigation, Writing—original draft M.F.; Resources M.S.; Validation D.S. All authors have read and agreed to the published version of the manuscript.

Funding: This project has received funding from the European Union's Horizon 2020 research and innovation programme under grant agreement No 862044. This work has been done using the INSPIRE Research Infrastructures and is part of a project that has received funding from the European Union's Horizon 2020 research and innovation programme under grant agreement No 730983. Martin Farkaš received funding from the internal project to support student scientific conferences SVK1-2023-005 and project SGS-2021-005: Research, development and implementation of modern electronic and information systems.

Institutional Review Board Statement: Not applicable.

Informed Consent Statement: Not applicable.

Data Availability Statement: The data presented in this study are available on request from the corresponding author.

Acknowledgments: The work was carried out in the Medipix collaboration. Full Authorlist of the PAN Collaboration: Giovanni Ambrosi (Istituto Nazionale di Fisica Nucleare, Italy), Philipp Azzarello[2], Mattia Barbanera[6], Benedikt Bergmann[3], Bruna Bertucci[1], Pavel Broulim[4], Petr Burian[3,4], Frank Raphael Cadoux[2], Michael Campbell[5], Mirco Caprai[6], Fabio Cossio[6], Matteo Duranti[6], Yannick Favre[2], Stefan Gohl[3], Maura Graziani[1], Tomoya Iizawa[2], Maria Ionica[6], Merlin Kole[2], Daniel La Marra[2], Edoardo Mancini[1], Petr Manek[3], Lorenzo Mussolin[1], Mercedes Paniccia[2], Stanislav Pospisil[3], Lukáš Pušman[4], Petr Smolyanskiy[3], Jerome Stauffer[2], Daniil Sukhonos[2], Pierre Alexandre Thonet[5], Nicola Tomassetti[1], Xin Wu[2], Pengwei Xie[2]. [1] Universita e INFN, Perugia, Italy; [2] Universite de Geneve, Switzerland; [3] Institute of Experimental and Applied Physics, CTU, Czech Republic; [4] University of West Bohemia, Czech Republic; [5] CERN, Switzerland; [6] Istituto Nazionale di Fisica Nucleare, Italy.

Conflicts of Interest: The funders had no role in the design of the study; in the collection, analyses, or interpretation of data; in the writing of the manuscript; or in the decision to publish the results.

Abbreviations

The following abbreviations are used in this manuscript:

LP	Low-Power
PAN	Penetrating Particle Analyzer
DAC	Digital to Analog Converter
ToT	Time over Threshold
ToA	Time of Arrival
PLL	Phase-Locked Loop

Appendix A. The Full Set of the Timepix3 Analog DAC Settings in HP and LP Mode

Table A1. Time stamping precision for Timepix3 Quad in normal mode and LP modes.

DAC	Normal	low-power
PreampOn	128	8
PreampOff	8	8
VPreamp	128	128
Ikrum	15	10 (LP20) and 5 (LP5)
Vfbk	164	128
DiscS1On	100	8
DiscS1Off	8	8
DiscS2On	128	8
DiscS2Off	8	8
Pixel	128	20
TpBufferIn	128	128
TpBufferOut	128	128
VtpCoarse	128	128
VtpFine	256	256
CpPLL	128	128
PLLVcntrl	128	128

References

1. Llopart, X.; Ballabriga, R.; Campbell, M.; Tlustos, L.; Wong, W. Timepix, a 65k programmable pixel readout chip for arrival time, energy and/or photon counting measurements. *Nucl. Inst. Meth. A* **2007**, *581*, 485–494. [CrossRef]
2. Stoffle, N.; Pinsky, L.; Kroupa, M.; Hoang, S.; Idarraga, J.; Amberboy, C.; Rios, R.; Hauss, J.; Keller, J.; Bahadori, A.; et al. Timepix-based radiation environment monitor measurements aboard the International Space Station. *Nucl. Instrum. Methods A* **2015**, *782*, 143–148. [CrossRef]
3. Kroupa, M.; Bahadori, A.; Campbell-Ricketts, T.; Empl, A.; Hoang, S.; Idarraga-Munoz, J.; Rios, R.; Semones, E.; Stoffle, N.; Tlustos, L.; et al. A semiconductor radiation imaging pixel detector for space radiation dosimetry. *Life Sci. Space Res.* **2015**, *6*, 69–78. [CrossRef] [PubMed]
4. Turecek, D.; Pinsky, L.; Jakubek, J.; Vykydal, Z.; Stoffle, N.; Pospisil, S. Small Dosimeter based on Timepix device for International Space Station. *J. Instrum.* **2011**, *6*, C12037. [CrossRef]
5. Granja, C.; Polansky, S.; Vykydal, Z.; Pospisil, S.; Owens, A.; Kozacek, Z.; Mellab, K.; Simcak, M. The SATRAM Timepix spacecraft payload in open space on board the Proba-V satellite for wide range radiation monitoring in LEO orbit. *Planet. Space Sci.* **2016**, *125*, 114–129. [CrossRef]
6. Gohl, S.; Bergmann, B.; Evans, H.; Nieminen, P.; Owens, A.; Posipsil, S. Study of the radiation fields in LEO with the Space Application of Timepix Radiation Monitor (SATRAM). *Adv. Space Res.* **2019**, *63*, 1646–1660. [CrossRef]
7. Proba-V Carrying Radiation Detector from CERN to Space. Available online: http://www.esa.int/Our_Activities/Technology/Proba_Missions/Proba-V_carrying_radiation_detector_from_CERN_to_space (accessed on 20 December 2023).
8. Baca, T.; Jilek, M.; Vertat, I.; Urban, M.; Nentvich, O.; Filgas, R.; Granja, C.; Inneman, A.; Daniel, V. Timepix in LEO Orbit onboard the VZLUSAT-1 Nanosatellite: 1-year of Space Radiation Dosimetry Measurements. *J. Instrum.* **2018**, *13*, C11010. [CrossRef]
9. Filgas, R. et al. RISEPix—A Timepix-based radiation monitor telescope onboard the RISESAT satellite. *Astron. Nachrichten* **2019**, *340*, 674–680. [CrossRef]
10. Poikela, T.; Plosila, J.; Westerlund, T.; Campbell, M.; De Gaspari, M.; Llopart, X.; Gromov, V.; Kluit, R.; Van Beuzekom, M.; Zappon, F.; et al. Timepix3: A 65K channel hybrid pixel readout chip with simultaneous ToA/ToT and sparse readout. *J. Instrum.* **2014**, *9*, C05013. [CrossRef]
11. Wong, W.S.; Alozy, J.; Ballabriga, R.; Campbell, M.; Kremastiotis, I.; Llopart, X.; Poikela, T.; Sriskaran, V.; Tlustos, L.; Turecek, D. Introducing Timepix2, a frame-based pixel detector readout ASIC measuring energy deposition and arrival time. *Radiat. Meas.* **2020**, *131*, 106230. [CrossRef]
12. Wu, X.; Ambrosi, G.; Azzarello, P.; Bergmann, B.; Bertucci, B.; Cadoux, F.; Campbell, M.; Duranti, M.; Ionica, M.; Kole, M.; et al. Penetrating particle analyzer (pan). *Adv. Space Res.* **2019**, *63*, 2672–2682. [CrossRef]
13. Burian, P.; Broulim, P.; Bergmann, B. Study of Power Consumption of Timepix3 Detector. *J. Instrum.* **2019**, *14*, C01001. [CrossRef]
14. Burian, P.; Broulim, P.; Jara, M.; Georgiev, V.; Bergmann, B. Katherine: Ethernet Embedded Readout Interface for Timepix3. *J. Instrum.* **2017**, *12*, C11001. [CrossRef]
15. Bergmann, B.; Pichotka, M.; Pospisil, S.; Vycpalek, J.; Burian, P.; Broulim, P.; Jakubek, J. 3D track reconstruction capability of a silicon hybrid active pixel detector. *Eur. Phys. J. C* **2017**, *77*, 421. [CrossRef]

16. Ruffenach, M.; Bourdarie, S.; Bergmann, B.; Gohl, S.; Mekki, J.; Vaillé, J. A New Technique Based on Convolutional Neural Networks to Measure the Energy of Protons and Electrons With a Single Timepix Detector. *IEEE Trans. Nucl. Sci.* **2021**, *68*, 1746–1753. [CrossRef]
17. Urban, M.; Nentvich, O.; Marek, L.; Hudec, R.; Sieger, L. Timepix3: Temperature Influence on Radiation Energy Measurement with Si Sensor. *Sensors* **2023**, *23*, 2201. [CrossRef] [PubMed]
18. Urban, M.; Doubravová, D. Timepix3: Temperature influence on X-ray measurements in counting mode with Si sensor. *Radiat. Meas.* **2021**, *141*, 106535. [CrossRef]

instruments

Article

Results and Perspectives of Timepix Detectors in Space—From Radiation Monitoring in Low Earth Orbit to Astroparticle Physics

Benedikt Bergmann [1,*], Stefan Gohl [1,2,*], Declan Garvey [1], Jindřich Jelínek [1,3] and Petr Smolyanskiy [1]

[1] Institute of Experimental and Applied Physics, Czech Technical University in Prague, 11000 Prague, Czech Republic; declan.garvey@cvut.cz (D.G.); jindrich.jelinek@etu.unige.ch (J.J.)
[2] Faculty of Mathematics and Physics, Charles University, 18000 Prague, Czech Republic
[3] Department of Nuclear and Particle Physics, University of Geneva, CH-1211 Geneva, Switzerland
* Correspondence: benedikt.bergmann@utef.cvut.cz (B.B.); stefan.gohl@utef.cvut.cz (S.G.); Tel.: +420-775-868-794 (B.B.)

Abstract: In space application, hybrid pixel detectors of the Timepix family have been considered mainly for the measurement of radiation levels and dosimetry in low earth orbits. Using the example of the Space Application of Timepix Radiation Monitor (SATRAM), we demonstrate the unique capabilities of Timepix-based miniaturized radiation detectors for particle separation. We present the incident proton energy spectrum in the geographic location of SAA obtained by using Bayesian unfolding of the stopping power spectrum measured with a single-layer Timepix. We assess the measurement stability and the resiliency of the detector to the space environment, thereby demonstrating that even though degradation is observed, data quality has not been affected significantly over more than 10 years. Based on the SATRAM heritage and the capabilities of the latest-generation Timepix series chips, we discuss their applicability for use in a compact magnetic spectrometer for a deep space mission or in the Jupiter radiation belts, as well as their capability for use as single-layer X- and γ-ray polarimeters. The latter was supported by the measurement of the polarization of scattered radiation in a laboratory experiment, where a modulation of 80% was found.

Keywords: space weather; scatter polarimeter; hybrid pixel detectors; Timepix; dE/dX spectrometer; low earth orbit; magnetic spectrometer; galactic cosmic rays; space instrumentation

Citation: Bergmann, B.; Gohl, S.; Garvey, D.; Jelínek, J.; Smolyanskiy, P. Results and Perspectives of Timepix Detectors in Space—From Radiation Monitoring in Low Earth Orbit to Astroparticle Physics. *Instruments* **2024**, *8*, 17. https://doi.org/10.3390/instruments8010017

Academic Editor: Pasquale Arpaia

Received: 30 October 2023
Revised: 12 February 2024
Accepted: 23 February 2024
Published: 29 February 2024

1. Introduction

In 1997, the Medipix collaboration was founded to evaluate hybrid pixel detector (HPD) technology, which was originally developed for particle tracking in high-energy physics for X-ray imaging [1]. Thus, single-photon-counting chips were developed, providing per-pixel information about the number of hits above predefined thresholds within a given time interval. Being mostly focused on the medical sector, these chips were called "Medipix". In 2006, within the Medipix2 collaboration and upon the request of the EUDET collaboration, the first Timepix was released. It does not only determine the number of hits above a threshold, but it could also be set to measure the time from the moment of triggering the pixel to the end of the acquisition, thereby enabling a measurement of electron drift times released by ionizing radiation in gaseous volumes for resolving particle trajectories in 3D [2–4]. In addition, Timepix provides per-pixel spectroscopy using the time over threshold mode. The success of early Medipix and Timepix chips then triggered follow-up collaborations that further improved the technology by adding features to the pixel–signal processing, e.g., analog charge summing modes and additional thresholds or an improved time resolution and a data-driven readout architecture for the Medipix and Timepix series, respectively [5]. In addition to their rich application potential on Earth [5,6], Timepix detectors [7,8] have become increasingly interesting for radiation monitoring in space science.

To date, most of the commonly used space radiation monitors rely on silicon diodes, thereby achieving particle (mainly electron and proton) separation through pulse height analysis, detector stacking, shielding, or electron removal by a magnetic field. The key advantage of HPDs is that in addition to the energy deposition measurement, particle signatures in the sensor are seen as tracks with a rich set of features. These track characteristics can be exploited for the identification of particle type, energy, and its trajectory. Determining these pieces of information on a single layer bypasses the need for sensor stacking or complex shielding geometries, so that HPD-based space radiation devices provide science-class data with a large field of view at an order of magnitude of lower weight and approximately half of the power consumption compared to commonly used space radiation monitors or science-class energetic particle detectors.

Since 2012, Timepix has been utilized in radiation environment monitors aboard the ISS [9–11], being the first Timepix (256×256 pixels, $55\,\mu$m pitch) used in open space is SATRAM (Space Application of Timepix Radiation Monitor) [12]. It is attached to the Proba-V, a satellite which was launched to low Earth orbit (LEO, 820 km, sun-synchronous orbit) in 2013 and has celebrated 10 years in orbit in May of 2023. During this time, it has been providing data for mapping out the fluxes of electrons and protons trapped in the Van Allen radiation belt, e.g., by in-orbit maps of the ionizing dose rate [13–15]. Over the years, different data analysis techniques have been successfully used for evaluation of the complex data set, incluing analytic categorization relying on the extraction of manually defined track features, as well as novel machine learning approaches [15–17]. The success of SATRAM initiated the development of advanced miniaturized space radiation monitors based on Timepix3 [18] and Timepix2 [19] technology. These are currently flown on the SWIMMR-1 (Space Weather Instrumentation, Measurement, Modelling and Risk) [20] mission (launched in 2023) and shall be used within the European Radiation Sensor Array (ERSA) [21].

The detector technology's science reach has been extended towards astroparticle physics application through the development of large area Timepix3 detectors (512×512 pixels, $55\,\mu$m pitch) for the demonstrator of the penetrating particle analyzer [22] (Mini.PAN), which is a compact magnetic spectrometer (MS) designed to measure the properties of cosmic rays in the $100\,\text{MeV/n}$–$20\,\text{GeV/n}$ energy range in deep space with unprecedented accuracy, thus providing novel results to investigate the mechanisms of origin, acceleration, and propagation of galactic cosmic rays and of solar energetic particles, as well as producing unique information for solar system exploration missions.

Nanosecond-precision-per-pixel time measurement provided by state-of-the-art Timepix detectors, together with a high spatial granularity, makes it possible to resolve the drift times not only in gaseous volumes, as originally intended, but also in thin semiconductor sensors, thereby segmenting their detection volume into a 3D grid of voxels with the dimensions $\sim(55 \times 55 \times 60)\,\mu\text{m}^3$ [23,24]. While the 3D reconstruction of interactions in the sensors improves the particle separation capability and impact angle determination, it also provides the means of using the detectors as a single-layer Compton camera for direction-sensitive hard X- or γ-ray detection [25–27]. The availability of sensors of different nuclear charge (e.g., silicon, CdTe/CZT, or GaAs) and thickness ($\sim$100 µm to 5 mm) allows for the optimization of detection efficiency and the ratio of Compton and photopeak signal across a broad energy band, thereby making it worthwhile to evaluate the capabilities of Timepix as baseline or complementary detectors in missions dedicated to the study of sources of hard X-rays or γ-rays in the not yet well-explored energy range from 0.1–2 MeV, which constitutes the low energy part of the so-called "MeV gap" [28]. Additionally, the inherent sensitivity of the Compton camera to photon polarization [29] provides a handle to advance the understanding of astrophysical γ-ray sources and environmental conditions [30].

The present manuscript elaborates on the capabilities of Timepix-type detectors in the space environment. We will describe the already well-established use as a single-layer radiation detector in near Earth orbits using the example of SATRAM while outlining their application potential as a tracker module in a compact magnetic spectrometer or as a compact Compton camera for direction- and polarization-sensitive X- and γ-ray detection

2. Materials and Methods

2.1. Timepix Series

Among the different readout ASICs developed in the Medipix collaborations, the Timepix series, namely the Timepix [7], Timepix2 [19], Timepix3 [8], and Timepix4 [31] ASICs, were dedicated to single-particle detection and tracking:

- Timepix was developed within the Medipix2 collaboration [32]. It segments the sensor into a square matrix of 256×256 pixels with a pixel pitch of 55 μm and purely relies on a frame-based readout scheme (dead-time > 11 ms). Each of the 65,536 pixels can be set to either of the three modes of signal processing: time-over-threshold (ToT), time-of-arrival (ToA, resolution > 10 ns), and hit counting.
- Timepix2, while still relying on the frame-based readout, provides additional features, e.g., a simultaneous measurement of ToA and ToT and an adaptive gain ToT mode for improved spectroscopy at high-energy deposition [33].
- The key improvements of Timepix3 are a time resolution below 2 ns and the data-driven mode. The latter provides an almost dead-time-free detector operation by reading out only the pixels, which are actually triggered by an ionizing particle, while all other pixels remain active (per-pixel dead time: $\sim$475 ns). Pixel hit rates up to 80 MHits s^{-1} can be sent off a chip at a bandwidth of 5.12 Gbps.
- Timepix4 comes with an increased pixel matrix featuring 512×448 pixels with a pitch of 55 μm (resulting in an area of $\sim$7 cm^2) [31]. Similar to Timepix3, it offers frame-based and data-driven readout schemes, but with 8 × higher maximal hit rate. The time binning is improved to 195 ps. The readout bandwidth can be up to 164 Gbps.

The ASICs can be coupled to different sensor materials by means of flip chip bump bonding. Currently available and tested sensor materials include silicon, CdTe, CZT, and high-resistivity chromium compensated GaAs:Cr with thicknesses ranging from 100 μm to 2 mm. Improvements in growth techniques facilitate the availability of thick sensors with low defect density (CdTe, CZT, and GaAs:Cr), which profit from a higher γ-ray detection efficiency and single-layer tracking performance.

2.2. The Space Application of Timepix Radiation Monitor (SATRAM)

The first application of a Timepix in open space was SATRAM (Space Application of Timepix Radiation Monitor) onboard the Proba-V satellite (see Figure 1) launched in May 2013. The satellite is orbiting Earth in a Sun-synchronous orbit at an altitude of 820 km with an inclination of 98.7°. The orbit duration is 101.21 min, and the local time at descending node is between 10:30 a.m. and 11:30 a.m. The SATRAM module is encapsulated in an aluminium alloy compartment. It has a thinned area above the sensor with a thickness of 0.5 mm. The module weighs 380 g, has a power consumption of 2.5 W, and dimensions of $55.5 \times 62.1 \times 197.1$ mm. The Timepix inside the module hosts a 300 μm thick silicon sensor. The threshold is globally set to 8 keV. The detector is operated in the ToT mode, with acquisition times for consecutive frames set to 20 s, 200 ms, and 2 ms to account for the different flux levels in orbit.

SATRAM's continued operation allowed for measurements of fluxes of electrons and protons trapped in the Van Allen radiation belts continuously during its ongoing mission. Current proton and electron separation relies on pattern recognition, together with the dE/dx information. Recent work started to use convolution neural networks (CNNs) to improve classification accuracy [16]. Based on the success of SATRAM, proposals to develop a miniaturized radiation monitor (MIRAM) [18] and a highly integrated Timepix-based radiation monitor (HITPix) have been funded by the European Space Agency (ESA). The major differences that set these detectors apart from other commonly used radiation monitors like ICARE [34,35] or the Standard Radiation Environment Monitor (SREM [36,37] are single-layer particle discrimination capabilities, which allow for the development of radiation monitors of small dimensions and low mass providing a large field of view.

Figure 1. Picture of SATRAM attached to Proba-V.

2.3. Pattern Recognition Tools and Particle Separation

Detectors of the Timepix family are sensitive to a broad variety of particle species: from X-rays and electrons with energies just above 3 keV up to particles in the GeV range. Due to detector segmentation and the charge transport properties of the semiconductor sensors, ionizing particles create imprints in the pixel screen (clusters or tracks), which are to some extent usable for particle identification.

A basic pattern recognition scheme was introduced in 2008 [38]. It defines six categories of events: dots, small blobs, curly tracks, heavy blobs, heavy tracks, and straight tracks, with each indicating different particle species and energy depositions (Figure 2). This methodology purely relies on the track morphology. Additionally, Timepix allows for the use of the energy information, with which properties like the deposited energy, cluster height (the energy of the highest energy pixel in a cluster), and the stopping power can be determined. Timepix can also provide timing information, but not while simultaneously measuring the energy. This ability was added in subsequent generations with Timepix2 and Timepix3. Together with increased energy resolution, the particle recognition capability was thus improved.

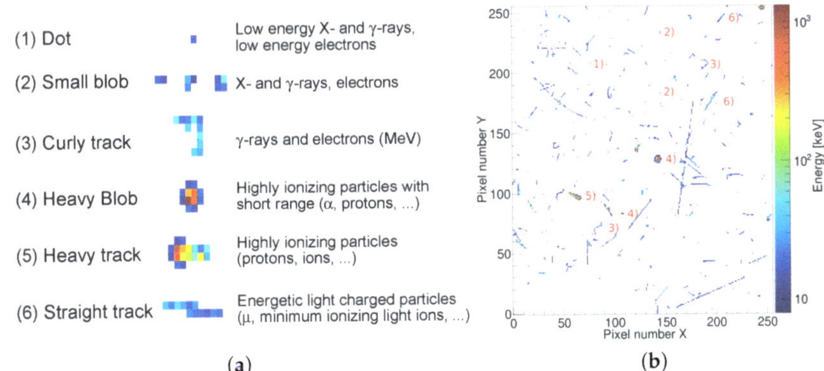

(a) (b)

Figure 2. (**a**) Illustration of the cluster shape classification scheme proposed by [38]. (**b**) A 20 s frame of the SATRAM response to the radiation field as measured in space. The different shapes seen in the pixel matrix can be categorized, and exemplary tracks are labeled according to the scheme in (**a**).

A first attempt to identify particle species in the open space radiation environment with Timepix technology is presented in [14] using SATRAM data. In low Earth orbit, electrons and protons are the most abundant particle species. In there, the track properties used for particle separation are the above-mentioned track morphology, cluster height, and stopping power (dE/dx). Monte Carlo simulations showed that for electrons, the cluster height is not higher than 300 keV, and the stopping power is not more than 10 MeV cm^2/g. The simulations were done for spectra that are expected in the radiation belts, i.e., for electrons with energies up to 7 MeV and protons with energies up to 400 MeV. While able to accurately determine electrons (correct classification in 98%), this method falls short in the identification of protons. High energy protons ($>$100 MeV) have a significant lower energy deposition and stopping power, which is on par with electrons. This is especially true for protons that pass through the detector perpendicularly, which create only short tracks that would often be misidentified as electrons.

In a follow-up work, neural networks were employed to improve previously achieved particle separation capability [16]. A fast Monte Carlo simulation of the detector response was used for the training of a CNN. Neglecting charge carrier transport, signal induction, and the behavior of the detector front end in the simulation, all particle tracks were only one pixel in width. While this is a valid approximation for the electron signatures, proton tracks seen in test beam measurements and in space data are usually two to four pixels wide. Still, it was shown that neural networks can be successfully used for particle identification in the case of electron- and proton-dominated space radiation data. Additionally, it was shown that the incident proton energy can be extracted, in particular at lower energy.

After implementing an improved detector response model accounting for charge sharing and induction to the previously used Monte Carlo simulation tool, another iteration of the NN was developed [15]. This feedforward neural network created in the TensorFlow framework [39] uses seven features to classify a cluster:

- The number of pixels in the cluster N;
- The deposited energy E_{dep} is defined as the sum of energies measured in each pixel of a cluster $E_{\text{dep}} = \sum_i^N E_i$;
- The maximal energy measured in a single pixel of the cluster $E_{\text{max}} = \max\{E_0, \ldots, E_N\}$;
- The linearity of the cluster, which is defined as the relative amount of pixel lying within a distance of one pixel from the longest line segment between two pixels of the cluster;
- The roundness of the cluster;
- The average number of neighboring pixels;
- The sum of the absolute values of cubic and quadratic terms of a third-order polynomial fit of the cluster.

The NN consists of an input layer, two hidden layers with seven neurons each, and one output layer. An overall testing accuracy of 90.2% was achieved, with protons being correctly classified in 89% and electrons in 91% of cases. In orbit, proton fluxes are still difficult to determine accurately, given that electron fluxes are often higher by about two orders of magnitude and electrons falsely identified as protons are of the same order of magnitude as the protons.

For the NN to work properly, it is required that clusters are well separated from each other. However, particle tracks inevitably overlap when frames have longer acquisition times and/or the fluxes are high. The NN is not able to recognize two or more tracks, let alone identify what particle species they are. To quantify this effect, occupancy has been introduced. It is calculated by the number of hit pixels divided by the number of available pixels. The result is expressed in percent. To ensure that the NN can work properly, only frames with a maximum occupancy of 20% were selected (low occupancy frames). For frames with higher occupancy (high occupancy frames), a different method had to be applied.

To estimate the electron fluxes of high occupancy frames, a statistical approach was chosen. The first step was to determine the mean energy of all clusters depending on the geographical position of the current measurement. The information was obtained by looking at the low occupancy frames in the region and calculating the mean energy for all particle tracks that were measured within the area. Typically, this local mean energy is higher in the SAA than the rest of the orbit due to the abundance of protons in that region. The estimation of the number of particles in the frame was then obtained by dividing the total measured energy in the frame by the local mean energy corresponding to the position of the satellite and then multiplying this by the fraction of electrons known from the previous low occupancy frame. Both methods are explained in detail in [15].

2.4. dE/dX Spectrum Unfolding

Spectrum deconvolution refers to the decomposition of a complex signal into its contributing spectrum components. There are many different iterative and statistical schemes that can be chosen for this process. In the present work, Bayesian unfolding has been chosen [40], which utilizes the probability formula:

$$p(A|B) = \frac{p(B|A)p(A)}{p(B)} \tag{1}$$

where p represents a generic probability function, $|$ is the given operator, and A and B are some arbitrary variables or system states. Despite the simplicity of the Bayesian formula, it is quite powerful and used in many areas of physics and statistics. The formula conveys the probability of A having a particular value or state given that B has a particular value or state, thereby essentially relating two otherwise unrelated states. The states A and B can be assigned some arbitrary distribution of two variables that will be referred to as the cause vector (x_C) and the effect vector (x_E), respectively. It can then be assumed that there exists an arbitrary probability distribution $p(x_E|x_C)$ given by the formula

$$p(x_C|x_E) = \frac{p(x_E|x_C)p(x_C)}{p(x_E)}. \tag{2}$$

The approximate values of $p(x_E|x_C)$ can be achieved through simulation. The remaining probability values for a specific experiment can be obtained through the Bayesian iterative deconvolution algorithm that is implemented using the library [41].

To use the Bayesian deconvolution algorithm, an incoming spectrum (x_C) is related to the measured spectrum (x_E) in a so-called "response matrix". Since previous works [42,43] have demonstrated the sensitivity of the dE/dX measurement to incident proton kinetic energy, we have chosen a response matrix relating the dE/dX spectra to an incoming monoenergetic omnidirectional electron or proton field (Figure 3). For each detected track, the stopping power was calculated as:

$$\frac{dE}{dX} = \frac{\sum_{i=0}^{N} E_i}{t_{\text{sensor}} \times \rho_{\text{Si}} \times \cos\theta} \tag{3}$$

with the per-pixel energy of a particle trace being E_i, the sensor thickness being $t_{\text{sensor}} = 300\,\mu\text{m}$, the density of silicon being $\rho_{\text{Si}} = 2.33\,\text{g\,cm}^{-3}$, and the reconstructed impact angle with respect to the sensor normal being θ.

Accounting for expected particles and energies, the response spectra were simulated in N_e electron primary energy bins and N_p proton primary energy bins with a flat distribution from 0 to 6 MeV and 0 to 400 MeV, respectively, using an in-house developed simulation tool based on Geant4 [44]. An omnidirectional particle environment was approximated by emitting $N_{\text{sim}} = 3 \times 10^6$ protons and 2×10^6 electrons with a $\cos^2(\theta)$ initial momentum direction distribution from the surface of a spherical source with a radius of $R_{\text{source}} = 20\,\text{cm}$. In order to speed up the simulation, and since we are only interested in particles arriving at the sensor, the emission angle range was restricted to point towards the sensor or its close surroundings, i.e., $\theta \in [0,3]$ degree.

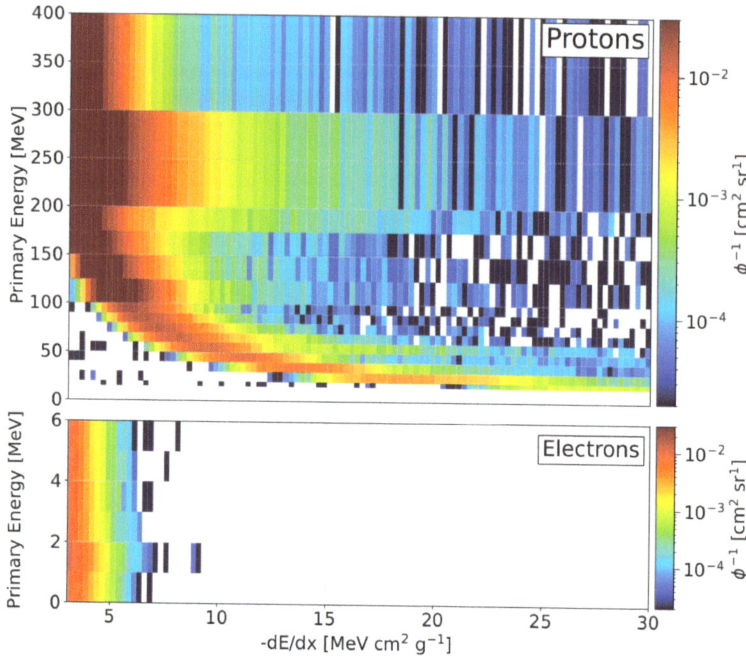

Figure 3. Graphical visualization of the response matrix $(p(x_C|x_E))$ used for the Bayesian deconvolution approach described in the text. It was obtained through simulation.

The response matrix was scaled with the particle flux through the sensitive volume, which was calculated for the simulated geometry as:

$$\frac{d\Phi_{\text{sim}}}{dA\,d\Omega} = \frac{N_{\text{sim}}}{A_{\text{surface}}\Omega_{\text{emission}}}, \tag{4}$$

$$\text{with } A_{\text{surface}} = 4\pi R_{\text{source}}^2 \tag{5}$$

$$\text{and } \Omega_{\text{emission}} = \int_0^{2\pi} d\varphi \int_0^{\frac{\pi}{60}} \cos(\theta)\sin(\theta)d\theta = \frac{1}{2}\left[1 - \cos\left(\frac{\pi}{60}\right)\right]. \tag{6}$$

The response matrix obtained likewise is presented in Figure 3. Asymmetric binning was chosen for the y axis to reflect the varying sensitivity towards spectral changes, which is higher at high stopping power (slow protons) and lower for higher energy protons. It can be seen that all electrons $\geq 1\,\text{MeV}$ were degenerate and all protons were asymptotically electron-like with increasing energy.

The presented methodology has been validated in clinical monoenergetic proton beams at the Danish Center for Particle Therapy, thereby finding proper incident energy reconstruction with angle-averaged incident energy resolutions of $\sigma_{125\,\text{MeV}} = 17\,\text{MeV}$, $\sigma_{175\,\text{MeV}} = 28\,\text{MeV}$, and $\sigma_{225\,\text{MeV}} = 42\,\text{MeV}$ at beam energies of 125 MeV, 175 MeV, and 225 MeV, respectively [45].

2.5. 3D Reconstruction of Particle Traces within the Semiconductor Sensor—Use as a Solid State Time Projection Chamber

Ionizing radiation interacting in the sensors creates free charge carriers, which start to drift towards the electrode of opposing charge. During this drift motion, currents are induced at the pixels in close vicinity. Due to the small pixel size compared to sensor thickness, detectable signals are only induced if the charge carriers are close to the pixel side (small pixel effect) so that the measured time corresponds approximately to the drift time across the sensor thickness.

The charge carrier (electrons e and holes h) drift along the z axis of the device can be described by

$$\vec{v_e} = -\mu_e \times \vec{E} \tag{7}$$

$$\vec{v_h} = \mu_h \times \vec{E} \tag{8}$$

where $\mu_{e/h}$ is the mobility of electron and holes, respectively. For planar silicon sensors in hole collection, a linear parameterization of the electric field can be used (see e.g., [23]):

$$\vec{E} = \frac{U_\mathrm{B}}{d}\vec{e_z} + \frac{2U_\mathrm{dep}}{d^2} \times \left(\frac{d}{2} - z\right)\vec{e_z}, \tag{9}$$

where d denotes the sensor thickness, U_B is the bias voltage, and U_dep is the depletion voltage. While U_dep depends on the quality of the sensor and should be determined individually, we can use $U_\mathrm{dep} = 40\,V \times \left(\frac{d}{300\,\mu m}\right)^2$ as a rule of thumb. For semi-insulation planar sensors (CdTe, CZT, and GaAs:Cr) with ohmic contacts, we assume a linear electric field across the sensor thickness:

$$\vec{E} = \frac{U_\mathrm{B}}{d}\vec{e_z}. \tag{10}$$

While the charge carrier drift motion can be described analytically, the induction process requires numeric calculation (iterative simulation). Therefore, the charge carrier drift motion and the amount of deposited charges were modeled, thereby creating lookup tables relating the energy deposition and measured time stamps to the interaction depth (see [24]).

The methodology was applied to silicon and CdTe sensors of thickness 500 μm [23] and 2 mm [24], respectively, thus finding z resolutions of $\sim$30 μm and 60 μm, respectively. Figure 4 shows typical event displays of 3D-reconstructed particle trajectories measured in relativistic particle beams.

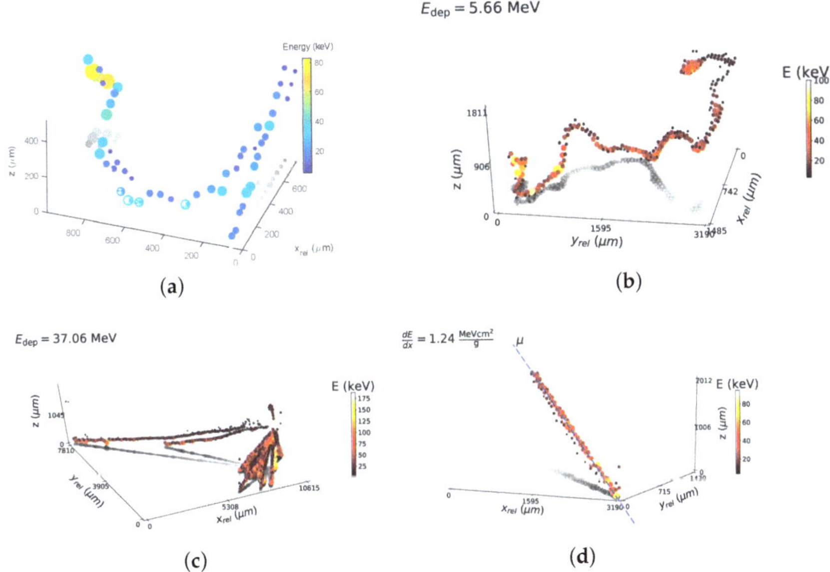

Figure 4. Event displays of 3D reconstructions of tracks: (**a**) 120 GeV/c pion passing through a 500 μm thick silicon sensor; (**b,c**) high-energy electron and fragmentation reaction measured during exposure

of a 2 mm thick CdTe sensor to a 180 GeV/c pion beam; (**d**) cosmic muon with a 3D line fit. Fitting the trajectory allows for a trajectory reconstruction with precision of <200 μm evaluated at a distance of 1 m. Reproduced from [23,24] (CC BY 4.0, no changes were made).

2.6. Single-Layer Compton Camera and Scatter Polarimetry

Fine pixelation and the (above-described) 3D reconstruction within the sensor permit Timepix3 utilization as a single-layer Compton camera. For this purpose, the Compton electron has to be detected together with the scattered X-ray photoelectron. With the Compton electron detected at $\vec{r}_e$ depositing energy E_e and the scattered photon detected at $\vec{r}'_\gamma$ with energy E'_γ, we can define Compton cones around the axis defined by the directional vector $\vec{r}_{\mathrm{dir}} = \vec{r}'_\gamma - \vec{r}_e$ with tips located at $\vec{r}_e$ [25]. The opening angle is given by

$$\cos \beta = 1 - m_e c^2 \times \left(\frac{1}{E'_\gamma} - \frac{1}{E_\gamma} \right) \tag{11}$$

with $m_e c^2 = 511$ keV being the electron rest energy, and $E_\gamma = E_e + E'_\gamma$ [25].

Polarized incoming radiation will create an asymmetry in the scattering angles evaluated, which can be described by the by the Klein–Nishina formula [29]

$$\frac{d\sigma}{d\Omega} = \frac{r_e^2}{2} \left(\frac{E'_\gamma}{E_\gamma} \right)^2 \left(\frac{E'_\gamma}{E_\gamma} + \frac{E_\gamma}{E'_\gamma} - 2 \sin^2(\beta) \cos^2(\varphi - \varphi_0) \right) \tag{12}$$

where $r_e = 2.818 \times 10^{-15}$ m is the classical radius of an electron, β is the angle of the outgoing photon with respect to the direction of the incoming photon, and $\varphi - \varphi_0$ is the angle between the scattering plane and polarization plane. The scattered photon preferentially flies perpendicularly ($\|\varphi_{\mathrm{pref}} - \varphi_0\| = 90° \implies \cos^2(\varphi_{\mathrm{pref}} - \varphi_0) = 0$) to the polarization of the incoming photon.

Concurrent detection of the Compton electron with the photoelectron therefore allows for the assessement of the X-ray polarization. Without loss of generality, we take the detector's x axis as reference and determine the angle φ as $\cos \varphi = \vec{r}_{\mathrm{dir}} \cdot \vec{e}_x / |\vec{r}_{\mathrm{dir}}|$, where $\vec{e}_x$ is the x axis unit vector. For partially polarized photon impact, the scattering angles φ will be distributed as

$$f(\varphi) = A(\mu \cos(2(\varphi - \varphi_0)) + 1), \tag{13}$$

where A is a scaling factor, μ is the modulation, and the phase φ_0 determines the polarization direction of the incoming X-rays with respect to the x axis. Here, we assumed that our detector is equally sensitive in all azimuthal directions φ. The degree of polarization is then

$$P = \frac{\mu}{\mu_{100}}, \tag{14}$$

where μ_{100} denotes the modulation response to a 100% polarized X-ray impact, which can be calibrated in simulation [46] or in a field of known polarization.

3. Results

3.1. Space Heritage—SATRAM and Its 10 Years of Operation as a Radiation Monitor

3.1.1. Measurement Stability—Noisy Pixel Appearance and Removal

In this section, the state of SATRAM in terms of radiation-induced effects on data quality shall be quantified. This will be done by looking at the amount of noisy pixels that may occur over time in a different quantity. Noisy pixels are defined as pixels exceeding the overall count rate at a statistically significant level. While most of the identified noisy pixels were recoverable and disappeared after resetting the detector configuration, some became permanently noisy. The latter were masked (removed from the data set) and, thus, are not considered in the analysis. Given that there are 65,536 pixels, the loss of a few hundred pixels is negligible for the overall data quality.

To determine if a pixel is noisy, one usually takes an arbitrarily high number of frames and counts how often a each pixel has sent a signal. In this study, the noisy pixel search was performed on the timescale of one week. The number of counts measured in each of the pixels within this time period were registered in a histogram. The resulting distribution was then fitted with a Gaussian distribution, therein obtaining the mean N_{mean} and the standard deviation σ. We defined the threshold for a pixel to be considered as noisy, N_{max}, as

$$N_{max} = N_{\mathrm{mean}} + 5 \times \sigma. \tag{15}$$

This procedure was done for every consecutive week from the beginning of available data in August 2014 until the 30 June 2023. Furthermore, this analysis was performed separately for the three different acquisition times. Naturally, more pixels will have sent a signal in longer frames than in shorter ones, which could have skewed the distributions. The resulting relative numbers of noisy pixels are shown in Figure 5.

Until the end of 2021, the number of noisy pixels were below 0.6%. In 2021, there were two periods, in spring and autumn, where an increased amount of noisy pixels was present over a few weeks. The same can be seen for the year 2022, but with a significantly higher number of noisy pixels. No definite explanation has been found for the increase nor for the periodicity.

In 2023, the detector seemed to have recovered. A careful inspection of the data has shown that most of the data was measured as expected. There were a few cases where the matrix became filled up to a large portion while being in the corresponding region of space where no high fluxes of radiation were expected. While the statistical method for noisy pixel determination used here is not suitable to detect this kind of behavior, these frames stood out and were excluded from analysis by comparison of their count rates with previous and subsequent frames.

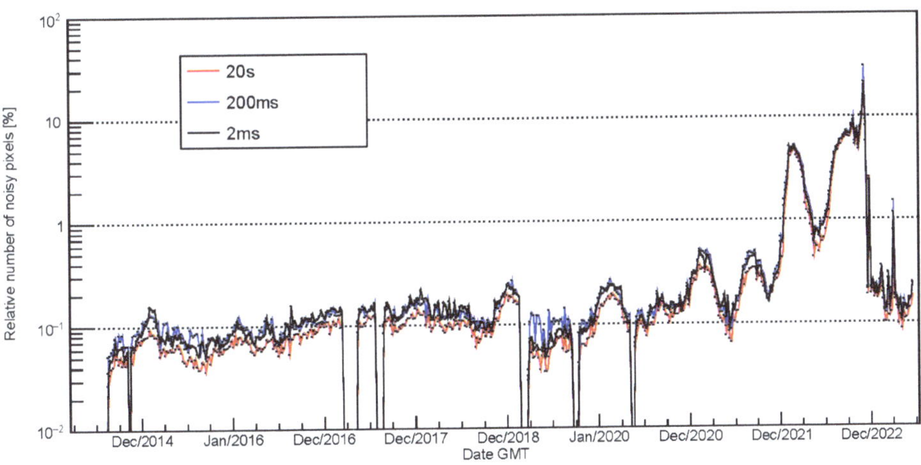

Figure 5. Relative number of pixel classified as noisy on a weekly basis from August 2014 to the end of June 2023 for the Timepix in SATRAM. The number of noisy pixels stayed below 0.6% until the end of 2021. In 2022, numbers were rising with a maximum of about 22% near the end of the year. The detector recovered in 2023.

The distribution of noisy pixels over the pixel matrix for the years 2015 and 2022 can be seen in Figure 6a and 6b, respectively. In 2015, most noisy pixels were concentrated in the lower left corner. This is a well-known firmware issue of SATRAM and was present already from the beginning. These pixels have been excluded from analysis in all previous studies. In 2022, the pixels in the lower left corner were still noisy and seemed to have worsened. Additionally, pixels along the edge were showing increased noise behavior. The edge pixels being noisy can be observed in many frames acquired in 2022. However, their position makes it easy to mask them and eliminate them from analysis. Their exclusion resulted in a reduction in the usable detector area by 21%.

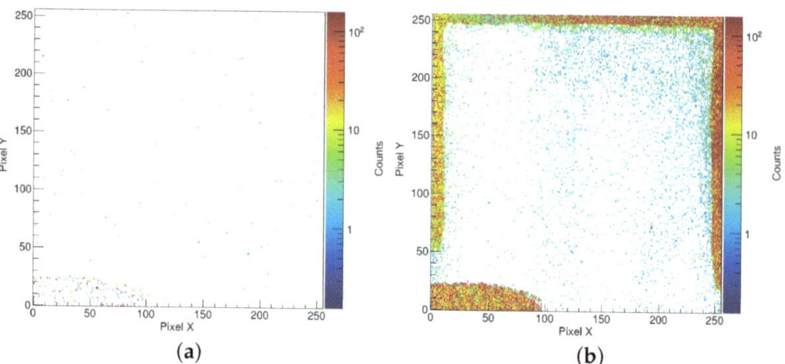

Figure 6. Noisy pixel distribution over the Timepix sensor for the years 2015 (**a**) and 2022 (**b**). The pixels in the lower left corner were damaged during launch of the Proba-V satellite. The color bar represents how often the pixels were considered noisy on a weekly basis and across all three acquisition times.

A follow up analysis, using the same method for noisy pixel detection but for a reduced number of pixels, was performed. The noisy pixels in the lower left corner and the edge pixels as seen in Figure 6b have been masked (excluded). The result is presented in Figure 7. The number of noisy pixels was greatly reduced to below 0.2%, except for a short period in late 2022, where about 10% of pixels were identified as noisy. This shows that by restricting the active area of the sensor through masking the problematic areas, SATRAM provides reasonable data during its entire 10 years of operation.

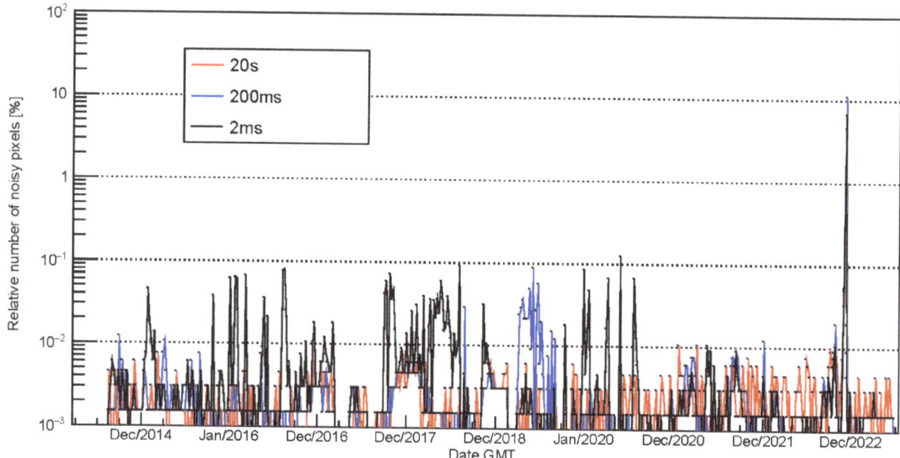

Figure 7. Same as Figure 5, but excluding noisy pixels from the lower left corner and pixels on the edges of the sensor, as seen in Figure 6b. The number of noisy pixels is greatly reduced.

3.1.2. Mapping Out Electron and Proton Fluxes in Orbit

Figure 8a,b show the fluxes of electrons and protons classified with the method described in Section 2.3. The majority of particles present in the radiation environment in LEO were protons (up to 400 MeV) and electrons (up to 7 MeV) that were trapped by the Earth's magnetic field. For electrons, three distinct structures are discernible, i.e., the northern polar horn, the southern polar horn, and the South Atlantic Anomaly (SAA). The northern and southern polar horns correspond to the points at which the satellite passed through the Earth's outer radiation belt. The SAA is present due to the satellite crossing the Earth's inner radiation belt. This crossing is possible due to the incline of the Earth's magnetic dipole combined with the deviation of the Earth's magnetic center with respect to the Earth's center of mass. While the outer radiation belt consists of electrons, the SAA is the only area in SATRAM's orbit, where a non-negligible flux of protons should be present. Thus, protons seen in the polar horns (in Figure 8b) were interpreted as misclassified electrons.

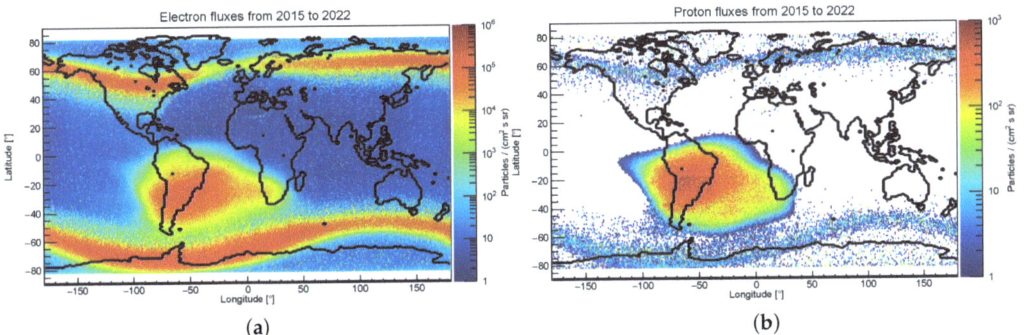

Figure 8. Electron (**a**) and proton (**b**) flux rates as measured with SATRAM at 820 km altitude in low Earth orbit averaged over the years from 2015–2022.

3.1.3. Measurement of the Proton Spectrum in the SAA

Figure 9a shows the dE/dx spectrum reconstructed for the central region of the SAA defined by *longitude* $\in [-70°, -25°]$ and *latitute* $\in [-40°, -12°]$. To avoid track overlap, which could result in an improper dE/dx determination, only tracks measured within the shortest frames (2 ms acquisition time) were used for the analysis. Within such frames, the matrix occupancy was consistently below 10%. A total of 22,784 frames were found during 2015–2018 operation, thus giving in a total effective measurement time of $t_{\text{meas.}} = 45.568$ s.

By applying the unfolding methodology defined in Section 2.4 to the measured dE/dx spectrum, we obtain the spectral-resolved differential flux equation:

$$\Phi_{\text{tot}}^{\text{unfold.}}(E) = \frac{N}{\Delta E \, d\Omega \, dt \, dA}, \tag{16}$$

where N denotes the number of particles measured within a bin of width ΔE per unit solid angle $d\Omega$, area dA, and time dt. It is shown as the blue curve in Figure 9b. Since no prior selection of the particle signature was performed, $\Phi_{\text{total}}^{\text{unfold.}}$ was thus defined as a linear combination of the differential proton (p) and electron (e) fluxes:

$$\Phi_{\text{total}}^{\text{unfold.}}(E) = \Phi_{p}^{\text{unfold.}}(E) + \Phi_{e}^{\text{unfold.}}(E). \tag{17}$$

As shown in Figure 3, the dE/dx response to the electrons is incident energy spectrum independent. We can thus estimate the "electron background" by applying Bayesian unfolding to the dE/dx spectrum simulated with an arbitrary incident electron energy spectrum. Thus, for our convenience, we utilized the simulation results of the above-described response matrix determination. In this way, we obtained an unfolded simulated fluence $\Phi_{\text{sim.}}^{\text{unfold.}}$, which is of proper spectral shape, but still needs to be scaled to resemble the electron flux rate from the measurement. This was accomplished using the relation

$$\Phi_e^{\text{unfold.}}(E) = \frac{N_e^{\text{meas.}}/dt}{N_e^{\text{sim.}}} \times \Phi_{\text{sim.}}^{\text{unfold.}},\qquad(18)$$

where $N_e^{\text{meas.}}/dt$ is the flux rate of particles identified as electrons in the measurement, and $N_e^{\text{meas.}}$ denotes the number of electrons detected in the simulation. The electron contribution obtained in this way is depicted in Figure 9b (in orange color). It becomes visible for the energy region above $\sim$100 MeV and dominates the count rate above $\sim$140 MeV.

The differential proton flux $\Phi_p^{\text{unfold.}}$ was then obtained by binwise subtraction of $\Phi_{\text{total}}^{\text{unfold.}}$ and $\Phi_e^{\text{unfold.}}$. An additional 20% error was added to account for possible systematic errors due to inaccurate detector response modeling. In Figure 9c, the result of the present work is compared with proton spectra measured with the EPT at different locations within the SAA [47]. Four geographic bins were close to the center, and a fifth bin was located at the edge of the SAA. While the overall spectral shapes of the EPT fell steeper with energy than the ones of the SATRAM, the deviations evaluated at each measured point were on a one sigma level over the entire energy range and across different locations. Considering that the data of the EPT and SATRAM were taken at different times, that the SATRAM bin averages the spectrum over a significantly larger geographic region, and the fact that the EPT has a limited field of view becoming narrower with higher proton energy, the agreement of our results with the EPT data is satisfactory.

3.2. Large Area Timepix3 Detectors as Tracking Modules in a Magnetic Spectrometer

The development of a penetrating particle analyzer (Mini.PAN) started in January 2020 in a collaboration formed by the Department of Nuclear and Particle Physics at the University of Geneva, the National Institute of Nuclear Physics at the Perugia Section, and the Institute of Experimental and Applied Phyics of the Czech Technical University. Mini.PAN employs position-sensitive (pixel and strip) detectors and (fast) scintillators to infer the particle type and velocity of GeV particles (and antiparticles) passing through the instrument's magnetic field by measuring their bending angles, charge deposition, and time of flight. Once in orbit, this device allows for a precise measurement of flux, composition, spectral characteristics, and directions of penetrating cosmic rays over the full solar cycle, thereby inherently providing the capability to search for antimatter. While such measurements exist within the heliosphere, Mini.PAN is designed as a compact instrument with low mass operating at low power. Thus, it counld be used in deep space or on smaller satellites. Figure 10a shows the developed pixel module featuring four Timepix3 detectors in a 2×2 geometry (quad) giving a $7.92\,\text{cm}^2$ area with 262,144 pixels at a pixel pitch of $55\,\mu\text{m}$. Figure 10 shows the pixel module integrated into the demonstrator.

Figure 11 shows the stopping power spectra measured with the developed device at different angles within a relativistic hadron beam (90% pions) measured at the CERN SPS. The distributions are modeled as the convolution of a Landau curve describing the physics the particle energy loss in the sensor smeared with a Gaussian whose width indicates the detector's energy resolution. Rotation of the device allows for the study of this resolution at different deposited energy. We find $\sigma(83.2\,\text{keV}) = 4.1\,\text{keV}$ (5%), $\sigma(97.1\,\text{keV}) = 6.4\,\text{keV}$ (6.6%), and $\sigma(180\,\text{keV}) = 13\,\text{keV}$ (7.2%).

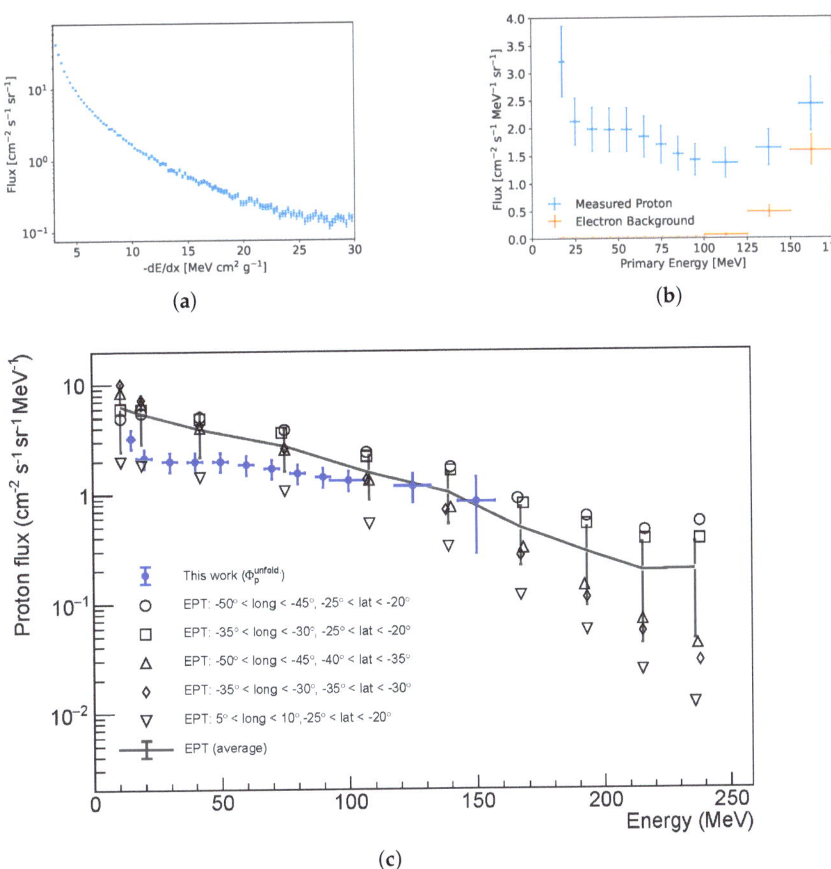

Figure 9. (**a**) dE/dx spectrum measured in the SAA used as input for the unfolding methodology. (**b**) Results of the unfolding methodology described in Section 2.4 for the dE/dx spectrum of (**a**) (blue line). The electron contribution determined in simulation (orange line, see text for details). (**c**) The energy-dispersive differential proton flux after electron background subtraction (blue markers). The SATRAM data measured within the SAA bin (*longitude* $\in [-70°, -25°]$, *latitute* $\in [-40°, -12°]$) are compared with a previous analysis of the EPT team [47] measuring at different geographic locations (different markers). The solid line resembles the average of these measurement points. Since the errors for the different locations are the same, these are drawn representatively for the averaged flux for improved visualization.

A detailed study of the temperature influence on the device performance and different operational parameters providing low power operation have been presented [48]. In the best case, a power consumption of $\sim$2 W for the entire quad was achievable.

3.3. Capabilities of Timepix3 as a Compton Camera and Scatter Polarimeter

The capabilities of a single-layer Timepix3 for use as a Compton polarimeter was studied in a laboratory experiment. The experimental setup is shown in Figure 12a. X-rays from a Hamamatsu microfocus tube were collimated onto a relatively large plastic target (dimensions: $2 \times 2 \times 2\,\text{cm}^3$) placed at a distance of 21.5 cm to the collimator. The tube voltage was set at $U_{\text{tube}} = 75\,\text{kV}$ with a tube current of $I_{\text{tube}} = 60\,\mu\text{A}$. A 1 mm thick pixelated silicon sensor (55 μm pixel pitch) attached to Timepix3, reverse biased at 400 V, was used to detect the scattered X-rays.

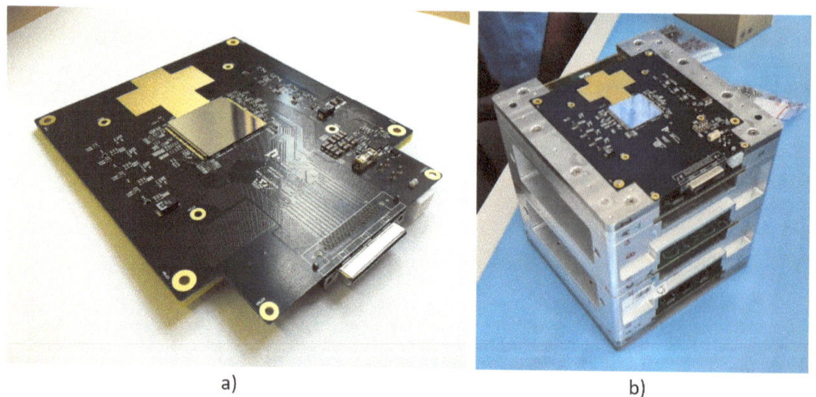

Figure 10. (**a**) Picture of the Timepix3 2 × 2 module developed for the use in demonstrator of a penetrating particle analyzer; (**b**) Timepix3 quad integrated into the Mini.PAN front end.

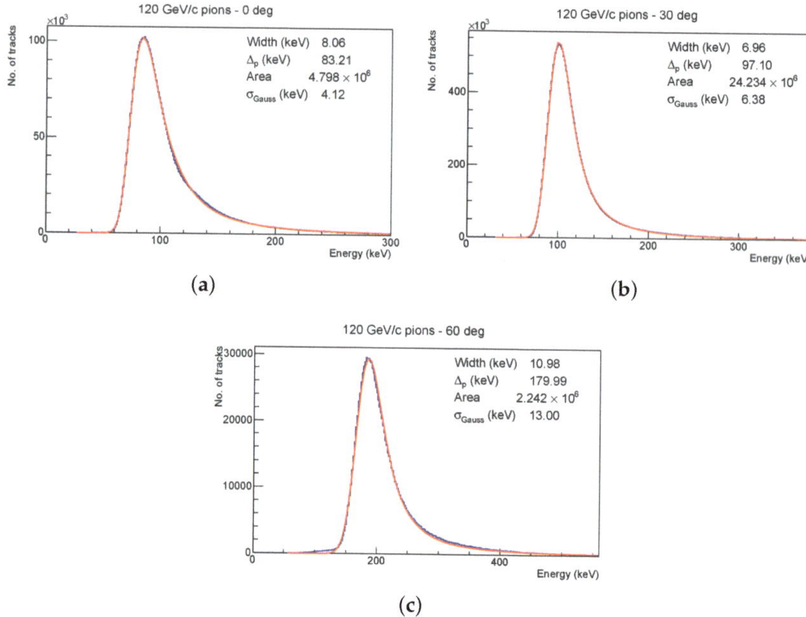

Figure 11. Energy deposition spectra measured with the Timepix3 quad module in a 180 GeV/c pion beam fitted with a Landau curve convolved with a Gaussian: (**a**) at perpendicular particle impact; (**b**) at 30 degrees; and (**c**) at 60 degrees impact angle with respect to the sensor normal.

The detector was placed at 16 cm from the target in a way that the X-rays of the highest degree of polarization (scattering off the target at 90°) could be recorded. Using the detector in data-driven operation, we searched for coincidentally detected pairs of clusters using a floating time window of $\Delta t = 65$ ns (drift time of holes across the whole thickness of the sensor). We refer to a set of coincidentally detected clusters as a "coincidence group". Coincidence groups larger than two clusters were omitted from the analysis. The cluster with higher energy E'_γ in each coincidence pair was assumed to be a photoelectron deposited by the scattered photon, while the lower energies E_e were assigned to Compton electrons. The histogram of energies of the clusters within coincident groups is shown in Figure 12b separately for clusters labeled as Compton electrons and as photoelectrons. The incoming photon energy was reconstructed by summing the two energy measurements $E_\gamma = E'_\gamma + E_e$. Using Equation (11), we calculated cosine of the scattering and selected only pairs with $-1 \le \cos\beta \le 1$.

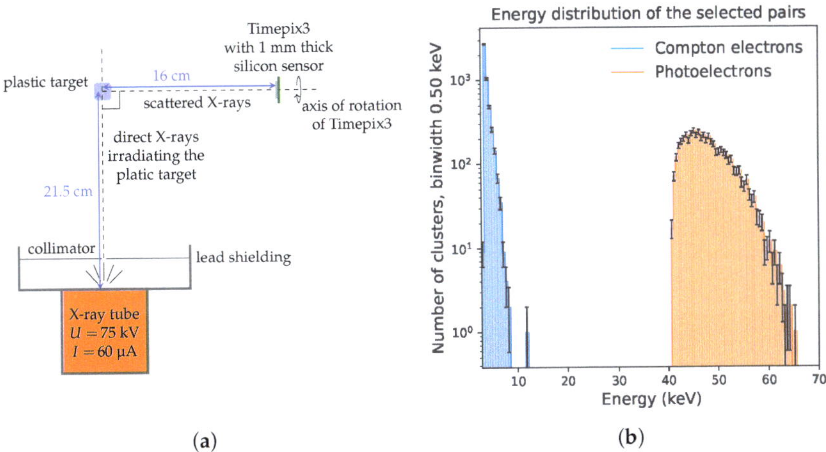

Figure 12. (**a**) Experiment design: A collimated beam from a Hamamatsu microfocus X-ray tube hits a plastic target to create polarized X-rays, which are absorbed in a 1 mm thick Timepix3 detector. The detector was placed at 90° to the axis defined by the tube and scattering target. A tube voltage $U_\text{tube} = 75$ kV was used at the tube current $I_\text{tube} = 60$ μA. (**b**) Energy histogram of the selected pairs of clusters. Compton scattering clusters with energies $\lesssim 3.5$ keV could not be detected due to per pixel detection threshold.

We further applied a cut on the pixel plane distance between the coincident clusters $d = \sqrt{\Delta x^2 + \Delta y^2}$, thereby restricting the range to $1.0\,\text{mm} < d < 10.0\,\text{mm}$. Figure 13 shows the measured scattering angle distributions fitted with Equation (13) to determine the modulation μ and phase shift φ_0. Overall, a modulation of 80% was found. To demonstrate that the seen modulation was in fact an effect observable in the laboratory frame and not inherent to the technology, the detector was rotated around the axis defined by the target and detector. The observed phase shifts φ_0 were consistent with the angle offsets.

Figure 14 shows the Compton camera reconstruction using simple back projection. Relative 3D coordinates were calculated as described in Section 2.5 using the timestamp measured by the Compton cluster within the coincidence pair as the time reference (t_ref). Furthermore, each cone was assigned a weight that favored cluster pairs with a higher energy of Compton electrons E_e (less uncertainty in $\cos\beta$), a larger absolute time of arrival difference $\|\Delta t\|$ (being close to either 0.0 ns or 65 ns), and a greater distance d (less uncertainty in cone axis vector).

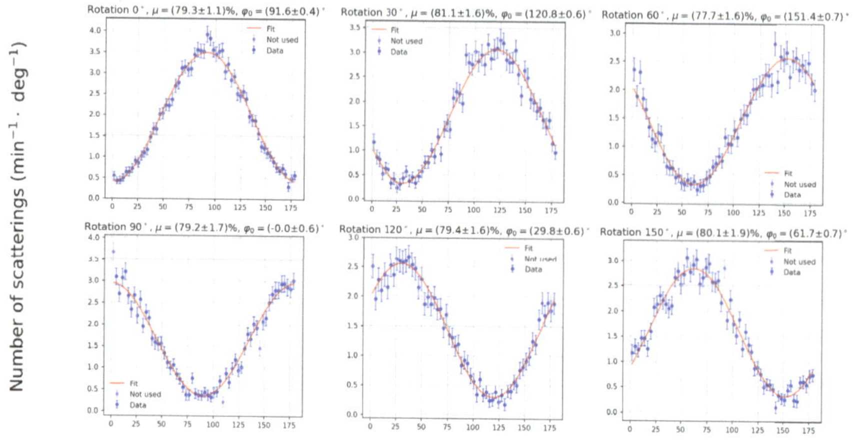

Figure 13. Modulations measured within the detector plane presented at different angles around the axis defined by the target and detector.

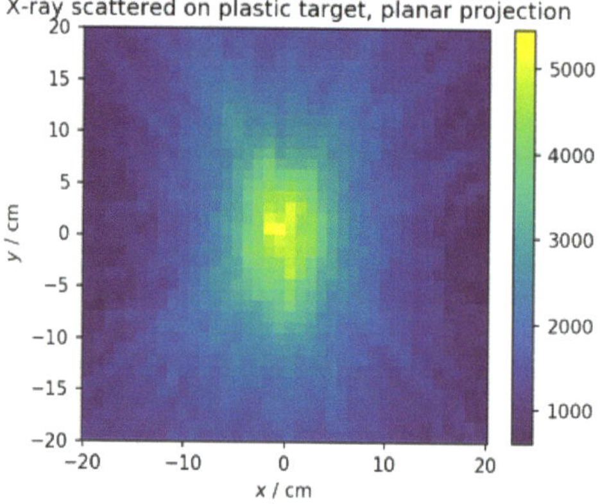

Figure 14. Application of the single-layer Compton camera reconstruction to the measured data.

4. Discussion

4.1. Timepix-Based Radiation Monitors

In contrast to commonly used space radiation monitors, Timepix-based devices provide the capability to separate different particle classes with a single-layer detector. This allows for the development of competitive low mass ($\sim$100 g) radiation detectors, which inherently provide an almost 4π field of view. In the present contribution, we have outlined these capabilities through the example of SATRAM, which has been operated in open space for more than 10 years.

While Timepix can be considered as a noise-free individual particle detector, due to single-event effects appearing in chip registers, individual pixels can "lose" their configuration and become noisy until their configuration is reset. Long-term irradiation additionally results in electronics baseline shifts, which could affect the noise behavior of the entire sensor. We have studied the appearance of noise patterns in the measured data by searching for outlier pixels with unphysically high count rates. For the first 8 years of operation, the number of such pixels was consistently on the level of 0.6%. During 2022 operation, the number of noisy pixels reached values of up to 22% and recovered in 2023. It was found that the noise level increase in particular affected pixels at the edges of the sensor, which were subsequently excluded to reduce the relative amount of noisy pixels to 0.1% (with the exception of a short period in late 2022 with up to 10%). While the effectively used detector area had to be reduced by 21% for 2022 operation, the overall data quality and scientific reach of the detector was not affected.

The current limitation of SATRAM compared to, e.g., with the Energetic Particle Telescope, is its insensitivity to resolve the spectral characteristics of incoming radiation. Within in the present work, we have overcome this problem—at least for protons—using a novel spectrum unfolding methodology. For the first time, we were able to present a proton energy spectrum measured with a single-layer device in low Earth orbit. The obtained result is consistent with a previous measurement performed with the science-class instrument EPT [47]. A comprehensive discussion and comparison with state-of-the art radiation belt models like AP-8 [49] or AP-9 [50] is out of the scope of this work. Energy-selective detection of electrons in the LEO radiation environment still remains unsolved, and should be addressed in future development, e.g., by implementing multidetector devices with sensors of differing stopping power or by adding electron stopping filters to the backside of the sensor.

A drawback of Timepix is that measurements are taken in frames of predefined acquisition times. Thus, at changing radiation fluxes, the overexposure of frames can occur and lead to track overlap and the misclassification of events. The typical mitigation strategy is the adjustment of frame times with the consecutive selection of frames with acceptable occupancy for analysis. While adaptive techniques for frame time adjustment are presented in [10], a computationally inexpensive approach had to be chosen for SATRAM. Measurements were collected in a predefined sequence of frames with acquisition times of 20 s, 0.2 s, and 2 ms. The overexposure issue has been addressed with the design of next generation Timepix ASICs. For example, Timepix2 provides "online" monitoring of the frame occupancy with automatic frame termination once a preset amount of columns is triggered; Timepix3 implements a data-driven mode, where only pixels triggered by radiation are read out, while all others remain active. The latter, however, comes with the possibility of high measured data rates. Considering typically limited resources for data storage and downlink, this imposes the requirement for an onboard data compression. Therefore, methodology and algorithms are needed that can analyze the data at low computing power. Development going in this direction has been started.

4.2. Towards Astroparticle Physics Application

Highly spatially segmented detectors with decent time resolutions are also a valuable asset for astroparticle physics instrumentation. In contrast to the space weather and radiation dosimetry studies, where small detectors are beneficial, astrophysical observations usually require detectors of a large area to cope with low flux rates.

4.2.1. From Mini.PAN to Pix.PAN

A Timepix3 quad detector was developed for application in Mini.PAN, which is a two-sector magnetic spectrometer proposed for an in situ spectrum-resolved measurement of the galactic cosmic ray fluxes. The developed detectors have an effective area of $7.92\,\text{cm}^2$, segmented in 262.144 pixels of $55 \times 55\,\mu\text{m}^2$. In the current instrument design, they are mainly supplemental detectors adding high flux capabilities, an additional charge, and position measurement. Their limited spatial ($dx = {}^{55}/\sqrt{12}\,\mu\text{m} = 16\,\mu\text{m}$) and temporal resolution ($\sim 2\,\text{ns}$), prevents them from being used as a standalone tracker (requirement: $dx < 7\,\mu\text{m}$) or as a segmented time-of-flight module (requirement: $dt < 200\,\text{ps}$). As outlined in [51], these issues can be overcome by the latest generation of Timepix-series chips, Timepix4, thus inherently providing a time granularity of $<200\,\text{ps}$ combined with an adapted sensor design using a "pitch adapter" to create rectangular pixels of $13.75 \times 1760\,\mu\text{m}^2$ in area. The small pitch in bending direction is sufficient for measuring the curvature of particles in the range up to $10\,\text{GeV}/c$ with the baseline Mini.PAN Halbach magnets of $0.5\,\text{T}$ [22]. The lax requirement in the nonbending direction further makes it possible to save power by switching off 7/8 of the pixels. The production and testing of the novel sensor design has been started.

The Pix.PAN design relies on three tracking stations, with each consisting of a stack of two Timepix4 quads [51]. While the synchronization of 24 detectors at picosecond precision requires careful electronics design, relying on a single detector technology represents a significant simplification compared to Mini.PAN. The high-rate capability of Timepix4 will allow for application in harsh radiation environments, such as the Jovian radiation belts.

4.2.2. Compton Scatter Polarimetry

At last, we have presented a simple laboratory experiment demonstrating the capability of Timepix3 to be used as a single-layer Compton camera and scatter polarimeter. Therefore, we profited from the capability of reconstructing the locations of the interaction of ionizing radiation within thick sensors in 3D, which was enabled by nanosecond-scale drift time measurement. We measured the modulation for X-rays from a microfocus tube (tube voltage: 75 kV) scattered at 90 degrees in a plastic target to be $\mu_{\text{meas.}} = 80\%$. This represented an improvement of $\sim$29% compared to previous work [46] using Timepix in a similar experiment (finding a modulation of 62%).

To further understand the detectors capability, a simulation in Geant4 [44] with X-rays hitting a $14.08 \times 14.08 \times 1.0\,\text{mm}^3$ silicon sensor was carried out. Simulated X-ray beams were monoenergetic, nondispersive, had a uniform spatial distribution, and were arriving at an angle of 90° to the sensor plane. Three types of beams were simulated: unpolarized, 100% polarized with polarization direction at 0°, and 100% polarized with the polarization vector oriented at 30° to to the sensor's x axis. Only events with the photon interacting twice in the sensor were selected. The same cuts on $\cos\beta$ and distance d were made as for the experimental data. Interactions with an energy deposit $E_{\text{dep}} < 3.5\,\text{keV}$, resembling the per pixel energy threshold, were omitted. We found that for 100% polarization in the incoming photon energy range from 45 to 75 keV, a modulation $\mu_{100} > 92\%$ could be achieved. We can assess the performance of the device according to the minimum detectable polarization (MDP) at a 99% confidence level describing the degree of linear polarization detectable within a given acquisition time. Neglecting the background, it can be estimated as [46]

$$\text{MDP}_{99\%} = \frac{4.29}{\mu_{100} \times \sqrt{N_{\text{det}}}}, \tag{19}$$

where N_{det} is the number of detected scatter electron–photon pairs, and μ_{100} is the modulation measured at 100% polarized radiation. We can solve Equation (19) for N_{det} to estimate the minimal amount of detected scatter events:

$$N_{\text{det}} = \frac{4.29^2}{\mu_{100}^2 \times \text{MPD}_{99\%}^2}. \tag{20}$$

With $\mu_{100} = 92\%$ (from the simulation) and $\text{MPD}_{99\%} = 10\%$, we find $N_{\text{det}} = 2000$. Further simulation studies implementing proper detector responses shall be the topic of future work and should focus on improving event selection criteria or obtaining ground truth data samples for machine learning techniques. The possibility to use different sensor materials of various thicknesses with the same Timepix readout ASIC hereby allows for the selection of sensors optimized for the desired photon energy range. The larger area and lower power density per unit area of Timepix4 will further enhance the applicability for space research. Thus, future work should study the capabilities of thick CdTe/CZT (studied up to 5 mm [52,53]) devices for measurement in the hard X-ray band, terrestrial γ flashes, or γ-ray bursts.

5. Conclusions

In summary, we have demonstrated that Timepix-family detectors' capability of single-layer particle tracking and particle species separation allows for the production of competitive radiation monitors with one order of reduction in mass and covering almost the entire solid angle range. The application of novel methdology utilizing the Timepix3 time resolution for the reconstruction of the z coordinate provides 3D reconstruction of particle tracks, which in "thick" sensors enables their use as a single-layer Compton camera and scatter polarimeter. Here, the possibility to combine the readout ASIC with sensors of different materials provides a means of optimization for different X- and γ-ray bands. While currently, Timepix3 detectors are an integral part of particle spectrometers for measurements of galactic cosmic ray properties (Mini.PAN), Timepix4 could be a baseline technology of future magnetic spectrometers, thereby adding high rate capability and electronics simplification (Pix.PAN).

Author Contributions: Conceptualization, B.B.; Formal analysis, D.G., S.G. and J.J.; Funding acquisistion, B.B.; Investigation, J.J.; Methodology, B.B., S.G., D.G. and J.J.; Project administration, B.B.; Resources, P.S.; Software, S.G., D.G. and J.J.; Supervision, B.B. and S.G.; Validation, S.G., D.G. and J.J.; Visualization, S.G., D.G. and J.J.; Writing—original draft preparation, B.B.; Writing—review and editing, S.G., D.G., J.J. and P.S. All authors have read and agreed to the published version of the manuscript.

Funding: The authors B.B., D.G., and P.S. are grateful for the funding received by the Czech Science Foundation (GACR) under grant number GM23-04869M. The development of the Timepix3 Quad module for Mini.PAN was funded by the European Union's Horizon 2020 research and innovation programme under grant agreement No. 862044.

Data Availability Statement: All SATRAM data are available through the online tool https://satram. utef.cvut.cz/ (accessed on 26 February 2024). Other presented data and the reconstruction codes used are available upon request.

Acknowledgments: Successful launch and operation of SATRAM would not have been possible without Stanislav Pospíšil, Carlos Granja, and Alan Owens.

Conflicts of Interest: The authors declare no conflicts of interest.

Abbreviations

The following abbreviations are used in this manuscript:

ASIC	Application-Specific Integrated Circuit
CdTe	Cadmiumtelluride
CZT	Cadmiumzinctelluride
GaAs:Cr	Chromium-Compensated Galliumarsenide
CNN	Convolution Neural Network
ESA	European Space Agency
EPT	Energetic Particle Telescope
HITPix	Highly Integrated Timepix radiation monitor
HPD	Hybrid pixel detector
ICARE	Influence sur les Composants Avancés des Radiations de l'Espace
LEO	Low Earth Orbit
MIRAM	Miniaturized Radiation Monitor
MS	Magnetic Spectrometer
MPD	Minimum Detectable Polarization
NN	Neural Network
PAN	Penetrating Particle Analyzer
SAA	South Atlantic Anomaly
SATRAM	Space Application Timepix Radiation Monitor
SPENVIS	Space Environment Information System
SREM	Standard Radiation Environment Monitor
SWIMMR	Space Weather Instrumentation, Measurement, Modelling and Risk

References

1. Campbell, M. 10 years of the Medipix2 Collaboration. *Nucl. Instruments Methods Phys. Res. Sect. A Accel. Spectrometers Detect. Assoc. Equip.* **2011**, *633*, S1–S10. [CrossRef]
2. Campbell, M.; Heijne, E.; Llopart, X.; Colas, P.; Giganon, A.; Giomataris, Y.; Chefdeville, M.; Colijn, A.; Fornaini, A.; van der Graaf, H.; et al. GOSSIP: A vertex detector combining a thin gas layer as signal generator with a CMOS readout pixel array. *Nucl. Instruments Methods Phys. Res. Sect. A Accel. Spectrometers Detect. Assoc. Equip.* **2006**, *560*, 131–134. [CrossRef]
3. Bamberger, A.; Desch, K.; Renz, U.; Titov, M.; Vlasov, N.; Wienemann, P.; Zwerger, A. Resolution studies on 5GeV electron tracks observed with triple-GEM and MediPix2/TimePix-readout. *Nucl. Instruments Methods Phys. Res. Sect. A Accel. Spectrometers Detect. Assoc. Equip.* **2007**, *581*, 274–278. [CrossRef]
4. George, S.; Murtas, F.; Alozy, J.; Curioni, A.; Rosenfeld, A.; Silari, M. Particle tracking with a Timepix based triple GEM detector. *J. Instrum.* **2015**, *10*, P11003. [CrossRef]
5. Ballabriga, R.; Campbell, M.; Llopart, X. Asic developments for radiation imaging applications: The medipix and timepix family. *Nucl. Instruments Methods Phys. Res. Sect. A Accel. Spectrometers Detect. Assoc. Equip.* **2018**, *878*, 10–23. [CrossRef]
6. Jakůbek, J. Semiconductor Pixel detectors and their applications in life sciences. *J. Instrum.* **2009**, *4*, P03013. [CrossRef]
7. Llopart, X.; Ballabriga, R.; Campbell, M.; Tlustos, L.; Wong, W. Timepix, a 65k programmable pixel readout chip for arrival time, energy and/or photon counting measurements. *Nucl. Instruments Methods Phys. Res. Sect. A Accel. Spectrometers Detect. Assoc. Equip.* **2007**, *581*, 485–494. [CrossRef]
8. Poikela, T.; Plosila, J.; Westerlund, T.; Campbell, M.; De Gaspari, M.; Llopart, X.; Gromov, V.; Kluit, R.; Van Beuzekom, M.; Zappon, F.; et al. Timepix3: A 65K channel hybrid pixel readout chip with simultaneous ToA/ToT and sparse readout. *J. Instrum.* **2014**, *9*, C05013. [CrossRef]
9. Stoffle, N.; Pinsky, L.; Kroupa, M.; Hoang, S.; Idarraga, J.; Amberboy, C.; Rios, R.; Hauss, J.; Keller, J.; Bahadori, A.; et al. Timepix-based radiation environment monitor measurements aboard the International Space Station. *Nucl. Instruments Methods Phys. Res. Sect. A Accel. Spectrometers Detect. Assoc. Equip.* **2015**, *782*, 143–148. [CrossRef]
10. Kroupa, M.; Bahadori, A.; Campbell-Ricketts, T.; Empl, A.; Hoang, S.M.; Idarraga-Munoz, J.; Rios, R.; Semones, E.; Stoffle, N.; Tlustos, L.; et al. A semiconductor radiation imaging pixel detector for space radiation dosimetry. *Life Sci. Space Res.* **2015**, *6*, 69–78. [CrossRef] [PubMed]
11. Kroupa, M.; Campbell-Ricketts, T.; Bahadori, A.A.; Pal Chowdhury, R.; Empl, A.; George, S.P.; O'Brien, T.P. Extravehicular Electron Measurement Based on an Intravehicular Pixel Detector. *J. Geophys. Res. Space Phys.* **2019**, *124*, 8271–8279. [CrossRef]
12. Granja, C.; Polansky, S.; Vykydal, Z.; Pospisil, S.; Owens, A.; Kozacek, Z.; Mellab, K.; Simcak, M. The SATRAM Timepix spacecraft payload in open space on board the Proba-V satellite for wide range radiation monitoring in LEO orbit. *Planet. Space Sci.* **2016**, *125*, 114–129. [CrossRef]
13. Granja, C.; Polansky, S.; Vykydal, Z.; Pospisil, S.; Turecek, D.; Owens, A.; Mellab, K.; Nieminen, P.; Dvorak, Z.; Simcak, M.; et al. Quantum imaging monitoring and directional visualization of space radiation with timepix based SATRAM spacecraft payload in open space on board the ESA Proba-V satellite. In Proceedings of the 2014 IEEE Nuclear Science Symposium and Medical Imaging Conference (NSS/MIC), Seattle, WA, USA, 8–15 November 2014; pp. 1–2. [CrossRef]
14. Gohl, S.; Bergmann, B.; Evans, H.; Nieminen, P.; Owens, A.; Posipsil, S. Study of the radiation fields in LEO with the Space Application of Timepix Radiation Monitor (SATRAM). *Adv. Space Res.* **2019**, *63*, 1646–1660. [CrossRef]
15. Gohl, S.; Bergmann, B.; Kaplan, M.; Němec, F. Measurement of electron fluxes in a Low Earth Orbit with SATRAM and comparison to EPT data. *Adv. Space Res.* **2023**, *72*, 2362–2376. [CrossRef]
16. Ruffenach, M.; Bourdarie, S.; Bergmann, B.; Gohl, S.; Mekki, J.; Vaillé, J. A New Technique Based on Convolutional Neural Networks to Measure the Energy of Protons and Electrons With a Single Timepix Detector. *IEEE Trans. Nucl. Sci.* **2021**, *68*, 1746–1753. [CrossRef]
17. Furnell, W.; Shenoy, A.; Fox, E.; Hatfield, P. First results from the LUCID-Timepix spacecraft payload onboard the TechDemoSat-1 satellite in Low Earth Orbit. *Adv. Space Res.* **2019**, *63*, 1523–1540. [CrossRef]
18. Gohl, S.; Bergmann, B.; Pospisil, S. Design Study of a New Miniaturized Radiation Monitor Based on Previous Experience with the Space Application of the Timepix Radiation Monitor (SATRAM). In Proceedings of the 2018 IEEE Nuclear Science Symposium and Medical Imaging Conference Proceedings (NSS/MIC), Sydney, NSW, Australia, 10–17 November 2018; pp. 1–7. [CrossRef]
19. Wong, W.; Alozy, J.; Ballabriga, R.; Campbell, M.; Kremastiotis, I.; Llopart, X.; Poikela, T.; Sriskaran, V.; Tlustos, L.; Turecek, D. Introducing Timepix2, a frame-based pixel detector readout ASIC measuring energy deposition and arrival time. *Radiat. Meas.* **2020**, *131*, 106230. [CrossRef]
20. RAL Space SWIMMR-1. Available online: https://www.ralspace.stfc.ac.uk/Pages/SWIMMR-1-launches-in-boost-to-UK-space-weather-forecasting-capabilities.aspx (accessed on 26 October 2023).
21. Filgas, R. (Institute of Experimental and Applied Physics, Czech Technical University, Prague, Czechia). Private communication, 2023.
22. Wu, X.; Ambrosi, G.; Azzarello, P.; Bergmann, B.; Bertucci, B.; Cadoux, F.; Campbell, M.; Duranti, M.; Ionica, M.; Kole, M.; et al. Penetrating particle ANalyzer (PAN). *Adv. Space Res.* **2019**, *63*, 2672–2682. [CrossRef]
23. Bergmann, B.; Pichotka, M.; Pospisil, S.; Vycpalek, J.; Burian, P.; Broulim, P.; Jakubek, J. 3D track reconstruction capability of a silicon hybrid active pixel detector. *Eur. Phys. J. C* **2017**, *77*, 421. [CrossRef]

24. Bergmann, B.; Burian, P.; Manek, P.; Pospisil, S. 3D reconstruction of particle tracks in a 2 mm thick CdTe hybrid pixel detector. *Eur. Phys. J. C* **2019**, *79*, 165. [CrossRef]

25. Turecek, D.; Jakubek, J.; Trojanova, E.; Sefc, L. Single layer Compton camera based on Timepix3 technology. *J. Instrum.* **2020**, *15*, C01014. [CrossRef]

26. Amoyal, G.; Schoepff, V.; Carrel, F.; Michel, M.; Blanc de Lanaute, N.; Angélique, J. Development of a hybrid gamma camera based on Timepix3 for nuclear industry applications. *Nucl. Instruments Methods Phys. Res. Sect. A Accel. Spectrometers Detect. Assoc. Equip.* **2021**, *987*, 164838. [CrossRef]

27. Wen, J.; Zheng, X.; Gao, H.; Zeng, M.; Zhang, Y.; Yu, M.; Wu, Y.; Cang, J.; Ma, G.; Zhao, Z. Optimization of Timepix3-based conventional Compton camera using electron track algorithm. *Nucl. Instruments Methods Phys. Res. Sect. A Accel. Spectrometers Detect. Assoc. Equip.* **2022**, *1021*, 165954. [CrossRef]

28. Carenza, P.; la Torre Luque, P.D. Detecting neutrino-boosted axion dark matter in the MeV gap. *Eur. Phys. J. C* **2023**, *83*. [CrossRef]

29. Caroli, E.; Moita, M.; Da Silva, R.M.C.; Del Sordo, S.; De Cesare, G.; Maia, J.M.; Pàscoa, M. Hard X-ray and Soft Gamma Ray Polarimetry with CdTe/CZT Spectro-Imager. *Galaxies* **2018**, *6*, 69. [CrossRef]

30. Weisskopf, M.C. An Overview of X-Ray Polarimetry of Astronomical Sources. *Galaxies* **2018**, *6*, 33. [CrossRef]

31. Llopart, X.; Alozy, J.; Ballabriga, R.; Campbell, M.; Casanova, R.; Gromov, V.; Heijne, E.; Poikela, T.; Santin, E.; Sriskaran, V.; et al. Timepix4, a large area pixel detector readout chip which can be tiled on 4 sides providing sub-200 ps timestamp binning. *J. Instrum.* **2022**, *17*, C01044. [CrossRef]

32. Medipix Collaboration. Available online: https://medipix.web.cern.ch/home (accessed on 18 July 2023).

33. Bergmann, B.; Smolyanskiy, P.; Burian, P.; Pospisil, S. Experimental study of the adaptive gain feature for improved position-sensitive ion spectroscopy with Timepix2. *J. Instrum.* **2022**, *17*, C01025. [CrossRef]

34. Boscher, D.; Bourdarie, S.A.; Falguere, D.; Lazaro, D.; Bourdoux, P.; Baldran, T.; Rolland, G.; Lorfevre, E.; Ecoffet, R. In Flight Measurements of Radiation Environment on Board the French Satellite JASON-2. *IEEE Trans. Nucl. Sci.* **2011**, *58*, 916–922. [CrossRef]

35. Boscher, D.; Cayton, T.; Maget, V.; Bourdarie, S.; Lazaro, D.; Baldran, T.; Bourdoux, P.; Lorfèvre, E.; Rolland, G.; Ecoffet, R. In-Flight Measurements of Radiation Environment on Board the Argentinean Satellite SAC-D. *IEEE Trans. Nucl. Sci.* **2014**, *61*, 3395–3400. [CrossRef]

36. Evans, H.; Bühler, P.; Hajdas, W.; Daly, E.; Nieminen, P.; Mohammadzadeh, A. Results from the ESA SREM monitors and comparison with existing radiation belt models. *Adv. Space Res.* **2008**, *42*, 1527–1537. [CrossRef]

37. Sandberg, I.; Daglis, I.A.; Anastasiadis, A.; Bühler, P.; Nieminen, P.; Evans, H. Unfolding and validation of SREM fluxes. In Proceedings of the 2011 12th European Conference on Radiation and Its Effects on Components and Systems, Seville, Spain 19–23 September 2011; pp. 599–606. [CrossRef]

38. Holy, T.; Heijne, E.; Jakubek, J.; Pospisil, S.; Uher, J.; Vykydal, Z. Pattern recognition of tracks induced by individual quanta of ionizing radiation in Medipix2 silicon detector. *Nucl. Instruments Methods Phys. Res. Sect. A Accel. Spectrometers Detect. Assoc. Equip.* **2008**, *591*, 287–290. [CrossRef]

39. Tensorflow. Available online: https://www.tensorflow.org/ (accessed on 9 September 2023).

40. D'Agostini, G. Improved iterative Bayesian unfolding. *arXiv* **2010**, arXiv:1010.0632.

41. Brenner, L.; Verschuuren, P.; Balasubramanian, R.; Burgard, C.; Croft, V.; Cowan, G.; Verkerke, W. RooUnfold: ROOT Unfolding Framework. 2020. Available online: https://gitlab.cern.ch/RooUnfold/RooUnfold (accessed on 9 September 2023).

42. Kroupa, M.; Bahadori, A.A.; Campbell-Ricketts, T.; George, S.P.; Zeitlin, C. Kinetic energy reconstruction with a single layer particle telescope. *Appl. Phys. Lett.* **2018**, *112*, 134103. [CrossRef]

43. Kroupa, M.; Bahadori, A.A.; Campbell-Ricketts, T.; George, S.P.; Stoffle, N.; Zeitlin, C. Light ion isotope identification in space using a pixel detector based single layer telescope. *Appl. Phys. Lett.* **2018**, *113*, 174101. [CrossRef]

44. Agostinelli, S.; Allison, J.; Amako, K.; Apostolakis, J.; Araujo, H.; Arce, P.; Asai, M.; Axen, D.; Banerjee, S.; Barrand, G.; et al. Geant4—A simulation toolkit. *Nucl. Instruments Methods Phys. Res. Sect. A Accel. Spectrometers Detect. Assoc. Equip.* **2003**, *506*, 250–303. [CrossRef]

45. Garvey, D. Advanced Methodology for Radiation Field Decomposition with Hybrid Pixel Detectors. Master's Thesis, Faculty of Nuclear Sciences and Physical Engineering, Czech Technical University in Prague, Prague, Czechia, 2023. Available online: https://dspace.cvut.cz/handle/10467/108693 (accessed on 22 February 2024).

46. Michel, T.; Durst, J.; Jakubek, J. X-ray polarimetry by means of Compton scattering in the sensor of a hybrid photon counting pixel detector. *Nucl. Instruments Methods Phys. Res. Sect. A Accel. Spectrometers Detect. Assoc. Equip.* **2009**, *603*, 384–392. [CrossRef]

47. López Rosson, G.; Pierrard, V. Analysis of proton and electron spectra observed by EPT/PROBA-V in the South Atlantic Anomaly. *Adv. Space Res.* **2017**, *60*, 796–805. [CrossRef]

48. Farkas, M.; Bergmann, B.; Broulim, P.; Burian, P.; Ambrosi, G.; Azzarello, P.; Pušman, L.; Sitarz, M.; Smolyanskiy, P.; Sukhonos, D.; et al. Characterization of a Large Area Hybrid Pixel Detector of Timepix3 Technology for Space Applications. *Instruments* **2024**, *8*, 11. [CrossRef]

49. Jordan, C.E. RADEX INC BEDFORD MA, Contract No F19628-89-C-0068. Available online: https://apps.dtic.mil/sti/citations/tr/ADA223660 (accessed on 26 October 2023).

50. Ginet, G.P.; O'Brien, T.P.; Huston, S.L.; Johnston, W.R.; Guild, T.B.; Friedel, R.; Lindstrom, C.D.; Roth, C.J.; Whelan, P.; Quinn, R.A.; et al. AE9, AP9 and SPM: New Models for Specifying the Trapped Energetic Particle and Space Plasma Environment. In *The Van Allen Probes Mission*; Fox, N., Burch, J.L., Eds.; Springer: Boston, MA, USA, 2014; pp. 579–615. [CrossRef]

51. Hulsmann, J.; Wu, X.; Azzarello, P.; Bergmann, B.; Campbell, M.; Clark, G.; Cadoux, F.; Ilzawa, T.; Kollmann, P.; Llopart, X. Relativistic Particle Measurements in Jupiter's Magnetosphere with Pix.PAN. *Exp. Astron.* **2023**, *56*, 371–402. [CrossRef]

52. Smolyanskiy, P.; Bergmann, B.; Burian, P.; Cherlin, A.; Jelínek, J.; Maneuski, D.; Pospíšil, S.; O'Shea, V. Characterization of a 5 mm thick CZT-Timepix3 pixel detector for energy-dispersive γ-ray and particle tracking. *Phys. Scr.* **2023**, *99*, 015301. [CrossRef]

53. Useche, S.J.; Roque, G.A.; Schütz, M.K.; Fiederle, M.; Procz, S. Characterization of a 5mm thick CdTe photon-counting detector: Advancing nuclear decommissioning with Compton camera technology. In Proceedings of the 2023 IEEE Nuclear Science Symposium, Medical Imaging Conference and International Symposium on Room-Temperature Semiconductor Detectors (NSS MIC RTSD), Vancouver, BC, Canada, 4–11 November 2023; p. 1. [CrossRef]

MDPI AG

Grosspeteranlage 5

4052 Basel

Switzerland

Tel.: +41 61 683 77 34

Instruments Editorial Office

E-mail: instruments@mdpi.com

www.mdpi.com/journal/instruments